Basic Electronics for the Petroleum Industry

FOURTH EDITION

written by

Will L. McNair

edited by

Ron Baker

Published by

PETROLEUM EXTENSION SERVICE

Continuing & Extended Education

The University of Texas at Austin

2004

Library of Congress Cataloging-in-Publication Data

McNair, Will L.
Basic electronics for the petroleum industry / Will L. McNair
p. cm.
ISBN 0-88698-199-9
1. Electronics. 2. Electricity. I. Title.
TK7816.M385 2003
621'.381—dc21

2003005019

First Edition published 1943. Fourth Edition 2004
Printed in the United States of America

Catalog No. 1.41040
ISBN 0-88698-199-9

Contents

Figures

Tables

Preface

Basic Electronics for the Petroleum Industry begins with simple electrical concepts and moves to more complex topics as the book progresses. This book is the outgrowth of a series of editions that began in 1965 and was last updated in 1980. It has evolved from a single book to individual units, and is now back to a single book that reflects current technology.

A study of basic electronics is best done with a combination of text materials, self-study, classroom study, and lab exercises. Creative work with construction of these circuits enhances the student's comprehension and practical knowledge of electricity. A number of electronic stores can provide components, integrated circuit chips, wiring boards, and other hardware to convert principles into reality.

Will L. McNair, P.E.

Acknowledgments

Over 35 years of contributions by various individuals and organizations have gone into this book. Former contributors to earlier editions include George L. Coenen, PETEX electronics instructor, Bruce Whalen, PETEX publications coordinator, David R. Skinner, consultant for PETEX, and representatives from companies that include Amerada Hess Corporation, Amoco, Atlantic Richfield, ITT Snyder, Shell Oil, Phillips Petroleum, Mobil Oil, and Continental Oil Company, some of which no longer exist at all or have changed names because of mergers.

Publication of this most recent book was achieved under the guidance of the publication staff of PETEX, including managing editor, Kathryn Roberts; layout artist, Debbie Caples; and proofreader, Doris Dickey; as well as the guidance and encouragement of PETEX director Ron Baker and Kilgore training manager, Bill Brantley.

Particular recognition is given to Sue Marie Myers who worked diligently in preparing the manuscript, organizing the book's structure, and performing initial editing of the work from rough notes and markups.

About The Author

Will L. McNair is a graduate electrical engineer from Mississippi State University with over 20 years of experience as a design and development engineer with companies that include General Electric, Varo, and Baylor Company. After his experience with these companies, he has been a consultant and technical trainer for companies worldwide, focusing on electrical applications for the petroleum industry. He has taught electrical and electronic courses in seminars and schools to over 5,000 personnel within the past 20 years. These subjects ranged from basic electricity to sophisticated solid-state topics such as SCR motor drives, variable frequency drives, programmable logic controllers, SCADA, and many others.

During this period, he headed the Energy Training Division at Texas A&M University for two years, has written three other textbooks and been published in various industry magazines, including the *Oil & Gas Journal*, *World Oil*, *Drilling Contractor*, and others. He has contributed to several patents related to solid-state technology and developed several advanced products for the oil and gas industry.

He is a registered engineer in Texas and member of IEEE, TSPE, Society of Petroleum Engineers, IADC, NFPA, and ISA.

1 Basic Mathematics

To understand electrical relationships in circuits and components, you must also understand mathematics. Mathematics—math, for short—helps you visualize how electricity and electrical *circuits* work. The math that follows is basic and is offered mainly as a refresher to those who may be out of practice in working with numbers.

ADDITION

Addition unites, or puts together, two or more things that are often represented by a symbol called a *number*. (Numbers are also called *integers* and *digits*.) The number obtained by adding two or more numbers is the *sum*, or total. The character for addition is the plus sign (+).

The first step in addition is to arrange the numbers to be added in vertical columns, such as those in Table 1. In numbers with more than one digit, the column on the right is the ones column. This column contains a one-digit number: 1, 2, 3, 4, and so on up to 9. The next column to the left of the ones column is the tens column. It is the tens column because the digits, or numbers, in this column are 10 or multiples of 10, as 20, 30, 40, and so on up to 90. The next column to the left is the hundreds column. The digits in this column are 100 or multiples of 100 up to 900.

TABLE 1
Numbers Arranged in Vertical Columns of Hundreds, Tens, and Ones

Hundreds	Tens	Ones
100	10	1
200	20	2
300	30	3
400	40	4
500	50	5
600	60	6
700	70	7
800	80	8
900	90	9

This columnar arrangement continues to the left in thousands, ten thousands, millions, billions, and so on. For example, the number 8,649 can also be expressed as having nine ones, four tens, six hundreds, and eight thousands.

Regardless of the sizes of the numbers, to add them correctly, the last digit to the right in each number should be placed directly under the other in a vertical column. For example, to add 14, 7, 121, and 2,390, they should be arranged as—

```
   14
    7
  121
2,390.
```

Notice that the numbers line up—are flush—with the right-hand ones column. Put another way, the numbers are flush right. Also notice that as long as the numbers are flush right, they may be written in any order. For example, the previous four numbers can be written as—

```
    7
2,390
   14
  121
```

or in any other order and the correct answer can be obtained. Remember, however, that the numbers must be flush with the right-hand column.

Always begin by adding the numbers in the right-hand, ones column first. If the sum can be expressed by one digit, from 0 to 9, write it under the column. If, on the other hand, the sum contains more than one digit, (10 or more), write the right-hand digit of the sum under the column added, and add (or *carry*) the remaining digits to the next column on the left. For example, if the sum of the column is 23, write 3 under the column, and add 2 to the sum of the digits of the next column to the left. The common expression for this action is, "put down 3 and carry the 2."

As an example, add 8,145, 234, and 756. The solution is—

$$\begin{array}{r} 8{,}145 \\ 234 \\ +\ \ 756 \\ \hline 9{,}135. \end{array}$$

In this case, 5, 4, and 6 add up to 15; so, write 5 under the ones column and carry, or add, 1 to the tens column. In the tens column, 1, 4, 3, and 5 are 13. Write 3 under the tens column and add 1 to the hundreds column. Thus, 1, 1, 2, and 7 are 11. Write 1 under the hundreds column and add 1 to the thousands column. Finally, add 1 to the thousands column to obtain the correct answer of 9,135.

SUBTRACTION

Subtraction is removing, or taking, one number from another number. The number that is subtracted is the *subtrahend*, and the number from which it is taken is the *minuend*. The number that represents the result of subtraction is the *remainder*. The minus sign (–) indicates subtraction. As an example, subtract 43 from 576. This action can also be stated as 576 minus 43, or 576 – 43. In any case, the solution is—

$$\begin{array}{rl} 576 & \text{minuend} \\ -\ \ 43 & \text{subtrahend} \\ \hline 533 & \text{remainder.} \end{array}$$

Notice again that the numbers line up (are flush) with the right-hand ones column.

If a number in the subtrahend is larger than the number in the minuend, then an operation termed *borrowing* occurs. For example, subtract 78 from 106. The solution is—

$$\begin{array}{r} 106 \\ -\ \ 78 \\ \hline 28. \end{array}$$

In this case, to subtract the 8 in the ones column, 1 is borrowed (taken from) 10, which leaves 9. The 1 in front of the 6 makes it 16. Then, 8 from 16 is 8, and 7 from 9 is 2. Thus, the answer is 28. This operation can be shown as—

$$\begin{array}{r} \not{1}\not{0}^{9}\ {}^{1}6 \\ -\ \ 78 \\ \hline 28. \end{array}$$

Cross out the 10 to leave 9, and put 1 in front of 6 to make it 16. Incidentally, when subtracting, you can check the answer by adding the remainder to the subtrahend. In this case, 28 + 78 = 106.

MULTIPLICATION

Multiplication is adding one of two given numbers as many times as there are units in the other number. Multiplication shortens the process of addition. Thus, multiplying 3 times 8 gives an answer equal to the sum of 3 + 3 + 3 + 3 + 3 + 3 + 3 + 3, which is 24. The number that is multiplied, or increased, is the *multiplicand*. The *multiplier* is the number by which the multiplicand is multiplied. It indicates how many times the multiplicand is to be increased. Thus, 451 times 25 means to add 451 to itself 25 times. The result obtained by multiplying one number by another is the *product*.

The multiplication sign (×) is called the times sign. For example, 3 × 4 is read as 3 times 4. When letters or symbols are used in an equation, the multiplication sign is often omitted. For example, if I is multiplied by R, it is usually written as IR instead of I × R. Also, sometimes the symbol • is used between the multiplicand and the multiplier. For example, 2 × 2 can also be written as 2 • 2. Using • instead of × avoids confusion when an unknown quantity in an equation is indicated by the letter x. For instance, while $x \times 2$ might be confusing, especially when writing it with pen or pencil on paper, $x \bullet 2$ will not be.

As an example problem, multiply 232 by 3. The solution is—

$$\begin{array}{rl} 232 & \text{multiplicand} \\ \times\ \ \ 3 & \text{multiplier} \\ \hline 696 & \text{product.} \end{array}$$

In multiplication, as in addition, numbers must sometimes be carried to obtain the correct answer. For example, what is 396 × 822? The solution is—

$$\begin{array}{r} 396 \\ \times\ \ 852 \\ \hline 792 \\ 1980 \\ 3168 \\ \hline 337{,}392. \end{array}$$

In this case, 2 × 396 is 792, 5 × 396 is 1,980, and 8 × 396 is 3,168. Adding the products of the individual numbers yields a final product of 337,392. Also, notice how the products line up. For example, the 0 in 1,980 is placed under the 9 in 792, and the 8 in 3,168 is placed under the 8 in 1,980. Proper alignment is essential to obtain the correct answer.

DIVISION

Division is the reverse of multiplication. It finds how many times one number is contained in another number. For example, to divide 8 by 4 means to find how many times 4 is contained in 8, or, put another way, what number is multiplied by 4 to result in 8? In this case, the number is 2.

When the division sign (÷) appears between two numbers, it indicates that the first number is to be divided by the second. For example, 16 ÷ 4 means to divide 16 by 4, or, to find the number that, if multiplied by 4, equals 16. The answer is 4. The number to be divided is the *dividend*, and the number that does the dividing is the *divisor*. The result obtained by division is the *quotient*.

Another way of writing numbers to be divided is to use the sign $\overline{)}$, wherein the dividend is to the right of the $\overline{)}$, the divisor is inside the $\overline{)}$, and the quotient on top, as in

$$\begin{array}{r} 3 \\ 6\,\overline{)\,18}. \end{array}$$

This format reads, "18 divided by 6 yields 3," or "6 into 18 is 3." You can check the answer by multiplying the quotient by the divisor; if it equals the dividend, then it is correct. In this example, 3 × 6 = 18.

As another example, divide 345 by 3, which, remember, is another way of determining what number times 3 equals 345. The answer is 345 (the dividend) ÷ 3 (the divisor) = 115 (the quotient). Check the answer by multiplying the quotient 115 by the divisor 3, which equals the dividend 345. This problem is an example of short division. It is called short division because one number, 3 in this case, goes into the other number, 345 in this case, in one simple step.

On the other hand, long division requires more calculation. For example, divide 42,899 by 86. The solution is—

$$\begin{array}{rl} & 498 \\ 86 & \overline{)\,42899} \\ & \underline{-\,344} \\ & 849 \\ & \underline{-\,774} \\ & 759 \\ & \underline{-\,688} \\ & 71. \end{array}$$

Because long division requires more steps, it has made a modern hand-held calculator a must-have item. However, as a review, let's manually solve the example problem of long division. First, determine how many times 8 goes into 42. Because 8 × 5 is 40, perhaps 5 is the first number in the quotient. However, 5 × 86 = 430, which is larger than 428. This larger number is unacceptable because we must subtract this number from the first three numbers in the dividend. So, 4 is the first number in the quotient because 4 × 86 = 344. Next, determine that 428 – 344 = 84 and bring down the 9 from the dividend. Eight goes into 84 nine times, and 9 × 86 = 774, which is written below 849. Now, subtract 774 from 849, which is 75. Bring down the 9, determine that 8 goes into 75 eight times, and multiply 8 times 86 to come up with 688. Finally, subtract 688 from 759, which yields 71.

Thus, in long division, 42,899 divided by 86 results in a quotient of 498 with a remainder of 71. A remainder results when one number (86 in this case) does not divide evenly into another number (42,899 in this case). The remainder can be expressed as a fraction so that the answer to the example is $498\,{}^{71}\!/_{86}$. The remainder can also be expressed as a decimal number. Simply divide 71 by 86 to obtain the decimal number 0.8256. So, another way of expressing the answer is to write 498.8256, which, incidentally, is how most hand-held calculators express the solution, although perhaps to more decimal places, as in 498.82558.

As you just read, numbers in fractional form, such as ${}^{25}\!/_{5}$ also indicate division. This fraction is the same as writing 25 ÷ 5. In the fraction, the dividend 25 is the *denominator* and the divisor 5 is the *numerator*. The solution, or quotient, is 5.

EXPONENTS

In electricity and electronics, very large numbers and very small decimal fractions are sometimes used. These numbers and fractions can be expressed more compactly as powers of 10. For example, 1,000 can be expressed as 10^3, which is a short way of writing 10 × 10 × 10. (10 × 10 = 100; and 100 × 10 = 1,000.) Thus, when a number is raised to a power, it is a way of stating that the number is multiplied by itself by the number it is raised to.

The number to which a number is raised is an *exponent*. For example, the 4 in 10^4 is the exponent. Another way to state it is, "10 raised to the fourth power." And, an even shorter way is, "10 to the 4." However, a number raised to the power of 2 is usually stated as the number squared. For example, 10^2

is usually stated as, "10 squared," and not, "10 to the second power," or "10 to the 2." Also, any number raised to 3 is usually stated as the number cubed. For example, 10^3 is usually read as, "10 cubed."

If a number has no written exponent, an exponent of 1 is understood—that is, $10 = 10^1$, for example. The number 100 can be represented as 10 x 10, or 10^2; 1,000,000 as 10^6. Notice that 1,000,000 is 1 followed by 6 zeros and that it can also be written as 10^6. Thus, the exponent indicates how many zeros the number contains. For example, 10^8 is a short way of writing the number 1 followed by 8 zeros, or 100,000,000. Finally, be aware that any number raised to the 0 power is equal to 1. Thus, 10^0 equals 1, as does $1{,}000^0$.

Raising a number to a negative power, such as 10^{-1}, is a short way to state a fraction or decimal of the number. For example, 10^{-1} is the same as $\frac{1}{10}$ or 0.1. (Incidentally, when writing decimals, mathematical convention requires placing a zero in front of the decimal place. Thus, .1 is always written as 0.1.) Other examples of numbers raised to negative powers are 10^{-2}, which is the same as $\frac{1}{100}$ or 0.01; and 10^{-3}, which is the same as $\frac{1}{1000}$ or 0.001. Notice that 10^{-3} equals $\frac{1}{1000}$, which is 1 over 1,000 or 1 followed by three zeros. The exponent indicates the number followed by the same number of zeros in the fraction. For example, 10^{-6} is the same as $\frac{1}{1000000}$. However, in a decimal, the exponent reflects the number of places the number is carried to. For example, 10^{-6} is 0.000001, or the decimal point followed by 6 numbers, only five of which are zeros.

Numbers besides 10 can also be written more compactly by using decimal points and exponents. The number 3,140,000, for example, is equal to $3.14 \times 1{,}000{,}000$. It is more compactly written as 3.14×10^6. In this case, the decimal point is moved six places to the left, and the exponent of 10 is 6. Another example is the decimal number 0.0002. This number is equal to 2×0.0001, or 2×10^{-4}. The decimal point is moved four places to the right, and the exponent of 10 is –4.

Multiplication of Exponents

To multiply exponents, add them. This statement in equation form is—

$$a^m \times a^n = a^{m+n} \qquad \text{(Eq. 1)}$$

where

a = base number

m, n = exponents.

For example, let's multiply 1,000 by 100. Rather than carrying out conventional multiplication, it is easier to use exponents and add them. Thus—

$$\begin{aligned} 1{,}000 \times 100 &= 10^3 \times 10^2 \\ 10^{3+2} &= 10^5 \\ 10^5 &= 100{,}000, \text{ or 1 followed by five zeros.} \end{aligned}$$

Similarly, to multiply 1,200 by 210—

$$\begin{aligned} 1{,}200 \times 210 &= (12 \times 10^2) \times (21 \times 10^1) \\ &= (12 \times 21) \times 10^3 \\ &= 252 \times 10^3 \\ &= 252{,}000 \text{ (252 followed by three zeros).} \end{aligned}$$

Division of Exponents

To divide exponents, subtract them. This statement in equation form is—

$$a^m \div a^n = a^{m-n} \qquad \text{(Eq. 2)}$$

where

a = base number,

m, n = exponents.

For example, divide 1,000,000 by 1,000—

$$\begin{aligned} 1{,}000{,}000 \div 1{,}000 &= 10^6 - 10^3 \\ &= 10^{6-3} \\ &= 10^3 \\ &= 1{,}000, \text{ or 1 followed by 3 zeros.} \end{aligned}$$

Similarly, to divide 252,000 by 210—

$$\begin{aligned} \frac{252{,}000}{210} &= \frac{252 \times 10^3}{21 \times 10^1} \\ &= \frac{252 \times 10^2}{21} \\ &= \frac{252}{21} \times 10^2 \\ &= 12 \times 10^2 \\ &= 1{,}200, \text{ or 12 followed by 2 zeros.} \end{aligned}$$

Raising to a Power

To raise a number to a power using exponents, multiply the exponents. This statement in equation form is—

$$(a^m)^n = a^{m \times n} \qquad \text{(Eq. 3)}$$

where

a = base number

m, n = exponents.

For example, to find the square of 100—

$$(100)^2 = (10^2)^2 = 10^{2\times 2} = 10^4 = 10{,}000.$$

Extracting Roots

The *root* of a number is the opposite of raising a number to a power. For example, 9^2 is 81, and the square root of 81 is 9. To extract a root using exponents, divide the exponent by the index of the root to be taken. The equation is—

$$\sqrt[n]{a^m} = a^{m \div n} \qquad \text{(Eq. 4)}$$

where

a = base number
m, n = exponents.

For example, to find the square root of 10,000—

$$\sqrt[2]{10{,}000} = \sqrt[2]{10^4} = 10^{4\div 2} = 10^2 = 100.$$

Although the text uses the symbol $\sqrt[2]{\ }$ to indicate a square root, usually the symbol $\sqrt{\ }$ is used alone (without the ^{2}) to indicate a square root. For example, the square root of 16 is written as $\sqrt{16}$. However, to indicate other roots, such as cube root and fourth root, the symbol is used with the appropriate numbers. For example, the symbol for the cube root of a number is $\sqrt[3]{\ }$. For the fourth root, the symbol is $\sqrt[4]{\ }$, and so on.

EQUATIONS

An *equation* is similar to a set of balanced scales (fig. 1). Just as objects of equal weight balance a scale, for an equation to be an equation, it must also be in balance. That is, an equation states that one thing or a group of things on one side of the equation is equal to another thing or another group of things on the other side of the equation. For example, 2 + 3 = 1 + 4. This equation states that 2 plus 3 is equal to (is the same thing as) 1 plus 4. It is a true equation because 2 + 3 = 5 and 1 + 4 = 5. The equation can also be written as 2 + 3 = 5, 1 + 4 = 5, or 5 = 5.

For an equation to be true, both sides must be equal in value. The symbol = means equal to. On the other hand, the symbol ≠ means not equal to. Thus, the following is not an equation: 2 + 3 ≠ 2 + 2.

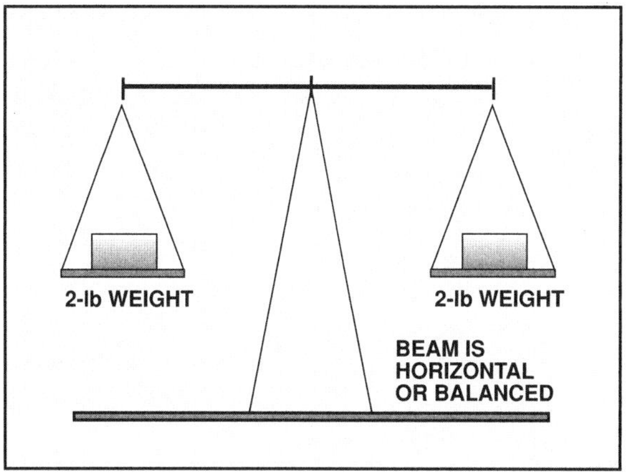

Figure 1 ***Beam balance balanced by two 2-pound weights.***

In other words, 2 + 3 ≠ 2 + 2 means that 2 + 3 (5) is not equal to 2 + 2 (4). Another symbol is ≈, which stands for approximately equal to. For example, 0.0009 ≈ 0.001. In this case, 0.0009 is approximately equal to 0.001; rounding 0.0009 down to 0.001 makes it almost, but not quite, equal.

Use of Symbols in Equations

Symbols, such as ϕ, or letters, such as x, are often used in equations to represent numbers. Symbols and letters make the equation more compact and easier to write. For example, in electrical work, a basic equation is Ohm's Law. This law equates current to the ratio of voltage and resistance. Thus, the equation can be written as—

$$\textit{Current (amperes)} = \frac{\textit{Voltage (volts)}}{\textit{Resistance (ohms)}}. \qquad \text{(Eq. 5)}$$

This equation can, however, be written more compactly and easily by letting I represent current in amperes (amps), E represent voltage in volts, and R represent resistance in ohms. Now, the equation can be written as—

$$I = \frac{E}{R} \qquad \text{(Eq. 6)}$$

where

I = current, amperes
E = voltage, volts
R = resistance, ohms.

Using symbols or letters also makes an equation fit not just one, but many situations. For instance, if George Ohm had written his equation as 2 amps = 10 volts/5 ohms, the equation would be valid for only these values. That is, it would be valid only in dealing with a circuit having 10 volts placed across a resistance of 5 ohms. Only in that specific case would the current be equal to 2 amps. So, instead of using specific values, Ohm's law uses letters, which can represent any number or value. Equation 6, $I = E/R$, is a general equation because many sets of values—not just one set—can be used in the equation.

Symbols or letters in an equation can stand for any number as long as the sum of the numbers on the left side of the equation equals the sum of the numbers on the right side of the equation. Remember: an equation is like a balance. One side must equal the other.

Rearranging Equations

Often, to solve an equation when the unknown value is on the right side of the equal sign, it must be rearranged. For example, if $6 = 12/x$, and you wish to know the value of x, you must rearrange the equation so that x is on the left side. The operation, in this case, is known as "solving for x." When solved for x, the equation reads as $x = 12/2$. To understand how this rearrangement was achieved, consider that rearranging equations involves performing mathematical operations to both sides of the equation. Several operations can be performed on an equation without destroying it, including addition, subtraction, multiplication, and division. These operations move unknown factors to the left side of the equation so that the equation can be easily solved. Also, keep in mind that more than one operation is sometimes required to rearrange an equation.

The first operation that rearranges an equation is adding the same number to both sides of it. To perform the operation, assume that $A = 2$, $B = 8$, and $C = 4$. A basic equation can be formed from the numbers or the letters, as $2 = 8/4$ or $A = B/C$. Without destroying the equation, the same value can be added to both sides, as $2 + 2 = 8/4 + 2$ or $A + A = B/C + A$. Notice that the equation has not been invalidated because the same number or letter is added to both sides of it. In the first case, 2 is added to 2 on the left side of the equation and 2 is added to $8/4$ on the right side of the equation. Similarly, in the second case, A is added to A on the left side of the equation and A is added to B/C on the right side. Because the same operation—addition in this case—has been done to both sides of the equations, they are still valid.

The next step is to *collect the terms*, which means to perform the mathematical operations. So, in the first equation, $2 + 2 = 2 + 2$, which is $4 = 4$. In the second equation, $2A = B/C + A$. Now, it is simply a matter of substituting the assumed values (the numbers), as $2 + 2 = 8/4 + 2$, which solves as $4 = 2 + 2$ and $4 = 4$.

The second operation that can be used to rearrange an equation is to subtract the same number or value from both sides of the equation. Assume that $A = 3$, $B = 9$, and $C = 2$. A basic equation can be stated in numbers or in letters. In numbers, let's say it is $3 = 9/3$; in letters, let's say it is $A = B/A$. The same value, 2 or C in this case, can be subtracted from both sides, which results in $3 - 2 = 9/3 - 2$ and $A - C = B/A - C$. Collecting the terms results in $1 = 3 - 2$, which is $1 = 1$. Substituting the values of A, B, and C in the symbolic equation, the result is $3 - 2 = 9/3 - 2$, which also comes out to $1 = 1$. Again, because the same operation occurs on both sides of the equation, it is not destroyed.

The third operation that can be used to rearrange an equation is to multiply both sides of the equation by the same number. Assume that $A = 6$, $B = 24$, $C = 4$, and $D = 2$. A basic equation can be stated in numbers as $6 = 24/4$, or in abstract symbols as $A = B/C$. Both sides can be multiplied by 2 or D, as in $6 \times 2 = 24/4 \times 2$ or $A \times D = B/C \times D$. The terms collect as $12 = 6 \times 2$ and $12 = 12$. And, substituting the assumed values in the symbolic equation, the result is $6 \times 2 = 24/4 \times 2$, which comes out to $12 = 6 \times 2$ and $12 = 12$. Once again, the equation is not destroyed.

The fourth operation that can be used to rearrange an equation is to divide both sides of the equation by the same number. Assume that $A = 24$, $B = 4$, $C = 6$, and $D = 2$. A basic equation can be stated in numbers or abstract symbols. In numbers it is $24 = 4 \times 6$; in symbols, it is $A = BC$. Both sides of the equation can be divided by the same value. In this case, let's say 2 or D. Dividing by 2 or D works out to $\frac{24}{2} = \frac{4 \times 6}{2}$ in numbers and $A/D = BC/D$ in symbols. Collecting the terms comes out to $12 = 24/2$ and $12 = 12$. Substituting the assumed values in the symbolic equation, the result is $\frac{24}{2} = \frac{4 \times 6}{2}$. Collecting the terms comes out to $12 = 24/2$, which is $12 = 12$.

So, to rearrange, or solve, equations, you can move values from one side of the equation to the other. And, in some cases, you can leave out some of the steps. You can move a term from one side of the equation to the other if you change the term's sign. That is, if you move a positive (+) number from one side to the other, it becomes a negative (–) number. (In the same way, a negative number becomes a positive number.) To show this action, assume that $A = 3$, $B = 5$, $C = 2$, and $D = 10$. These numbers and symbols can be put into equation form as $3 + 5 + 2 = 10$ and $A + B + C = D$.

In the numeric equation, let's move +2 from the left to the right side, where it becomes –2. The equation now reads 3 + 5 = 10 – 2, which works out to 8 = 8. In the equation with symbols, let's move $+C$ to the other side, where it becomes $-C$. The equation then reads $A + B = D - C$. In both cases, we moved one value from the left side to the right side and changed its sign from positive (+) to negative (–). Now, let's add –2 to the numerical equation and $-C$ to the symbol equation, as in 3 + 5 = 10 – 2, which works out to 8 = 8. In the equation with symbols it comes out as $A + B = D - C$. When we substitute numbers for the symbols, the result is 3 + 5 = 10 – 2 or 8 = 8.

Also, in equations that contain multiplication and division, when you move a multiplied or divided term from one side to the other, you change the operation—that is, if the equation contains division on the right side, moving it to the left changes it to multiplication and vice versa. For example, assume that $A = 2$, $B = 4$, and $C = 8$. Now suppose that $AB = C$. If we wish to solve for A, we can move B to the right side, where it is divided into C. This operation makes the equation now appear as $A = C/B$. Substituting numbers for letters, the solution is $2 = 8/4$ or 2 = 2. The operation did not negate the equation.

Another short cut is *cross-multiplying*, whereby a number is removed from the denominator of one side of an equation and placed in the numerator of the other side of the equation. Assume that $A = 2$, $B = 4$, $C = 3$, and $D = 6$. The basic equation can be stated in numbers as $2/4 = 3/6$ and in letters as $A/B = C/D$. We can now cross-multiply—that is, multiply the numerator by the denominator on each side of the equation. Thus, $2 \times 6 = 4 \times 3$, which works out to 12 = 12. In the equation with letters, $AD = BC$. Substituting the numbers for the letters results $2 \times 6 = 4 \times 3$, which works out to 12 = 12.

A third shortcut is to invert both sides of a fractional equation. Assume again that $A = 2$, $B = 4$, $C = 3$, and $D = 6$. As usual, the basic equation can be stated in numbers and in letters. In numbers, it is $2/4 = 3/6$, or $1/2 = 1/2$; in letters, it is $A/B = C/D$. Inverted, the equations are $4/2 = 6/3$ (2 = 2) and $B/A = D/C$.

Note that when the numerator or denominator contains two or more factors separated by a plus or a minus sign, the whole unit must be handled as a group. To explain, assume that $A = 3$, $B = 5$, $C = 4$, and $D = 6$. Then, assume that 1 is added to one side of the equation and 2 to the other side, so that the result is $\frac{3}{5+1} = \frac{4}{6+2}$ or $\frac{A}{B+1} = \frac{C}{D+2}$. To solve these equations, cross-multiply both sides of the equations to obtain $3 \times (6 + 2) = 4 \times (5 + 1)$ and $A(D + 2) = C(B + 1)$. [Note that $A(D + 2)$ is the same as $A \times (D + 2)$.] Collecting the terms results in 18 + 6 = 20 + 4, which is 24 = 24. Similarly, substituting numbers for the letters gives a result of 3(6 + 2) = 4(5 + 1), which equals $3 \times 8 = 4 \times 6$ to yield a final result of 24 = 24.

As a final example of rearranging (solving) an equation, let's look at Ohm's Law, which, you recall, in equation form is $I = E/R$. This equation can be solved for one unknown if two values are given. Let's say that $E = 10$ and $R = 2$. In this case, $I = 10/2$, which equals 5 amps. However, if $I = 2$ and $E = 10$ and R is unknown, then the equation reads $2 = 10/R$.

To determine R, the equation should be rearranged—that is, the equation needs to be solved for R. The first step is to multiply both sides of the equation by R to make it read $R \times I = E/R \times R$. Multiplying both sides of the equation by R allows the Rs on the right side of the equation to *cancel out*. That is, the R on the right is the same as $R/1$, so the equation can now be written as $R \times I = E/R \times R/1$. The Rs cancel each other, leaving the equation to read $R \times I = E$ (E is the same as $1E$). Finally, because R is the desired quantity, I is moved from the equation's right side to the left side. Recall that in the case of multiplication and division, moving an expression from one side to the other of an equation changes the expression's mathematical operation. That is, because I is multiplied on the left, moving it to the right means it is divided. So, the equation finally reads as $R = E/I$. In other words, the equation has been solved for R.

With the equation rearranged, the known values of 10 for E and 2 for I are substituted and the indicated division performed. The quotient that results is equal to the unknown, R. Thus, $R = 10/2$, and R is equal to 5 ohms.

SUBSCRIPTS

Frequently, a letter representing a quantity is annotated with a small number or letter on its lower right side. These small numbers are *subscripts*. For example, in electrical circuits, resistors are often identified as R_1, R_2, R_3, or R_4, which means that four resistors are in the *circuit*. (A circuit is a complete electrical path from one terminal of a source of electricity to the other.) Each is identified with the letter R and a subscripted number—1 through 4 in this case.

BINARY NUMBERS

Numbers and characters in digital, or binary, form have made it possible to develop digital electronic devices such as computers, programmable logic controllers, and similar digital equipment. Digital electronic devices cannot recognize conventional numbers such as 1, 2, 3, 4, 5, and so on. They can only recognize the presence or absence of a binary signal that is abbreviated as 1 or 0. A binary number is so called because it consists only of two numbers—0 and 1.

The binary numbering system is a base 2 system—that is, only two numbers, 0 and 1, form all numbers. By contrast, the conventional system is a base 10 system in that 10 numbers, from 0 to 9, form all numbers. (It is also called the decimal system from the fact that *deci* means 10 in Latin.) In the binary system, 1 is the largest value while 0 is the lowest. Digital devices use only 1 and 0 to make discrete decisions or show their states.

A binary number can be converted to decimal and decimal to binary. Perhaps the easiest method is to use a table, such as Table 2, which counts through four binary digits. [Incidentally, binary digits are termed *bits*, which comes from combining b(inary) and (dig)it. So, four binary digits are also 4 bits.] In any case, binary numbers larger than those in the table, which run from 2^0 to 2^3, can be achieved by adding more columns whose values are 2^4, 2^5, 2^6, 2^7, 2^8, and so on. The second row of numbers across the top of the table converts 2 raised to various powers. For example, 2^3 equals 8.

With the table, it is easy to determine equivalent numbers in each system. For example, the binary number 1010 is 10 in the decimal system. Likewise, the decimal number 9 is 1001 in the binary system.

TABLE 2

Decimal and Binary Equivalent Numbers

2^3	2^2	2^1	2^0	
8	4	2	1	**Decimal Value**
0	0	0	0	0
0	0	0	1	1
0	0	1	0	2
0	0	1	1	3
0	1	0	0	4
0	1	0	1	5
0	1	1	0	6
0	1	1	1	7
1	0	0	0	8
1	0	0	1	9
1	0	1	0	10
1	0	1	1	11
1	1	0	0	12
1	1	0	1	13
1	1	1	0	14
1	1	1	1	15

OCTAL NUMBERS

Octal numbers are used in a digital device called a programmable logic controller, or PLC, for short. PLCs are used in process control systems to run and manage the variables that occur in such systems. For example, a PLC could maintain a constant level of liquid in a tank or a vessel by reading level sensors and sending corrective signals to devices to adjust the liquid flowing into and out of the vessel.

As the name implies, octal numbers are a base 8 system—that is, the system consists of eight numbers: 0, 1, 2, 3, 4, 5, 6, and 7. A subscript of 8 following an octal number designates it as such. So, a number such as 571_8 indicates that the number is octal and not decimal.

We convert binary numbers to octal numbers in two steps. First, combine the binary number in groups of three bits. For example, the binary number 110111101001001111 is regrouped as 110 111 101 001 001 111. Second, the equivalent decimal number for each group is determined. The easiest way to make the conversion is to use a table, such as Table 3. The first row has the grouped binary numbers; the second row shows the decimal equivalents.

Table 3 shows that the octal number for the example binary number is 675117_8.

TABLE 3

Binary Number in Groups of Three (Top) and Decimal Equivalents (Bottom)

110	111	101	001	001	111
6	7	5	1	1	7

HEXADECIMAL NUMBERS

Many electronic display systems, such as a computer screen, use hexadecimal numbers to electronically convert binary numbers for displaying characters and objects. The hexadecimal system is a base 16 system—that is, it uses 16 symbols, which are 0, 1, 2, 3, 4, 5, 6, 7, 8, 9, A, B, C, D, E, and F. A hexadecimal (hex, for short) number is indicated with a subscript of 16. For example, the number $42A9D_{16}$ is a hex number.

Combining the binary in groups of 4 bits as shown in Table 4 facilitates converting binary numbers to hex numbers.

TABLE 4
Hexadecimal and Binary Equivalent Numbers and Letters

2^3	2^2	2^1	2^0	Hex Value
8	4	2	1	
0	0	0	0	0
0	0	0	1	1
0	0	1	0	2
0	0	1	1	3
0	1	0	0	4
0	1	0	1	5
0	1	1	0	6
0	1	1	1	7
1	0	0	0	8
1	0	0	1	9
1	0	1	0	A
1	0	1	1	B
1	1	0	0	C
1	1	0	1	D
1	1	1	0	E
1	1	1	1	F

As an example of converting a binary number to a hex number, suppose the binary number is 11011110100100 1111. First, arrange the binary numbers in groups of four, then find the equivalent alphanumeric numbers in the table and write them under the groups, as—

0011	0111	1010	0100	1111
3	7	A	4	F.

So, the hexadecimal equivalent for the example binary number is $37A4F_{16}$.

BINARY CODED DECIMAL CODE

Binary coded decimal (*BCD*) *code* is used to store information in computers and is used in electronic input devices such as thumbwheel switches and seven-segment displays. It consists of the four-digit binary equivalent of the 10 decimal numbers from 0 to 9. To show a decimal number in BCD, we write the equivalent binary number for each digit. For example, the decimal number 72 becomes 0111 0010 in BCD, because the binary number for 7 is 0111 and the binary number for 2 is 0010. The space between the four digits signifies a BCD coded number. Table 5 shows BCD, binary, and decimal equivalents. Notice that all the BCD numbers have four digits.

TABLE 5
BCD, Binary, and Decimal Equivalent Numbers

BCD	Binary	Decimal
0000	0	0
0001	1	1
0010	10	2
0011	11	3
0100	100	4
0101	101	5
0110	110	6
0111	111	7
1000	1000	8
1001	1001	9

AMERICAN STANDARD CODE FOR INFORMATION INTERCHANGE

American standard code for information interchange (ASCII) code is used in PLCs to represent numbers, letters and symbols. It is also used to send and receive alphanumeric data such as the letters and symbols found on a computer keyboard. Each character or symbol uses up to 7, or sometimes 8, binary digits to represent it. Table 6 is a partial list of characters and symbols in the ASCII code. They can also be found in digital handbooks and on the worldwide web (the internet).

TABLE 6
ASCII Characters and the Equivalent Binary Numbers

Character	Binary No.	Character	Binary No.
@	100 0000	0	011 0000
A	100 0001	1	011 0001
B	100 0010	2	011 0010
C	100 0011	2	011 0010
D	100 0100	3	011 0011
E	100 0101	5	011 0101
F	100 0110	6	011 0110
G	100 0111	7	011 0111
H	100 1000	8	011 1000
I	100 1001	9	011 1001
J	100 1010	:	011 1010
K	100 1011	;	011 1011
L	100 1100	<	011 1100
M	100 1101	=	011 1101
N	100 1110	>	011 1110
O	100 1111	?	011 1111
P	101 0000	SP	010 0000
Q	101 0001	!	010 0001
R	101 0010	"	010 0010
S	101 0011	#	010 0011
T	101 0100	$	010 0100
U	101 0101	%	010 0101
V	101 0110	&	010 0110
W	101 0111	'	010 0111
X	101 1000	(	010 1000
Y	101 1001	)	010 1001
Z	101 1010	*	010 1010
[	101 1011	+	010 1011
\	101 1100	,	010 1100
]	101 1101	-	010 1101
^	101 1110	.	010 1110
_	101 1111	/	010 1111

As an example of using the table, note that the letter F has the binary ASCII code of 1000110 and that the + sign has the ASCII code of 0101011.

GRAY CODE

Gray code is useful in mechanical encoders because a slight change in location only affects one bit in the code. Thus, it is often used to transmit binary code from rotating equipment when tracking the position of the shaft. As the shaft rotates from one position to another, only one bit changes in the code to permit a more stable and precise monitoring of its position. Table 7 shows the Gray code number next to the equivalent binary number.

TABLE 7
Gray Code Numbers with Equivalent Binary Numbers

Gray Code	Binary
0000	0000
0001	0001
0011	0010
0010	0011
0110	0100
0111	0101
0101	0110
0100	0111
1100	1000
1101	1001
1111	1010
1110	1011
1010	1100
1011	1101
1001	1110
1000	1111

REVIEW QUESTIONS

1. What is the sum of 100, 1,000, and 10,000?
2. Divide 300 by 6.
3. Multiply 40 by 25.
4. Show the exponent of 4 to the number 89.
5. What is the binary equivalent number of 14?

The Nature of Electricity

Matter is anything that has *mass* or weight and occupies space. Mass is the quantity of matter that a substance contains. It is related to weight, but, unlike weight, mass is independent of external conditions such as buoyancy of the atmosphere or acceleration caused by gravity. For example, an object on earth may weigh 24 grams, but the same object on the moon weighs only 4 grams. Because the earth's gravity is six times that of the moon, the object weighs six times more on the earth than it does on the moon. However, if an object on earth has a mass of 24 grams, it has a mass of 24 grams whether it is on the moon or anywhere else in our universe.

Matter is electrical in nature. The three forms of matter are solids, liquids, and gases. Matter is made up of elements and combinations of elements. An *element* is an individual substance not made up of other elements or substances. For example, the element copper is not made up of other substances. An element consists of only one kind of atom. Atoms make up the smallest substance within an element and are the fundamental building blocks of our universe.

About 100 elements occur naturally, but chemists and physicists have created more in nuclear laboratories, so that 115 have been identified at this time. Future experiments may yield even more. In any case, elements combine chemically to form materials found in everyday life. For example, the elements sodium and chlorine combine to form sodium chloride, which is ordinary table salt.

As stated earlier, an *atom* is the smallest unit of an element. When the atoms of an element combine with themselves or with other elements, they form a *molecule*. For example, oxygen, which has the chemical symbol of O, naturally occurs as a molecule, the symbol of which is O_2. An oxygen molecule is abbreviated as O_2 to indicate that it is composed of two atoms of oxygen. Similarly, a molecule of sodium chloride, or table salt, is abbreviated as NaCl—sodium is Na and the chloride molecule is Cl. NaCl means that a sodium atom and a chloride molecule combine to form a molecule of table salt.

Sometimes elements merely mix together and do not combine chemically. Such a combination is a *mixture*. Examples of mixtures are natural gas, steel, brass, and air. Air, for example, is a mixture of several gases, such as nitrogen, oxygen, and carbon dioxide. These gases do not react chemically; so, they are a mixture.

ATOMS

As mentioned earlier, the atom is the smallest division of an element. An atom consists of a heavier part called the *nucleus* and very small particles revolving around it called *electrons*. The nucleus is the central region of an atom, composed of protons and neutrons and containing almost all of the mass of the atom. The mass of an electron is very small; indeed, an electron's mass is only 9.1×10^{-28} grams. 9.1×10^{-28} means 9.1 with the decimal point shifted 28 places to the left. Thus, the mass of an electron is very small indeed. Besides being very light in mass, an electron exhibits a quality that is called a *negative charge*. Usually, we simply say that electrons are negatively charged.

The nucleus of the atom contains particles called *protons* and *neutrons*. These particles are about 1,845 times heavier than the electron. The proton exhibits an opposite and equal relation to the electron and has a positive electrical charge. Neutrons have no electrical charge—that is, they are *neutral*. Other particles and forces exist in the nucleus, and nuclear scientists spend a great deal of time studying them.

When two or more atoms unite chemically and the properties of the separate elements are changed, a *compound* is produced. The smallest unit of a compound is the molecule, which, as mentioned previously, contains two or more atoms.

Atoms have different amounts of electrons, protons, and neutrons. Indeed, it is the different number of electrons, protons, and neutrons that account for the fact that different elements have different physical properties. For example, the difference between gold and mercury—at least on an atomic level—is that gold has 79 electrons and protons, and mercury has 80 electrons and protons. Moreover, each has a different number of neutrons in its nucleus. Of course, gold and mercury differ greatly in their outward appearance.

The copper atom has 29 electrons, which rotate around its nucleus in distinct groups called *rings*, or *shells*, much as planets rotate around the sun (fig. 2). The copper atom has four separate rings of electrons rotating around its nucleus. The inner ring has two electrons, the second has eight, the third has 18, and the fourth has one. The outer ring may not hold all the electrons it is capable of holding. Put another way, the outer ring may not be completely filled with electrons. This phenomenon is true for the copper atom, because the outer ring has only one electron, which leaves it unfilled. Consequently, the electron in the outer ring is so loosely bound to the nucleus that it can migrate from one atom to another and is called a *free electron*.

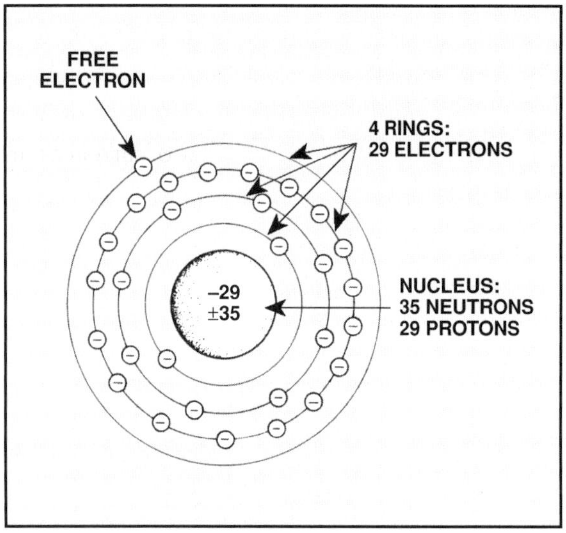

Figure 2 A copper atom has 29 electrons.

Positive and negative, or opposite, charges attract each other, and like charges (negative and negative or positive and positive) repel each other. Because electrons are negatively charged and the proton in the nucleus is positively charged, it would seem that the negatively-charged electrons would go to the positively-charged nucleus and remain there. However, the electrons rotate, or orbit, the nucleus at such great speed that centrifugal force holds them away from the nucleus, just as centrifugal force holds the earth away from the sun.

If an atom has enough negatively-charged electrons to balance the positively-charged protons, it is electrically neutral. Normally, atoms are just that: neutral. If, however, an atom loses an electron to a neighboring atom—if the electron becomes a free electron—the atom has one more positive charge than negative charges. In such a state, the atom becomes positively charged. Such an atom is in a stressed and abnormal condition. So, it exerts certain forces in an attempt to obtain a free electron from another atom and get back to its normal state. Heat, light, and electrical pressure can furnish the extra energy to cause this exchange of free electrons among atoms.

Some atoms have a large number of free electrons, and others have very few. Copper, for instance, has 1.64×10^{24} free electrons per cubic centimetre (cm^3). Copper is therefore a good *conductor*—that is, it has many free electrons that flow between the copper atoms. On the other hand, atoms that make up rubber compounds have only three or four free electrons per cubic inch (in^3). Such material is a *nonconductor*, or an *insulator*. In a nonconductor, very little flow of free electrons occurs. Thus, a conductor's free electrons are not tightly bound to the nucleus, while a nonconductor's electrons are tightly bound to the nucleus.

VOLTAGE

The electrical pressure, or *potential*, that causes electrons to move in a conductor is *electromotive force* (abbreviated *emf*). Often, the pressure that is emf is called *voltage* or, in some cases, potential. The letters *E* or *V* usually stand for voltage. The *volt* is the unit of measurement for voltage. A volt is the amount of electrical pressure that forces 1 ampere (amp) of current through a resistance of 1 ohm. Batteries and generators are two common sources of voltage.

The letter *E* usually abbreviates voltage that comes from a battery, generator, or transformer. The letter *V* usually abbreviates the voltage that exists across the *terminals* (a terminal is one of the electrical input or output points of a circuit or a component) of an electrical component. This voltage is often referred to as the *voltage drop* or *load voltage*. Also, *potential difference* is another term for voltage that flows from one terminal to another in a component.

CURRENT

Visualize a conductor, such as a copper wire, connected to the terminals of a battery (fig. 3). One end is attached to the battery's positive terminal and the other end to the negative terminal. The potential difference between the terminals causes the conductor's free electrons to move from the battery's negative terminal to the positive terminal. As these free electrons flow from one terminal to the other, they leave spaces, or *holes*, in the atoms. The holes also move, but in the opposite direction from the electrons. Thus, the holes leave the positive terminal of the battery and return to the negative side of the battery. Hole flow is exactly opposite from free electron flow.

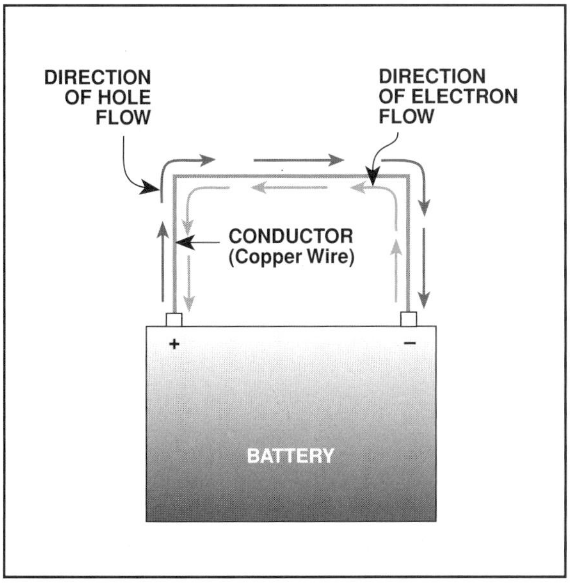

Figure 3 Battery with a conductor connects to + and – terminals. Note direction of electron flow and hole flow.

In electronics, conventional current flow is shown as hole flow and not electron flow. Consequently, *circuit diagrams* (*schematics*) for electronic devices show current flowing from the positive terminal to the negative terminal (fig. 4). Because current flows from the positive terminal of the voltage source, the positive terminal is the *hot terminal.* The *negative terminal* is the return terminal and is sometimes *grounded*—that is, it is connected to the earth or to a conducting body that serves in place of the earth. Be aware that current flows between the terminals of a battery through an external path, such as a length of conductive wire, in the same direction that it flows where no external path is provided, such as between the terminals of a single cell.

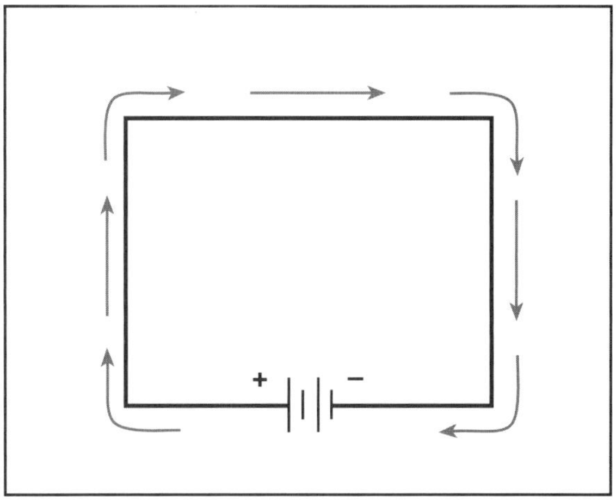

Figure 4 Circuit diagram showing current flowing from + to – terminal of battery.

The unit that indicates the amount of current flowing through a conductor is the *ampere*, which is often shortened to *amp*. An ampere is defined in terms of the rate of current flow. Specifically, 1 amp of current flow equals 6.28×10^{18} electrons (or holes) flowing past a point in one second. Current flow is similar to fluid flow in, say, barrels per hour (bbl/hr) or cubic metres per hour (m^3/hr). Since nothing happens instantly, most electrical occurrences are rated in reference to the time it takes for them to happen. The letter I is a symbol for current in amperes and represents the intensity of the current flow.

RESISTANCE

The amount of current passing through a conductor depends on the electrical pressure applied to it and on the nature of the conductor's material. Materials with a great number of free electrons make good conductors. Materials with characteristics that lie between being good conductors and being nonconductors are *semiconductors*.

A material's ability to conduct current is called *conductivity*, and its ability to oppose the flow of current through it is called *resistivity*. Most materials are characterized by their resistivity rather than their conductivity. All materials have some ability to oppose the flow of current, except at extremely low temperatures.

In practice, no perfect conductor exists—that is, all conductors have the property of resistance to some degree. Resistivity to current flow is a property of a material, just as color, hardness, or weight are properties of a material. *Resistance* is a measurement of resistivity, which is usually represented by the letter R and is measured in units called *ohms*.

INTRODUCTION TO OHM'S LAW

The relationship between voltage, current, and resistance states that the amount of current flow in an electrical circuit is directly proportional to the amount of voltage and inversely proportional to the amount of resistance. In equation form, this relationship known as Ohm's law is—

$$I = \frac{E}{R} \qquad \text{(Eq. 7)}$$

where

I = current, amperes
E = voltage, volts
R = resistance, ohms.

Rearranging this equation, or solving the equation for each value, results in—

$$E = IR \qquad \text{(Eq. 8)}$$

and

$$R = \frac{E}{I} \qquad \text{(Eq. 9)}$$

INTRODUCTION TO POWER LAW

When voltage forces current to flow through resistance, heat is produced. Voltage can also force current flow through a direct current (DC) motor and produce speed and torque from the motor. The product of voltage and current to produce useful work is referred to as *power*, which is usually represented by the letter P. Power in a DC circuit is expressed in units of *watts* or in large quantities, *kilowatts* (*KW*).

In equation form, power is—

$$P = EI \qquad \text{(Eq. 10)}$$

where

P = power, watts
E = voltage, volts
I = current, amps.

Because the letter V can also stand for voltage, the power equation can also be stated as—

$$P = VI \qquad \text{(Eq. 11)}$$

where

P = power, watts
V = voltage, volts
I = current, amps.

If a lot of power is involved, then the power equations can be expressed in kilowatts, or thousands of watts, as—

$$P = \frac{EI}{1{,}000} \qquad \text{(Eq. 12)}$$

or

$$P = \frac{VI}{1{,}000} \qquad \text{(Eq. 13)}$$

REVIEW QUESTIONS

1. Define matter.
2. What are the three states of matter?
3. What is an element? Name several.
4. What is the difference between a compound and a mixture?
5. What is an atom? A molecule?
6. What does the nucleus of an atom contain?
7. What causes the different physical properties of elements?
8. What is meant by a ring, or shell?
9. In what ring, or shell, is the free electron found?
10. Under what conditions is an atom electrically neutral?
11. When is an atom positively charged?
12. Define the following in terms of free electrons: good conductor, semiconductor, and nonconductor, or insulator.
13. What is an electric current? What unit is used to express the rate at which electric current flows?
14. What is conductance and resistance?
15. Explain the difference, in terms of free electrons, between materials having low resistance and those having high resistance.
16. What is the unit of resistance?
17. What is voltage? What unit is used to express voltage?

3 Galvanic Cells and Batteries

Electrical and electronic circuits require a source of voltage to function. Current does not flow unless a source of electrical pressure acts on the circuit components. Direct current (DC) sources include electrochemical devices such as *galvanic cells* and *batteries*. Although solid-state devices are another source of DC, they are discussed in another section. Many devices deliver voltage, including *electrochemical cells*, or batteries, *photovoltaic cells*, and *thermoelectric devices*. Each of these devices produces DC voltage that can deliver current to components and equipment.

GALVANIC CELLS

An electrochemical, or galvanic, cell is built by using the proper chemicals and two dissimilar materials immersed in, or surrounded by, the chemicals. A conductor or an electronic device is attached to each end of the dissimilar materials. The place where the connections are made is called a terminal. One terminal has a negative charge and the other has a positive charge. The negative terminal has an excess of electrons, and the positive terminal has a lack, or deficiency, of electrons. This situation sets up an energy difference between the two terminals, which is called potential or potential difference.

If an external path or a component is connected to the terminals, the potential causes the free electrons and corresponding holes to move through the external path or component. The holes flow from the positive terminal to the negative terminal of the cell. This movement of holes in one direction is referred to as *current*.

A device consisting of a single-positive and a single-negative terminal along with activating chemicals is referred to as a *cell*. A cell is the smallest unit of this type of device. A typical cell contains sulfuric acid as its *electrolyte* and two dissimilar terminals made of lead and a lead compound. This cell is a *lead-acid cell*. It produces a voltage of about two volts. (An electrolyte is a chemical that, when dissolved in water, dissociates into positive and negative *ions*, thus increasing its electrical conductivity. An ion is simply a positively- or negatively-charged atom.)

BATTERIES

A battery is made up of several galvanic cells connected in *series* to produce a voltage higher than that of one cell. (Connecting an electrical device or circuit in series means that current can travel only in a single path. Also, the current is the same throughout the device or circuit.) The kind of electrical charge at a cell or battery's terminal determines the terminal's *polarity*. A terminal's polarity is either positive or negative, depending on whether its electrical charge is positive or negative. That is, if the terminal's electrical charge is positive, the terminal is said to have positive polarity. Conversely, if the terminal's charge is negative, it has negative polarity.

A common lead-acid battery is an automobile battery. Typically, such a battery consists of six galvanic cells connected in series to produce a voltage at its terminals of 12 volts DC. One terminal is positive (+) while the other is negative (-).

Strictly speaking, a battery is formed only when two or more cells are connected. However, it is customary to call a single cell a battery. Although not strictly correct, this manual follows conventional usage and also refers to a single cell as a battery.

Construction of Batteries

Many years ago, humans discovered that if a person placed two different metals in a liquid so that they did not touch each other, and if the liquid reacted chemically with the metals, a potential difference, or voltage, was set up between the metals. This arrangement is a cell. For example, a lead-acid cell is made with a lead plate and another plate with lead peroxide pressed into its grids. The negative plate is the lead plate, and the positive plate is the lead peroxide plate. The liquid is a solution of sulfuric acid and water and is an electrolyte. This combination produces a potential difference (a voltage) of 2.1 volts at the terminals of the cell when fully activated.

When the same materials from a cell are used, the voltage is the same, even if the size of the material varies. It is the nature of the materials that determines the voltage of a cell, not the quantity or size of the material. For example, a small AAA (triple A) flashlight cell and a large number 6 dry cell both furnish 1.5 volts, yet the number 6 cell is many times bigger than a triple A battery. While the quantity of materials does not affect the amount of voltage, it does determine the amount of current (amperes) that a cell can furnish. The number 6 dry cell can furnish a greater current for a longer time than the small flashlight cell.

Earlier, the manual mentioned that, in order to make a battery cell, two dissimilar metals must be placed in a liquid (an electrolyte), and that the liquid must react chemically with the metals. However, as voltage flows through the electrolyte from one of the metals to the other, the active material in the liquid is chemically transformed into another compound. When the *chemical reaction* is complete—that is, when all the active material in the solution has been transformed chemically into another chemical, the battery is dead.

Also, sometimes the electrolyte is not a liquid. Instead, it is formed from a dry, solid material. So, some batteries, like flashlight batteries, are *dry cells*—the terminals are not in a liquid, but rather, in a dry electrolytic material. On the other hand, car batteries, in which the terminals are placed in a liquid solution of sulfuric acid, are *wet cells*.

Dry cells, such as flashlight batteries, are *primary cells* while those with liquid electrolyte are *secondary cells*. Primary cells cannot be recharged and are discarded. However, secondary cells, such as a lead-acid car battery, can be recharged from an electronic *battery charger* or another source of DC voltage and current. By passing a current from an external source through the battery in the reverse direction, the chemical action can be reversed and the active chemicals driven back into the solution.

The rating of a battery can be expressed in several ways, including its voltage, maximum continuous current over an 8-hour period, *ampere-hour rating* and *cold cranking amperes (CCAs)*. Typically, the larger the battery, the more amperes it delivers.

Batteries Connected in Series

When cells or batteries are connected in series, their voltages add up (fig. 5). The purpose in connecting batteries in series is to increase the voltage. When cells or batteries are connected in series, unlike potentials are connected—that is, a positive to a negative. When the batteries are properly connected, the terminals are opposite in polarity.

Batteries Connected in Parallel

A *parallel circuit* is a circuit with two or more paths through which the current flows. The total current equals the sum of the currents through the various paths. However, the voltage is the same throughout the circuit. So, connecting batteries in parallel increases the amount of current that can be taken from

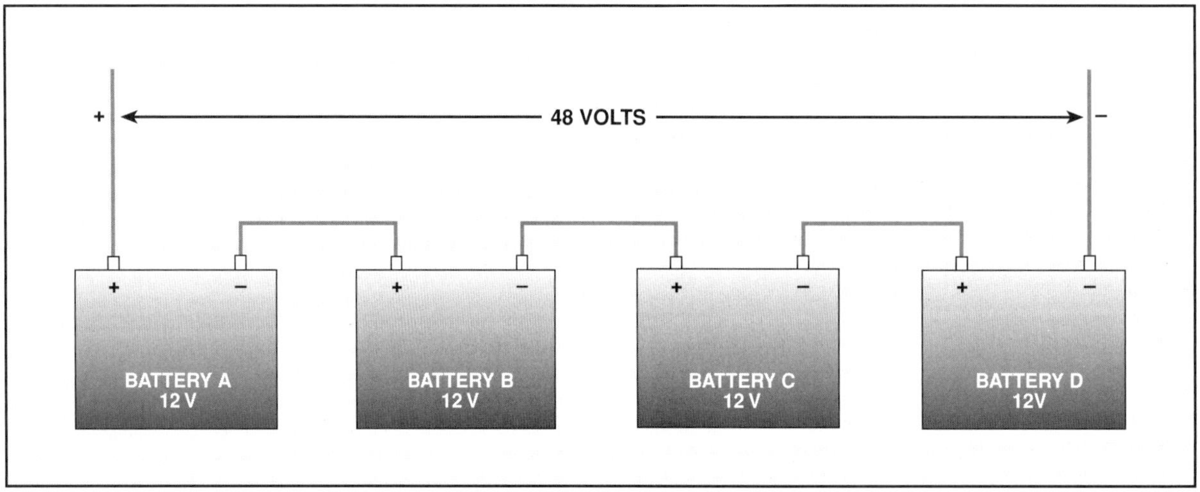

Figure 5 Four batteries connected in series. Note that current flows in only one path.

the battery (fig. 6). Batteries of equal voltages must be used in the parallel circuit.

As an example, a number 6 dry cell can furnish 1 ampere for 30 hours. Should a need occur for a current flow of 2 amperes for a period of 30 hours, two number 6 dry cells can be connected in parallel to meet this need. (Note that connecting batteries in parallel does not increase the voltage.) When two batteries of equal ratings are connected in parallel, the supply of chemical material available to be transformed into electrical energy doubles. Electrically, this doubling means that the batteries yield the same voltage but deliver more amperes for the electrical loads.

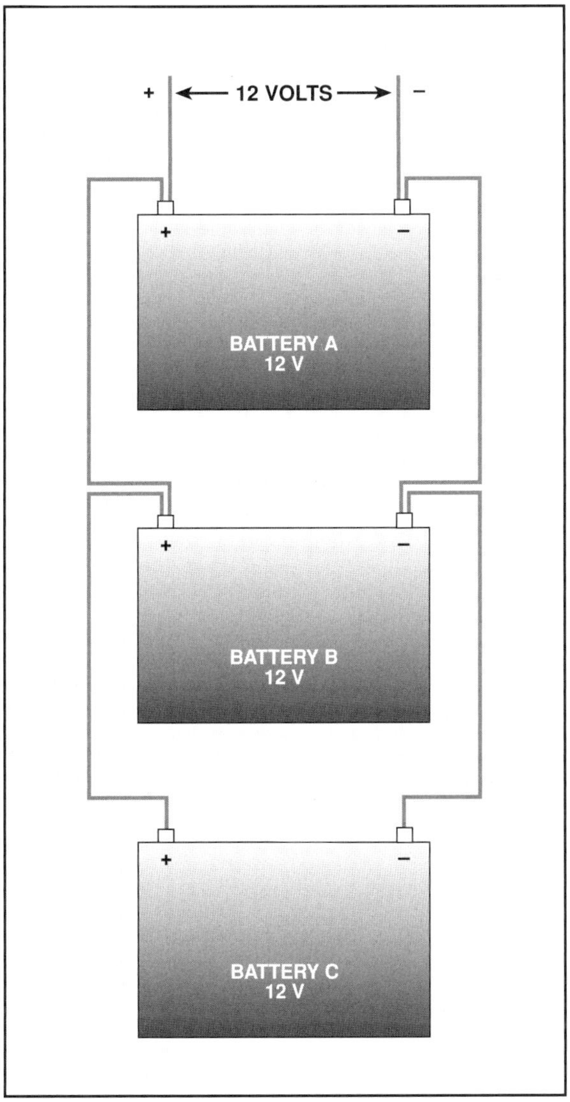

Figure 6 Three batteries connected in parallel. Note that current flows through two paths, one on the battery's positive side, and the other on the negative side.

A typical 12-volt car battery can supply 1 ampere for 100 hours. With two of these batteries connected in parallel, the voltage is still 12, but the system can furnish 2 amperes for 100 hours. The additional current from a second, or booster, battery can make it easier to start a car or truck in very cold weather.

A measurement taken from the negative terminal of one battery to its positive terminal yields +12 volts; the same measurement taken on the next battery also yields +12 volts. The difference between +12 volts and +12 volts is 0. So, no potential difference exists between the two terminals. Since no potential difference occurs, the terminals may be joined together without any current flow. No potential difference means no current flow.

On the other hand, batteries of unequal voltage cannot be connected because a potential difference exists at the terminal points. If two unequal voltage batteries were connected, heavy current would flow from one battery to the other. In other words, the higher voltage would discharge into the battery with the lower voltage.

Battery Ratings

Most commercially available lead-acid batteries are rated at 12 volts. However, other lead-acid batteries that are rated for 8 or 6 volts may be purchased. Within certain limits, batteries can be constructed in a variety of voltages in multiples of 2 volts.

The short-time current rating for a battery is typically rated in cold cranking amperes (CCAs). A typical battery's CCA can be hundreds of amperes for several seconds, which may be required to crank, or start, a cold engine. For a battery that is constantly being used, its maximum continuous current is its ampere-hour rating divided by 8 hours. An ampere-hour equals a flow of 1 ampere for one hour.

Charging Batteries

Batteries become discharged after the ampere-hour rating of use is consumed. To put it another way, if a battery is rated for 100 ampere-hours, and it is continuously delivering 10 amperes, it becomes fully discharged in about 10 hours. The battery will require recharging with an electronic battery charger.

To recharge a battery, the charger's positive lead is connected to the battery's positive terminal and the charger's negative lead is connected to the battery's negative terminal. The charger delivers the required ampere-hours to replace those that were used in operation. The charging rate can be either a low-ampere *trickle charge* or it can be a fast charge.

Battery Safety

Batteries can cause injury or damage if precautions are not observed when working around them. Since the batteries contain acid, avoid any contact with the skin or other material, such as clothing, that acid may damage. In wet cell batteries, the flammable hydrogen gas can escape from the battery vents. So, be sure no open flames, or anything that can cause a spark, is near the battery. Otherwise, an explosion or fire can result. Avoid any sparks at the battery terminals.

Overcharging or overdischarging can produce internal heat in batteries, which may produce high pressure and rupture of the battery cases. Avoid overcharging batteries and do not allow them to become overdischarged.

REVIEW QUESTIONS

1. Can a primary battery be recharged?
2. Name three electrical ratings of a battery.
3. State the components of a lead-acid battery.
4. Series batteries are used for what purpose?
5. Paralleled batteries are used for what purpose?
6. How should battery chargers be connected to batteries for recharging?
7. What type of gas is generated from a battery during operation and is it dangerous?

Ohm's and Power Laws

CIRCUITS

A *circuit* is a complete path in which current will flow. The path is from one battery terminal, or another source of voltage, back to the other battery terminal. The three elementary circuits are *simple*, *series*, and *parallel*. A simple circuit consists of a single path from one terminal of a source of electricity, through one electrical device, and back to the terminal of the electricity source. For example, one simple circuit consists of a battery, a light bulb, and a conductor such as copper wire. Current flows from the negative terminal of a battery, through the copper wire and light bulb, and back to the battery's positive terminal. A series circuit consists of a single path from the battery or source of voltage through two or more electrical devices back to the other terminal of the battery. A parallel circuit consists of two or more complete and separate paths for current flow.

Incidentally, circuits often contain *resistors*. Resistors are electrical devices that oppose the flow of electricity through a circuit. Besides the actual electrical component that is called a resistor, other devices in a circuit create resistance when they convert electrical energy into another form. For example, an incandescent lamp creates resistance as it converts electrical energy to light, and an electric heater creates resistance as it converts electric energy to heat.

OHM'S LAW

Ohm's law states that the strength or intensity of an electrical current (I) is directly proportional to the electromotive force (E) and inversely proportional to the resistance (R) of the circuit. Ohm's law relates voltage, which is the electrical pressure that causes the current flow; current, which is the rate at which electrons flow; and resistance, which is the opposition that a circuit offers to the flow of electrons.

Ohm's law can determine the amount of current that will flow in a circuit. In this case, use $I = E/R$ to find the current in amperes. To determine the amount of resistance, in ohms, in a circuit to limit current to a given value, use, $R = E/I$. Finally, to determine the voltage needed to create a flow of current through a given resistance, use $E = IR$. When two variables (voltage, current, or resistance) are known, Ohm's law provides an easy way to find the third by selecting the Ohm's law equation that has the unknown on the left side.

A drawing is available for finding the appropriate Ohm's equation to solve a problem. Called the Ohm's law wheel (fig. 7), it shows how to select the correct equation. For example, if values for E and R are known, then it shows that $I = E/R$. Similarly, if values for I and R are known, then the wheel shows that $E = IR$.

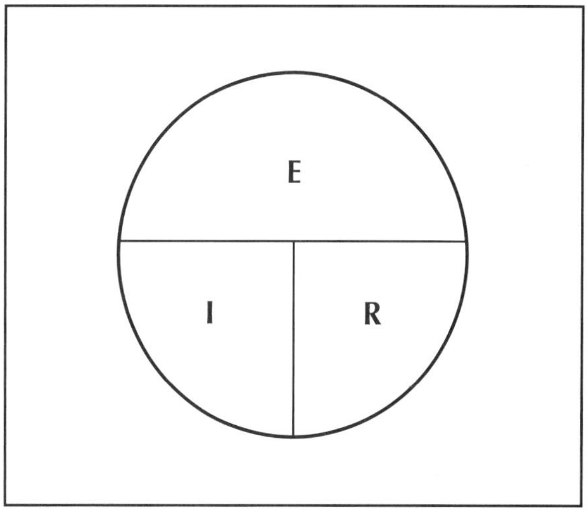

Figure 7 Ohm's law wheel

POWER LAW

As you recall, besides Ohm's law, the *power law* is available to determine the power a circuit is capable of producing. Like Ohm's law, the power law also consists of three variables; voltage, E or V, measured in volts; current, I, measured in amperes; and power, P, measured in watts. In equation form, the power law is $P = EI$, or $P = VI$. To determine the amount of power in watts produced in a circuit component or the supply, use $P = VI$ or $P = EI$. To determine the amount of current in amps required to produce the power, use $I = P/E$. To determine the voltage in volts needed to produce a given amount of power, use $E = P/I$. As long as two of the variables of voltage, current, or power are known, the power law provides an easy way to find the third by selecting the power law equation that has the unknown on the side. Just as for Ohm's law, a power law wheel (fig. 8) is available for finding the appropriate equation.

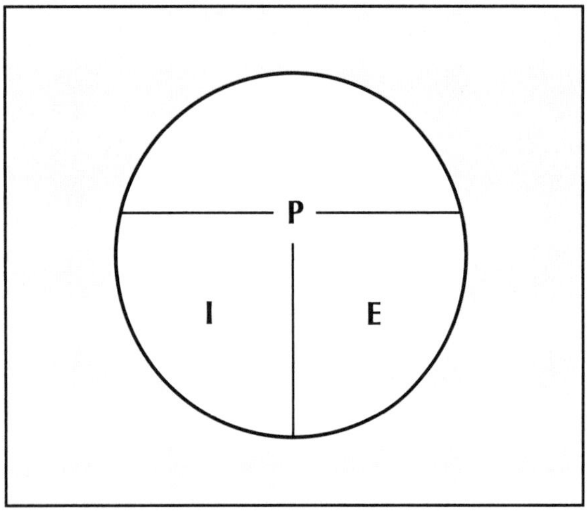

Figure 8 Power law wheel

OHM'S LAW EXAMPLES

Voltage

An electrical device is rated at a certain current and resistance. Often, it is necessary to know the voltage required to cause current to flow through the device. Ohm's law in the form $E = IR$—that is, where electrical pressure (E) equals current (I) times resistance (R)—can provide this answer. For example, what is the voltage needed to force 1.2 amperes through a device whose resistance is 80 ohms? To solve it, substitute the known values in the equation. Thus, $E = 1.2$ volts times 80 ohms, or $E = 1.2 \times 80$, which equals 96 volts. Therefore, 96 volts are required to force 1.2 amps through a device with 80 ohms of resistance.

Another example: a relay has a resistance of 100 ohms and requires 0.01 amperes to operate. What voltage is required? The solution is $E = IR$, where E = unknown, $I = 0.01$ amperes, and $R = 100$ ohms. Substituting known values in the equation yields $E = 0.01 \times 100 = 1$ volt.

Resistance

To calculate the resistance of a device, it is necessary to know the amount of current that flows through it when a certain voltage is applied. To obtain the answer, use Ohm's law in the form of $R = E/I$, where resistance (R) equals voltage (E) divided by current (I). For example, what is the resistance of a light bulb if the applied voltage in the circuit is 125 volts and the current through the bulb is 2 amperes? To solve it, substitute the known values in the equation so that $R = 125/2$, which means that $R = 62.5$ ohms.

Note that when a light bulb is not energized (it is turned off), the resistance of its *filament* is low: about 20 ohms for a 60-watt bulb. In this case, the resistance of the bulb can be measured with an *ohmmeter*. (An ohmmeter is an instrument with special *leads*, or wires, which, when placed in the circuit on both sides of a resistor, directly measures the resistance in ohms of an electrical circuit.) When the bulb is operating in a circuit, the filament becomes very hot: about 2,000°F (1,093°C). This change in temperature causes a large change in the filament's resistance. The way in which an ohmmeter works keeps it from being used in an energized circuit. So, in this case, you can use Ohm's law to calculate it. Unlike measuring resistance, however, instruments are available that can measure voltage and current in energized circuits.

Current

The current that flows in an electric circuit is equal to the voltage divided by the resistance. The equation is $I = E/R$, where, as usual, I = current in amperes, E = electrical pressure in volts, and R = resistance in ohms. Thus, if we wish to know how much current flows through an electric heater that has a resistance of 10 ohms if the heater is connected to a 120-volt power outlet, we simply substitute the known values in the equation. The equation then reads $I = 120/10$, which works out to 12 amperes.

Power

The power that produces heat in a resistive component such as a lamp is equal to the current times the voltage. The equation is $P = VI$ or $P = EI$, where P = power in watts, V or E = electrical pressure in volts, and I = current in amperes. So, if we want to know how much power is dissipated in an electric heater that is connected to a 120-volt power outlet with a current rating of 12 amperes, we simply substitute the known values in the equation. The solution is $P = 120 \times 12$, which equals 1,440 watts or 1.44 kilowatts.

SIMPLE CIRCUITS

A simple circuit consists of a single path from one terminal of the source of electromotive force, through one electrical device, and back to the other terminal of the source of electromotive force. For example, figure 9A shows a single path from one battery terminal going to a lamp and back to the battery's other terminal. This circuit is a simple circuit because it has only one device—in this case, a lamp. Current flow is from the positive terminal of the battery, through a copper wire to the light bulb, through the filament in the bulb, and through a wire back to the negative terminal of the battery.

As long as the circuit is not broken and the battery can furnish current, the light bulb glows. The filament of the bulb is tungsten wire that can withstand high temperatures without melting. The wire is also small in diameter and is relatively long because it is coiled. Most of the resistance of this circuit is concentrated in the filament. The connecting wires from the battery are made of copper and are large enough in diameter—they have a heavy *gauge*—to have low resistance. Wire gauges are stated in numbers and commonly range from 4 to 18. The smaller the gauge number, the larger is the diameter of the wire. For example, number 6 wire has a diameter larger than number 12 wire. The larger the wire's gauge, the less resistance to current flow it offers. Thus, with the heavy gauge wires in figure 9A, very little of the battery's voltage is applied to the ends, or terminals, of the filament.

On the other hand, the filament has high resistance. High resistance means that fewer free electrons exist in the filament's material. Also, the free electrons that exist in the material must travel at a higher velocity than they do in the copper conductors. The high-velocity electrons have collisions with the atoms in the filament's material, and the energy of motion (*kinetic energy*) of the electron is transformed into heat. These collisions are similar to a high-speed bullet striking a steel plate that it cannot penetrate. The bullet hits the plate with so much energy that it melts. The transformation of the energy of motion of electrons into heat in the filament causes the filament to become white hot and give off light. The process must be continuous for the light bulb to continue to glow. In other words, the current must flow continuously through the filament for the filament to continue to glow. Without current flow, no electron collisions occur, no heat is given off, and no light occurs. Keep in mind, though, that it is the resistance of the filament that limits and controls the amount of current that flows in the circuit.

Figure 9B is a schematic diagram of the simple circuit in figure 9A that uses symbols to represent each component. The diagram uses a battery symbol at

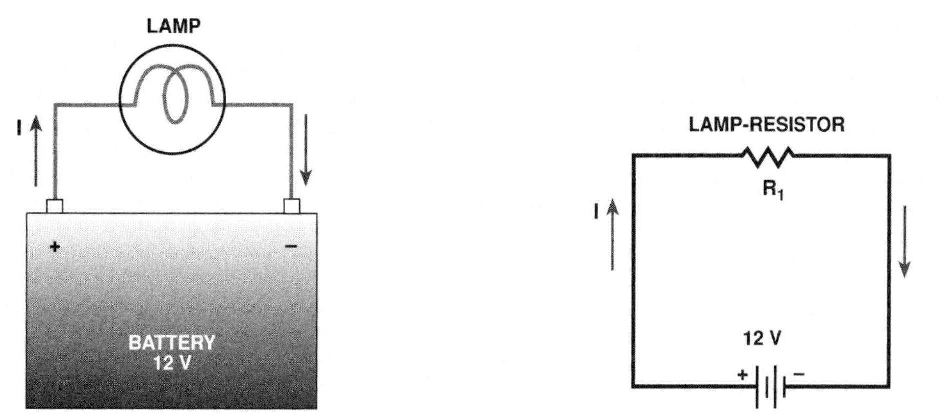

A. Simple Circuit with Physical Components

B. Simple Circuit Using Component Symbols

Figure 9 A simple circuit with a battery, lamp, and copper wire

bottom, and is labeled 12V, which abbreviates 12 volts. So, the schematic shows a 12-volt battery in the circuit as the source of electrical pressure, or electromotive force. Figure 9B also uses a symbol for a resistor, labeled R_1, which represents the lamp. The straight heavy lines represent the copper wire. Schematic diagrams normally present electrical circuits because they are simpler than a detailed drawing of the circuit's components as in figure 9A.

A symbol can represent virtually every electrical component in a circuit diagram. In general, a rectangular box, with descriptive terms beside it, represents fixed equipment such as control boxes, cabinets, and similar devices. Circular symbols represent round or rotating devices, such as lamps, coils, motors, and generators.

In circuit diagrams, components often are designated with an abbreviation and a reference designation such as that of the lamp-resistor, R_1, in figure 9B. Reference designations can be employed when using Ohm's law to solve for the current in a circuit, and when describing a circuit, troubleshooting it, maintaining it, and ordering replacement parts.

The current in the circuit shown in figure 9A–B can be determined using Ohm's law if the battery voltage is 12 volts DC and the lamp resistance is 6 ohms. Thus, $I = {}^{12}/_{6}$, which equals 2 amperes. Further, the *voltage drop* in the circuit, measured across the lamp, is $V = IR$, which is $2 \times 6 = 12$ volts. In other words, the voltage developed across the resistor (the voltage drop) is the same as that delivered from the battery (the *supply voltage*).

The power dissipated in the lamp in figure 9 can also be determined using the power law. In this case, it is $P = VI$, which works out to $P = 12 \times 2$, or 24 watts.

SERIES CIRCUITS

When more than one component is added in sequence to another, a series circuit is created (fig. 10). For example, if four resistive components are placed end-to-end and the resulting circuit load terminates at a DC supply, current flows in each component. The resulting current is the same in each component as well as the supply source. Current flows from the positive terminal of the battery, through each resistor in series, and returns to the negative terminal of the battery.

To solve for current in the series circuit in figure 10, the supply voltage must be known as well as the values of the resistors.

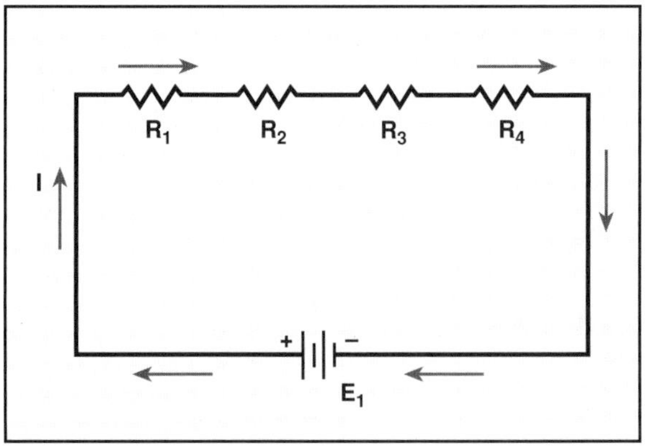

Figure 10 A series circuit

Resistance in Series Circuits

It is important to understand that the resistance used to determine the current in the circuit is not that of an individual resistor but the total circuit resistance. In equation form, the total circuit resistance is—

$$R_T = R_1 + R_2 + R_3 + R_4 \qquad \text{(Eq. 14)}$$

where

R_T = total resistance, ohms.

For example, if R_1 is 50 ohms, R_2 is 10 ohms, R_3 is 40 ohms, and R_4 is 20 ohms, the total resistance is 50 + 10 + 40 + 20 = 120 ohms.

Current in Series Circuits

With the known total resistance and voltage in the circuit, current can be calculated using Ohm's law. So, for the circuit in figure 10, if the total resistance of a circuit is 120 ohms and E_1 is 240 volts, then $I = {}^{240}/_{120}$. Thus, current (I) is equal to 2 amperes.

Voltage Drops in Series Circuits

The value of the individual voltage drops across each resistor can be determined and the results added to obtain total voltage drop through the circuit. To determine individual voltage drops in the circuit in figure 10, take the circuit's current and multiply it by the resistance (R) of each resistor. Thus, the voltage drop across resistor R_1 is $V_1 = IR_1 = 2 \times 50 = 100$ volts.

The corresponding voltage drops across the remaining resistors are—

$$V_2 = IR_2 = 2 \times 10 = 20 \text{ volts}$$
$$V_3 = IR_3 = 2 \times 40 = 80 \text{ volts}$$
$$V_4 = IR_4 = 2 \times 20 = 40 \text{ volts.}$$

Adding the sum of the total voltage drops yields the following result—

$$V_T = V_1 + V_2 + V_3 + V_4 = 100 + 20 + 80 + 40 = 240 \text{ volts.}$$

The sum of the voltage drops in this circuit, then, is equal to the value of the supply voltage. In any electrical series circuit, the sum of the voltage drops equals the supply voltage.

Power Dissipated In Series Circuits

The power dissipated by each component in a series circuit is obtained from the power law. The power law states that the power in watts (P) is equal to the product of the current (I) through a component, times the voltage (V) across a component, or $P = IV$. In the circuit in figure 10, where several resistors exist, the power (P_1) in the first resistor (R_1) is $P_1 = IV_1 = 2 \times 100 = 200$ watts.

The corresponding power that is dissipated in the remaining resistors is—

$$P_2 = IV_2 = 2 \times 20 = 40 \text{ watts}$$
$$P_3 = IV_3 = 2 \times 80 = 160 \text{ watts}$$
$$P_4 = IV_4 = 2 \times 40 = 80 \text{ watts.}$$

Adding the sum of the individual resistor's power dissipation, the result is $P_T = P_1 + P_2 + P_3 + P_4 = 200 + 40 + 160 + 80 = 480$ watts.

The amount of power delivered by the DC supply is also determined by using the power equation. In this case, $P_1 = IE_{supply} = 2 \times 240 = 480$ watts. The sum of the power dissipated in each component in this circuit, which is 480 watts, is equal to the power delivered by the supply source, which is also 480 watts. In any electrical series circuit, the sum of the individual power losses is equal to the power delivered by the supply source.

PARALLEL CIRCUITS

A parallel circuit is a circuit in which two or more complete and separate paths are available for the current to flow. For example, the circuit in figure 11A has three current paths. Current leaves the positive battery terminal, flows through each bulb, and back to the battery. In a series circuit, if one bulb burns out or is removed, all the other bulbs will fail to light because the burned out, or missing bulb, interrupts current flow. On the other hand, in a parallel circuit, one bulb can be removed or burned out,

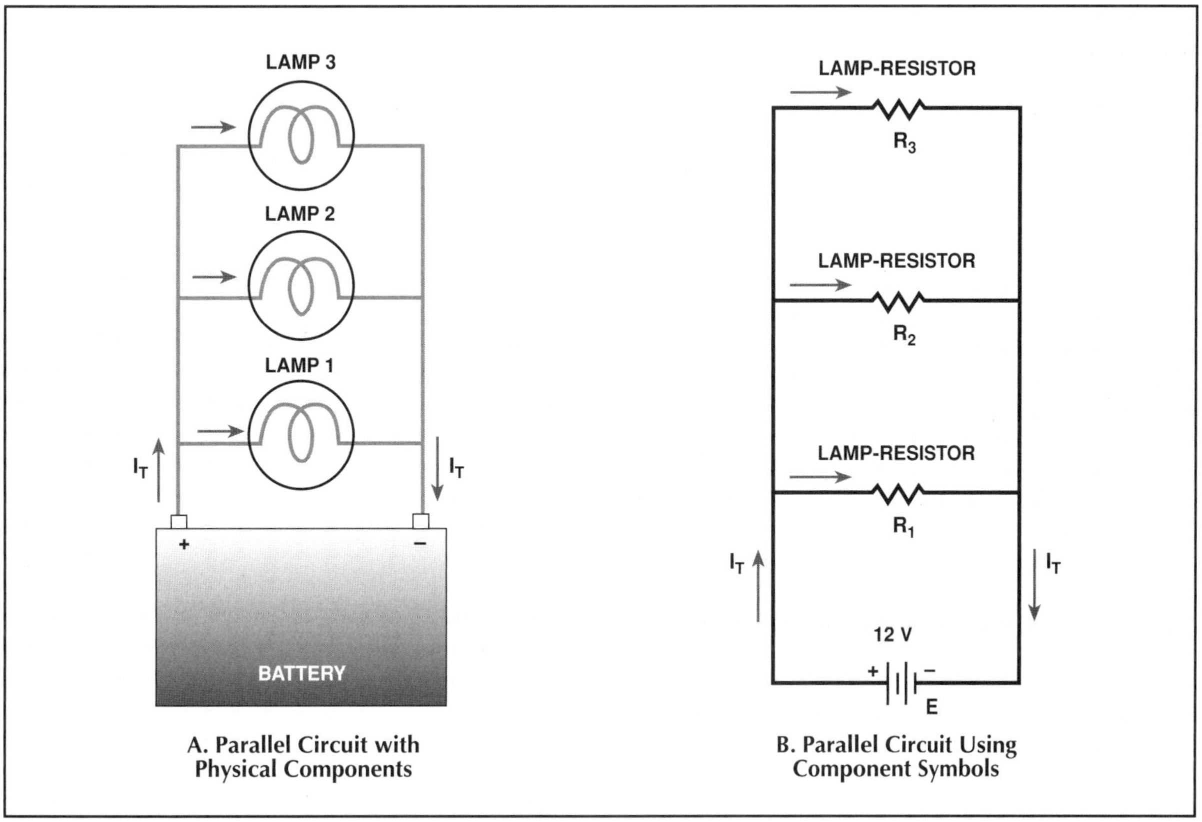

Figure 11 Parallel circuits

and the other bulbs will continue to burn. Current continues to flow through the alternate paths. Consequently, parallel circuits are widely used in the wiring systems of houses, automobiles, and other systems.

In a parallel circuit, the current leaving or returning to the battery divides through the number of paths used. The voltage across each branch of a parallel circuit is the same. The current through each branch of the circuit may or may not be the same, depending on whether the resistances of the branches are the same or different. One lamp could be a large bulb with a large amount of current flowing through it, while another might be a small bulb with correspondingly smaller current flowing though it. In any case, the three currents passing through the three bulbs in figure 11A add up to the current leaving and returning to the battery. In the parallel circuit, all lamps receive the full battery voltage.

Resistance and Parallel Circuit Currents

In the parallel circuit in figure 11B, the sum of the currents from the DC supply is equal to the sum of the individual branch, or parallel component, currents. In equation form, the total current is equal to—

$$I_T = I_1 + I_2 + I_3 \qquad \text{(Eq. 15)}$$

where

I_T = total current, amperes.

This relationship is true for any number of parallel circuit branches—that is, the sum of the currents (I_T) into parallel loads is equal to the supply current.

To determine the current in each individual branch using Ohm's law, the supply voltage (E) and the value of the branch resistance (R_1, R_2, R_3, etc.) must be known. With this information, the current is equal to $I_1 = {}^{E}/_{R_1}$. If only the supply current is to be calculated, and not the individual branch currents, the total circuit resistance, R_T, must be known. To determine total circuit resistance (R_T), the *reciprocal equation* is used. This equation is suitable for any number of parallel branches. The reciprocal equation is—

$$R_T = \frac{1}{\frac{1}{R_1} + \frac{1}{R_2} + \frac{1}{R_3} + \frac{1}{R_4} + etc.} \qquad \text{(Eq. 16)}$$

where

R_T = total resistance, ohms.

For example, suppose that the values for the lamp-resistors in figure 12 are—

R_1 = 50 ohms
R_2 = 10 ohms
R_3 = 40 ohms, and

that voltage (E_1) in the circuit is 240 volts DC. The current in branch 1 is $I_1 = {}^{E}/_{R_1} = {}^{240}/_{50} = 4.8$ amperes. The current in branches 2 and 3 are $I_2 = {}^{E}/_{R_2} = {}^{240}/_{10} = 24$ amperes and $I_3 = {}^{E}/_{R_3} = {}^{240}/_{40} = 6$ amperes. So, the total current from the supply is $I_T = I_1 + I_2 + I_3$, which equals 4.8 + 24 + 6 = 34.8 amperes.

If only the value of the supply current is needed, and not the individual branch currents, supply current can be determined with the resistance reciprocal equation for parallel circuits. So,

$$R_T = \frac{1}{\frac{1}{50} + \frac{1}{10} + \frac{1}{40}},$$

which works out to $\frac{1}{0.02 + 01 + 0.025} = \frac{1}{0.145} = 6.897$ ohms. Because resistance and voltage are now known, total current (I_T) can be determined with the equation $I_T = {}^{E}/_{R_T} = {}^{240}/_{6.897}$, which equals 34.8 amperes. The two methods of determining the supply current produced the same result of 34.8 amperes. The procedure to follow is determined by that which is desired in the analysis.

MALFUNCTIONING CIRCUITS

When something goes wrong with an electrical circuit, many people immediately blame the trouble on a *short*, which means a *short circuit*. In reality, two types of defective circuits are common. One is a short circuit and the other is an *open circuit*.

In one type of short circuit (fig. 12), defective insulation on the connecting wires allows bare wires to touch. Heavy current can then flow from the negative terminal of the battery to the point where the bare wires touch, and back to the positive terminal of the battery. The result is a short circuit. Very little resistance exists in this accidental circuit, so current flow is high; indeed, flow can be great enough to cause the wires to become red hot and burn in two.

The other type of faulty circuit is one in which one of the connecting wires has become broken or cut (fig. 13). All current flow ceases, so the lamp cannot burn. Such a fault is an open circuit.

When trouble occurs in a circuit, chances are good that it is either a short circuit or an open circuit. A short circuit blows fuses, trips circuit breakers, burns wires or components, and can create fire and smoke.

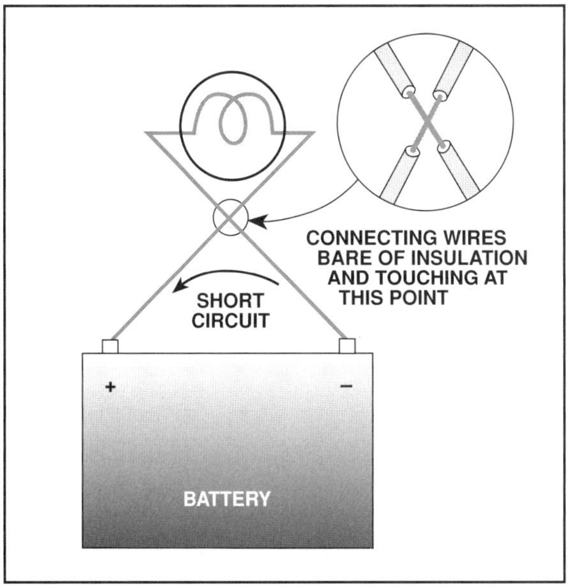

Figure 12 Short circuit

An open circuit shows up as equipment failure or malfunctioning components. Moreover, open circuits usually do not blow fuses or trip a circuit breaker.

SWITCHES IN CIRCUITS

A switch, such as the one in figure 14, is a handy mechanical device for opening or closing an electrical circuit. A switch starts and stops the action of a device in the electrical circuit. Figure 15 is a graphic representation of a switch in a simple circuit. Opening or closing the switch turns the bulb on or off. Figure 16 represents a switch in a series circuit. Operating the switch turns three bulbs, wired in series, on or off.

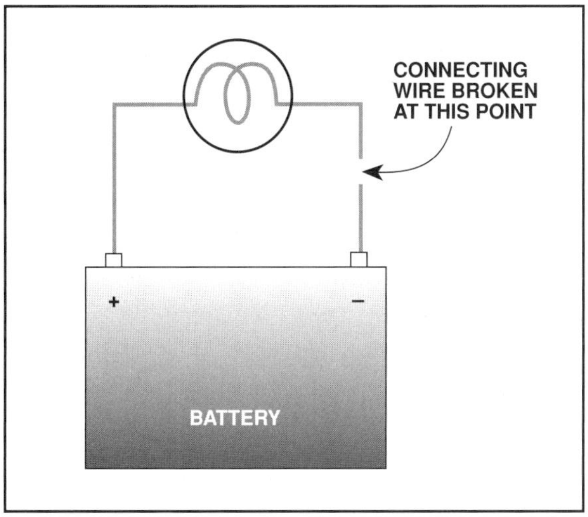

Figure 13 Open circuit

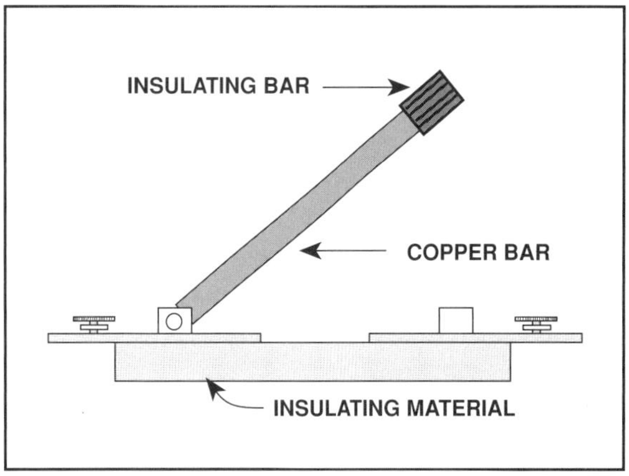

Figure 14 Simple switch

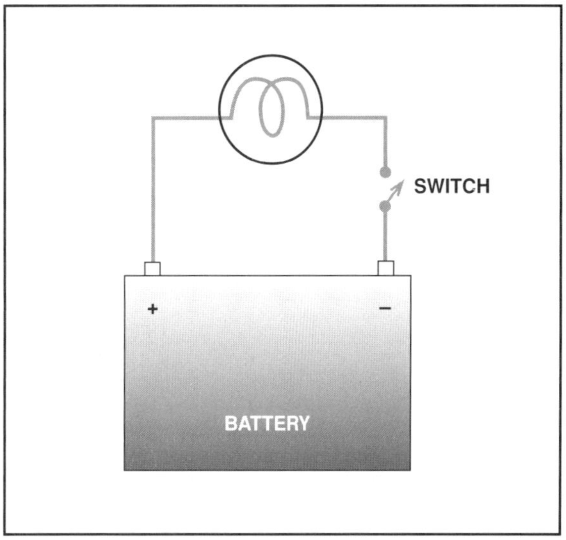

Figure 15 Switch in a simple circuit

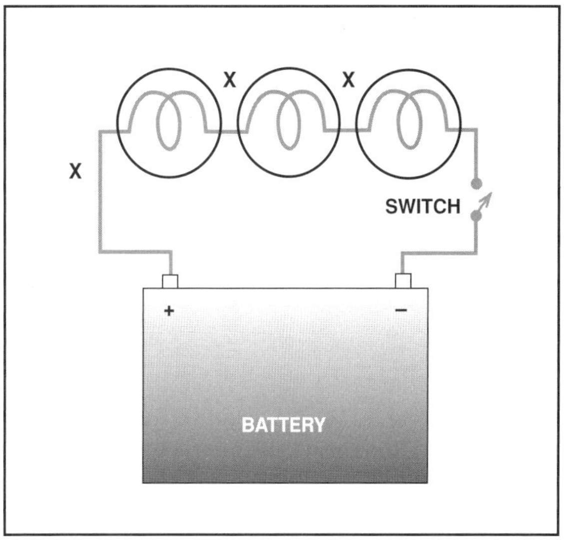

Figure 16 Switch in a series circuit

Further application of switches is shown in a parallel circuit (fig. 17). Switch 1 (S_1) is a *master switch* because it must be closed before the circuits controlled by switch 2 (S_2) and switch 3 (S_3) can be completed back to the battery. After S_1 is closed, S_2 and S_3 are used to turn on and off the lamps in their circuits. Put another way, the master switch can turn everything off, and the individual switches can turn on or off separate lamps or circuits.

The switches in figure 17 are the simplest of all and are called *single-pole, single-throw (SPST) switches*. In figure 17, they are in open position. Switches may be combined in one unit and operated by the same lever. For example, besides the single-pole, single-throw (SPST) switch, three others are the *single-pole, double-throw (SPDT)*; the

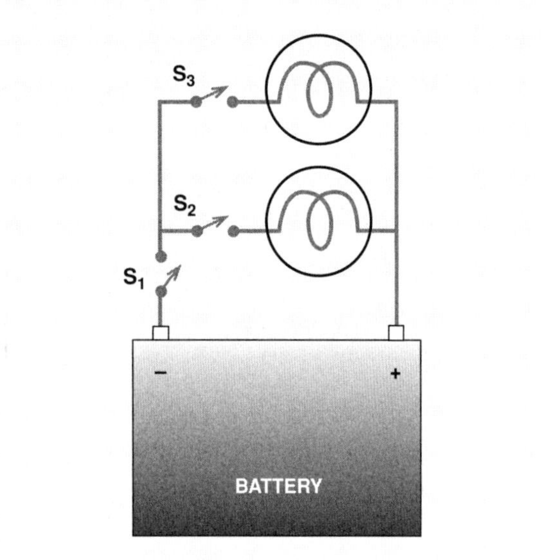

Figure 17 Switches in a parallel circuit

double-pole, single-throw (DPST); and the *double-pole, double-throw (DPDT)* (fig. 18). In the SPST switch, one side of the single switch is attached to the circuit and the switch is either open or closed on the other side of the circuit. In the SPDT switch, a single switch is open or closed on two sides of the circuit. In the DPST switch, two switches are combined and are operated on one side of the circuits. In the DPDT switch, two switches are combined and operate on both sides of the circuit.

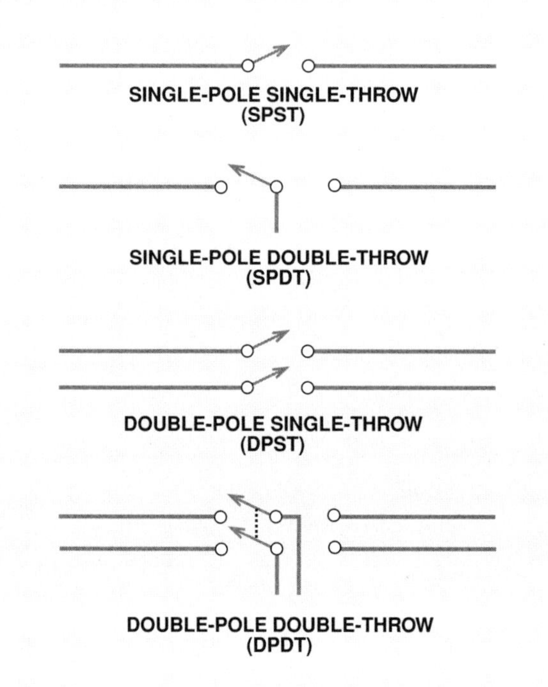

Figure 18 Four types of switches

REVIEW QUESTIONS

1. What is a circuit? Define a series circuit.
2. In an eight-bulb, 120-volt Christmas tree light string that is wired in series, what voltage is developed across each lamp?
3. Will two lamps connected in a series circuit burn with the same brilliance if they are of different sizes?
4. In a series circuit, what unit of electricity is the same through each unit?
5. When the voltages across several units in a series circuit are added, what is the sum equal to?
6. Draw a series circuit showing how a switch should be connected.
7. How many branches or paths does a parallel circuit have?
8. What is the same across each branch of a parallel circuit?
9. What is the sum of the currents flowing in the branches of a parallel circuit equal to?

5 Conductors and Insulators

CHARACTERISTICS OF MATERIALS

Various materials exhibit various electrical properties. For example, materials such as aluminum, copper, and silver have low resistance. Consequently, they make good conductors of electricity. Thus, they are often braided or drawn to make wire, or *cable*, for connecting electrical components. Connecting wire should have low resistance so that little of the available voltage is lost while the equipment is operating. Therefore, resistance of wires and cables should be as low as economically possible. Economics is an important factor. For example, silver has less resistance—is a better conductor—than copper or aluminum; however, silver is much more expensive, so it is seldom used.

Another consideration is that one conductor should not make contact with another conductor at a different voltage, or potential. A short circuit can develop, which is a malfunction. To prevent short circuits, insulating materials cover conductors at various points in the circuit. Rubber or plastic thick enough to be effective is one material that can coat wires and cables to prevent short circuits and the damage they cause.

Also, some wire and cable not only have plastic insulation for protection, but also an outer metal *cladding*, or coating. This cladding may be solid aluminum sheathing or braided copper. In either case, the cladding protects the cable or wire from mechanical damage—that is, it protects the cable when it is struck or crushed by an outside force such as a dropped tool, people stepping on it, or similar forces.

Figure 19 shows several insulated wires. The 300-ohm *ribbon line* at the top is often used to connect an antenna to a television set. The 72-ohm double-wire transmission line is used in various gauges to conduct alternating current in households, industrial plants, offices, and so on. *Coaxial cable* has many uses. It is called coaxial because several wire conductors are bundled around each other to form a single, although thick, wire. *Hookup wire* is used to wire electrical components together in equipment such as amplifiers and other electronic devices. *Magnet wire* is a copper conductor that is insulated with thin enamel or enamel-like coating. Prior to making a connection with it, this coating must be scraped away from the wire. Magnet wire gets its name from its use as windings in electromagnets, transformers, motors, and generators. Its insulation must be thin so that it can be wound several times around a core: thick insulation would not allow it to be wound an adequate number of times. So, instead of a plastic coating, varying grades of enamel or varnish are baked onto magnet wires to provide good insulating properties.

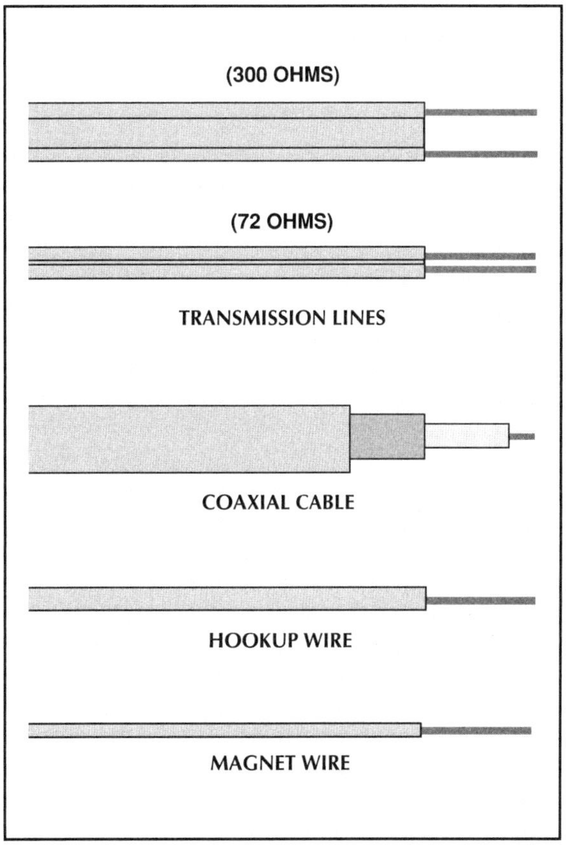

Figure 19 Examples of insulated wires

The insulation that covers a conductor depends on the temperature of the conductor and the voltage to be applied to the circuit. Consider, for example, spark-plug wires. In a spark-plug wire, the conducting wire is small in gauge because it conducts only about 0.001 ampere. The insulation, on the other hand, must be of good quality and be thick enough to withstand about 25,000 volts. Spark-plug wire insulation is much like a high-pressure pipe because a high-pressure pipe must also be made of high-quality materials and have relatively thick walls.

The conductors used in wiring and cable may be either solid or stranded. Solid wire is less costly, but, in large sizes, it is difficult to bend and flex. Stranded wire easily bends around corners in conduits or trays, so it is used where such flexibility is required.

Ribbon line can conduct high-frequency alternating currents from one point to another with minimum loss in signal strength. As previously mentioned, one popular use is to connect television antennas to the receiver. Ribbon line consists of two conductors covered with polyethylene insulation that is formed into a ribbonlike strip. Thin, flat, twin-lead transmission line is rated in terms of its characteristic surge *impedance*. Impedance is a measure of the total opposition to current flow in an *alternating current* (*AC*) circuit. (AC is current that flows in one direction in a circuit and then reverses flow. Reversal of current flow happens rapidly—for example, in many AC circuits in the United States, reversal happens 60 times in 1 second.) Because impedance is similar to resistance, it is measured in ohms. Thus, ribbon line is available with impedance values of 300 ohms, 150 ohms, and 75 ohms. The impedance of the wire is determined by the size of the wire and spacing between conductors.

Coaxial cable is constructed with a center wire that is covered with polyethylene insulation. The second conductor is copper braid that fits over the polyethylene insulation. Finally, a vinyl cover protects the braided conductor. Coaxial cable (*coax* for short) is used for the same purpose as twin-lead ribbon line but has the advantage of not being affected by outside electric and magnetic fields. It, too, is rated in terms of its characteristic impedance.

CURRENT-CARRYING CAPACITY OF CONDUCTORS

Earlier, this manual mentioned that wire is often sized by gauge, with lower gauge numbers having diameters larger than higher numbers. Another measure of wire diameter is *mil size*, which is based on the size of the cross-sectional area of the wire. (Mil size is further explained later in this section.) The gauge or circular mil size of wires used in circuits depends on the amount of current (amperes) that will flow in the circuit. If a circuit will carry large currents, then the connecting conductors must also be large in size. The size of the conductor varies directly not only with the current it must conduct, but also with the length of the conductor. The greater the current to be carried, and the longer the conductor is, the larger the conductor must be.

Current flowing through a wire is much like fluid flowing through a pipeline. If the volume of fluid passing through the line is large, then the pipe must be large. Further, if the pipeline is long, fluid pressure drops because of friction as the fluid moves against the pipe walls. Consequently, a large diameter pipe must be used to combat pressure losses.

A car's battery cables are somewhat like the large pipeline. They must be large because the starter circuit demands a high current; also, the voltage drop must be kept to a minimum. On the other hand, the car's headlamp circuit is more like a small pipeline; it is wired with smaller line because the headlights demand a smaller current.

As current passes through wire, heat is given off. This heating effect is proportional to both the resistance of the wire and the square of the current it is carrying. The electrical industry has established standards for the maximum current-carrying capacity of conductors for various types of service. The *National Electric Code*, NFPA 70, Article 310, is the primary standard used in selecting wire size and insulation for various applications.

Wire size is a measure of its cross-sectional area. As you now know, it can be given in terms of gauge, in which case it can be stated as *American wire gauge* (*AWG*). The wire's area increases as the AWG size decreases. On some wire tables, the smallest size is 40 AWG, which is so small it can carry only a small amount of current. AWG increases are indicated by double digit numbers lower than 40, such as 12 and 10, and single digit numbers, such as 4, 2, and 0. The largest AWG is quadruple zero, or 0000, which is usually stated as, "four ought," and is written as 4/0. AWG 4/0 can carry hundreds of amperes.

For wire or cable sizes beyond 4/0, some wire tables give them in *circular mils* (*cmil*) or *thousand circular mils* (*kcmil*). (A circular mil is the square of the wire's diameter in thousandths of an inch.) In countries where the metric system is used, wire size is expressed in square millimetres (mm^2).

As stated earlier, wire sizes in the United States are made in accordance with the AWG system. This system calculates the diameter and circular mil area for each gauge number. As the gauge number increases, the size of the wire decreases. Thus, No. 30 gauge wire is very small, with a circular mil area of 101. The next larger size wire is No. 29 gauge. It has a circular mil area of 127, which is 1.26 times 101 (the circular mil area of No. 30 gauge wire is 101). No. 28 gauge wire is 1.26 times 127, which is 160. This system continues to include all wire sizes.

A circular mil is the square of the wire's diameter measured in thousandths of an inch, or *milli-inches*. For example, if a stranded cable has 50 strands, each strand of which is 0.1 inches (in.) in diameter (*d*), the circular mil size is calculated by first determining that the strand's diameter (*d*) is equal to 100 milli-inches, because $0.1 \times 1{,}000 = 100$. Then, the circular mil size of a single strand is d^2, which is 100^2, or 10,000 cmils. This figure may also be expressed as 10 kcmils because $10{,}000 \div 1{,}000 = 10$. Finally, because the cable contains 50 strands, the cable's total size is 10×50, or 500 kcmils.

Table 8 lists the typical maximum current-carrying capacities for several sizes of copper wire. For example, a copper wire with an AWG of 12 can safely conduct 25.0 amperes of current (the shaded area in the table).

TABLE 8
Current-Carrying Capacity of Copper Wire

AWG Size	Current, Amperes
20	2.5
18	5.0
16	10.0
14	20.0
12	25.0
10	40.0
8	50.0
6	80.0
4	105.0

Keep in mind, however, that many factors come into play when engineers plan electrical installations. Therefore, a particular installation could require wires with more or less current capacity than shown in the table. For example, bare wires of a given size that transmit power in the open countryside can safely carry a great deal of current because they are well ventilated. The same size wire, however, could not carry nearly as much current when used as a winding in an unventilated transformer.

Length and Cross-Sectional Area of Conductors

Electrical circuits transform electrical energy into other forms of energy, such as heat or mechanical energy. Various kinds of conductors form circuits, and they all have the property of resistance. Four factors determine the resistance of a conductor—

1. the resistivity of the material,
2. the length,
3. the cross-sectional area (the gauge), and
4. the temperature.

As the length of a conductor is increased, its resistance increases, just as the resistance to fluid flow in a long pipe is greater than in a short one. Conversely, as the cross-sectional area of the conductor increases, the resistance decreases, just as the resistance to fluid flow is less in a large-diameter pipe than in a small-diameter pipe. The difference in resistivity is also similar to fluid flow through a slick plastic pipe versus a rough iron pipe. Flow through the plastic pipe is easier than through the iron pipe.

As you read earlier, because conductors are usually round, their cross-sectional area is usually measured in circular mils—that is, the conductor's diameter is measured in thousandths of an inch, multiplied by 1,000, and this number is squared. Thus, a wire 0.004 in. in diameter has a cross-sectional area of 16 circular mils, because $0.004 \times 1{,}000 = 4$, and $4^2 = 16$. Here the word mil means one thousandth of an inch, or 0.001 in. Wire or conductor that is 0.001 in. in diameter and 1 foot (ft) long is a *mil-foot*.

Conducting Materials

The resistance of a conductor also depends on the kind of material used to make the conductor. For example, a mil-foot of copper wire has a resistance of 10.4 ohms at 200°C (392°F), compared with 17.0 ohms for a mil-foot of aluminum wire. Copper is less resistant than aluminum because copper has more free electrons than aluminum. Table 9 lists the resistivities of several elements and alloys used for conductors. The relative merit of these materials as conductors can be seen from this table. Incidentally, *Constantan* is an alloy that is made of 55 percent copper and 45 percent nickel. *Nichrome* is the trademarked name of an alloy of nickel and chrome. Because of Nichrome's high resistance (600 ohms), it is often used for the elements in electric heaters.

TABLE 9
Resistivities of Conductors
(per mil-foot at 200°C)

Silver	9.9 ohms	Iron	75 ohms
Copper	10.4 ohms	Lead	132 ohms
Gold	14.0 ohms	Constantan	302 ohms
Aluminum	17.0 ohms	Mercury	577 ohms
Nickel	47.0 ohms	Nichrome	600 ohms

Copper and aluminum should never be joined together, or, as electrical engineers say, they should never be combined in a common termination. If they are connected, corrosion occurs at the point of the connection. The corrosion is *galvanic corrosion* because a small electric current flows from one conductor to the other. This electrical flow has nothing to do with the normal flow of electricity in the circuit. Rather, the flow is the result of a miniature galvanic cell, or battery, that is created when two different, or dissimilar, metals are joined. That is, an electric potential occurs between the dissimilar metals and electrons flow from one metal to the other, which, over time, builds up corrosion. The corrosion caused by the contact of dissimilar metals results in high resistance, loose connections, and undue heat because of voltage drops. Indeed, enough heat can be produced to cause a fire and other damage.

Temperature Effects

Temperature affects the resistance of a conductor and is an important variable in industrial applications of electric or electronic equipment. The resistance of pure metals (elements) increases with an increase in temperature and decreases with a decrease in temperature. Certain alloys, such as Constantan, show very little variation in resistance with changes in temperature. Such alloys are therefore used in circuits where resistance changes owing to temperature changes cannot be tolerated.

Liquids that conduct current (electrolytes) behave the opposite of metals: an electrolyte's resistance goes up when the temperature goes down. For example, a solution of sulfuric acid and water is the electrolyte in an ordinary car battery. Because an electrolyte's resistance increases as the temperature decreases, a car battery furnishes less current in cold weather than in warm weather.

The amount that resistance increases per degree Celsius per ohm is the *temperature coefficient.* This value ranges between 0.003 and 0.006 for pure metals. Table 10 gives the temperature coefficients per mil-foot for several conductors at a temperature of 20°C (68°F).

TABLE 10
Temperature Coefficients of Conductors
(per mil-foot at 20°C or 68°F)

Nickel	0.006	Aluminum	0.0039
Iron	0.0055	Silver	0.0038
Tungsten	0.0045	Nichrome	0.00044
Copper	0.00393	Constantan	0.000008

RELATIONSHIP EQUATION

The relationship of the four factors that determine the resistance of a conductor (specific resistance, length of the conductor, cross-sectional area of the conductor, and temperature), is shown in the equation—

$$R = x\left[\frac{l}{cm}\right] \qquad \text{(Eq. 17)}$$

where

R = resistance in ohms
x = resistance per mil-foot
l = length in feet
cm = area in circular mils.

In spite of equation 17's availability, practicing electricians and electronics personnel seldom use it because copper wire tables have been prepared that give the same information. It is easier to simply look it up in a table. Table 11 is an example of a commercially available copper wire table. It lists several AWG numbers for copper wire. It also gives the wire's diameter in mils, circular mils, and square inches (in.^{2}), as well as the resistance in

TABLE 11
Characteristics of Standard-Gauge Copper Wire Conductor

AWG No.	Diameter, mils (0.001 in.)	Cross-sectional area		Ohms per 1,000 feet at 25°C (77°F)	Pounds per 1,000 ft
		Circular mils	in.^{2}		
0000	460.0	212,000.0	0.1660000	0.0500	641.0000
000	410.0	168,000.0	0.1320000	0.0630	508.0000
00	365.0	133,000.0	0.1050000	0.0795	403.0000
0	325.0	106,000.0	0.0829000	0.1000	319.0000
1	289.0	83,700.0	0.0657000	0.1260	253.0000
2	258.0	66,400.0	0.0521000	0.1590	201.0000
3	229.0	52,600.0	0.0413000	0.2010	159.0000
4	204.0	41,700.0	0.0328000	0.2530	126.0000
5	182.0	33,100.0	0.0260000	0.3190	100.0000
6	162.0	26,300.0	0.0206000	0.4030	79.5000
7	144.0	20,800.0	0.0164000	0.5080	63.0000
8	128.0	16,500.0	0.0130000	0.6410	50.0000
9	114.0	13,100.0	0.0103000	0.8080	39.6000
10	102.0	10,400.0	0.0081500	1.0200	31.4000
11	91.0	8,230.0	0.0064700	1.2800	24.9000
12	81.0	6,530.0	0.0051300	1.6200	19.8000
13	72.0	5,180.0	0.0040700	2.0400	15.7000
14	64.0	4,110.0	0.0082300	2.5800	12.4000
15	57.0	3,260.0	0.0025600	3.2500	9.8600
16	51.0	2,580.0	0.0020300	4.0900	7.8200
17	45.0	2,050.0	0.0016100	5.1600	6.2000
18	40.0	1,620.0	0.0012800	6.5100	4.9200
19	36.0	1,290.0	0.0010100	8.2100	3.9000
20	32.0	1,020.0	0.0008020	10.0400	3.0900
21	28.5	810.0	0.0006360	13.1000	2.4500
22	25.3	642.0	0.0005050	16.5000	1.9400
23	22.6	509.0	0.0004000	20.8000	1.5400
24	20.1	404.0	0.0003170	26.2000	1.2200
25	17.9	320.0	0.0002520	33.0000	0.9700
26	15.9	254.0	0.0002000	41.6000	0.7690
27	14.2	202.0	0.0001580	52.5000	0.6100
28	12.6	160.0	0.0001260	66.2000	0.4840
29	11.3	127.0	0.0000995	83.4000	0.3840
30	10.0	101.0	0.0000789	105.0000	0.3040
31	8.9	79.7	0.0000626	133.0000	0.2410
32	8.0	63.2	0.0000496	167.0000	0.1910
33	7.1	50.1	0.0000394	211.0000	0.1520
34	6.3	39.8	0.0000312	266.0000	0.1200
35	5.6	31.5	0.0000248	325.0000	0.0954
36	5.0	25.0	0.0000196	423.0000	0.0757
37	4.5	19.8	0.0000156	533.0000	0.0600
38	4.0	15.7	0.0000123	673.0000	0.0476
39	3.5	12.5	0.0000098	848.0000	0.0377
40	3.1	9.9	0.0000078	1,070.0000	0.0299

ohms per 1,000 ft, and how much 1,000 ft of the wire weighs.

As an example of using the table, you can see that AWG No. 12 copper wire (the shaded area in the table) has a diameter of 81.0 mils (or 0.081 in.), a circular mill area of 6,530, a cross-sectional area of 0.00513 in.^{2}, a resistance of 1.62 ohms per 1,000 ft, and a weight of 19.8 pounds per 1,000 ft.

REVIEW QUESTIONS

1. How does the length of a wire affect its resistance?
2. What effect does a cross-sectional area have on the resistance of a wire?
3. How does temperature affect the resistance of a wire?

Properties and Characteristics of Magnetism

A close relationship exists between electricity and magnetism. Therefore, it is important to learn about magnetism and its relation to electronics.

MAGNETS AND MAGNETISM

A *magnet* is a piece of iron or steel that attracts other bits of iron or steel. *Magnetism* is the force that occurs when a magnet attracts iron. Magnetism also occurs when electricity moves through a conductor. Put another way, magnets and certain electrical devices exhibit magnetism. Magnetism is responsible for the operation of generators, magnetos, motors, voltmeters, relays, and many other important devices.

Magnetism can transform electrical energy into mechanical energy and mechanical energy into electrical energy. For example, an electric motor transforms electrical energy into mechanical energy because electricity turns a shaft in the motor. Then, when the shaft is connected to another device, such as a pump, it performs mechanical energy, or work. On the other hand, a generator transforms mechanical energy into electrical energy. For example, when a device that provides power, such as a diesel engine, is hooked up to a generator to make it turn, the generator makes, or generates, electricity.

HISTORY

Over 2,000 years ago, humans discovered that a type of naturally-occurring iron ore possessed the peculiar property of being able to attract small pieces of iron. The Greeks named this iron ore, which was a form of iron oxide, *magnetite*. Later, ancient explorers found that an elongated piece of magnetite, if suspended by a string, would always point in the same direction. People then began to navigate using magnetite as a compass. Because it led early explorers in their travels, they called it *lodestone*, which means way stone, in the sense that the stone led the way. Today, lodestone is important mainly because of its historical interest; nevertheless, it was the beginning of humankind's knowledge of magnetism. It was not until the 1800s, however, that scientists began to investigate and to discover the properties of magnetism.

EARTH AS A GIANT MAGNET

The earth is a huge magnet and is surrounded by a *magnetic field*. The earth has *magnetic poles* and *geographic poles* (fig. 20). Because the earth's magnetic axis does not quite coincide with the geographic axis, the magnetic and geographic poles are not at the same place on the surface of the earth. This fact is important to those who plot courses or find directions when traversing the earth. This difference in the location between the magnetic pole, or *magnetic north*, and the geographic pole, or *true north*, is the *angle of declination*, or *magnetic deviation*.

Magnetic deviation varies from place to place on the earth. Because declination varies, geographers

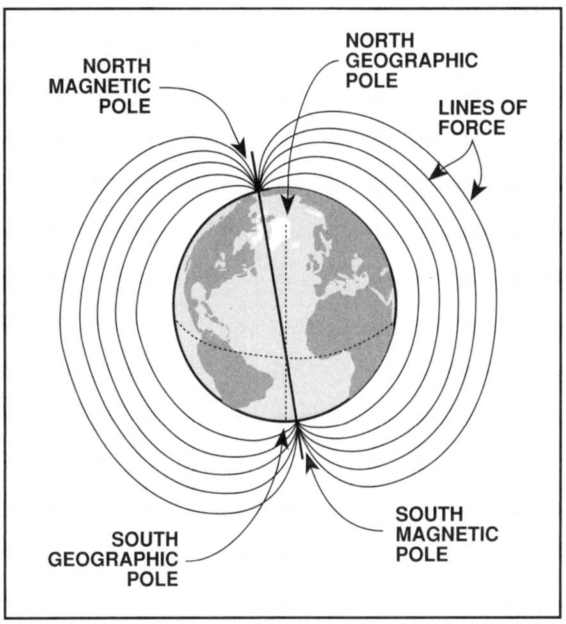

Figure 20 Earth's poles and magnetic lines of force

have developed and published maps that show the amount of declination for a particular location. For example, in Los Angeles, California, the angle of declination is 14° west of true north. So, to determine true north from this location, one must subtract 14° from the compass reading. On the other hand, in New York City, the angle of declination is 15° east of true north. So, add 15° to the compass reading to determine true north from this location.

Early users of the compass discovered that one end of the compass needle always pointed to the North Pole while the other end pointed to the South Pole. On some maps, geographers designated the *north magnetic pole* as the northward direction the compass always pointed, whereas, in fact, the magnetic pole is close to, but not the same as, the geographic North Pole. Geographers also termed the South Pole the *south magnetic pole*.

Magnetic lines of force are assumed to come out of the north pole of a magnet and to enter the magnet's south pole as closed loops (fig. 21). The compass needle aligns itself so that the earth's lines of force enter at its south pole and leave at its north pole. Because the north pole of the needle is defined as the end that points in a northerly direction, it follows that the magnetic pole in the vicinity of the north geographic pole is in reality a south magnetic pole and vice versa.

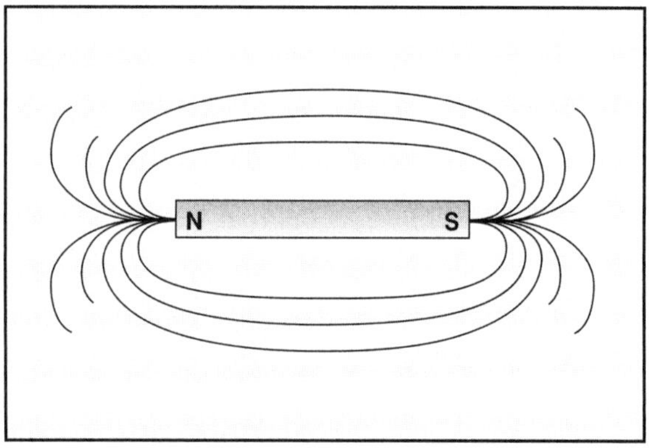

***Figure 21** Magnetic lines of force*

MAGNETIC PROPERTIES

Of all the materials on earth, only a few can be magnetized. Of these, iron, cobalt, and nickel are the most important. Iron exhibits a higher degree of magnetism than any other material. Consequently, it is the most important material used for electrical purposes. Indeed, iron is more important to the field of electricity than copper because copper cannot be magnetized, and materials other than copper can be used as conductors. Because *steel* contains iron, it is also an important material.

Iron or steel bars can be magnetized by—

1. placing them in a north-south direction and jarring them with a hammer.
2. rubbing them with a magnet.
3. inserting them in a current-carrying coil.

The third method is the best way to magnetize iron or steel.

NATURE OF MAGNETISM

In the 1830s, the German physicist Wilhelm Weber formulated a theory about the nature of magnetism. Weber's theory assumes that all metals contain molecular magnets. That is, each metal molecule is a tiny magnet. However, the molecular magnets that compose an unmagnetized bar of iron or steel are arranged at random (fig. 22A). With this arrangement, adjacent molecules neutralize each other's magnetism, and no external magnetic effect is produced. But, when an unmagnetized bar is subjected to a magnetizing force, the individual north and south poles of the molecules align themselves so that all the north poles are pointing in one direction and all the south poles are pointing in the other direction (fig. 22B). The presence of excessive heat or vibration can cause these molecules to become misaligned.

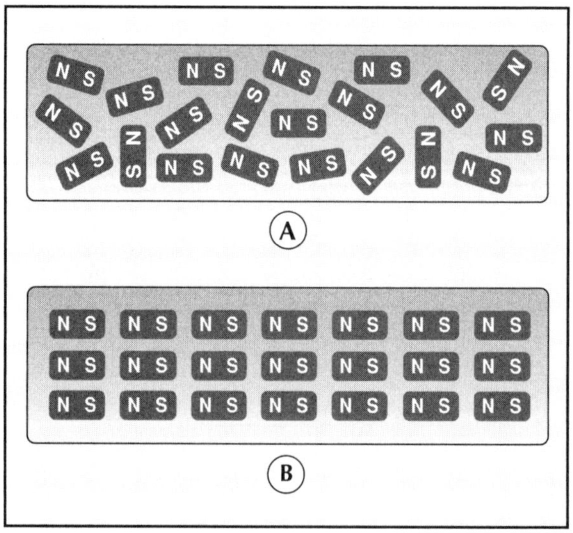

***Figure 22** A. Magnetic molecules arranged in a random manner. B. Magnetic molecules arranged in an organized manner*

Retentivity

Hard steel is difficult to magnetize. The molecules in hard steel offer more resistance to a change in their alignment than do the molecules in soft iron. A stronger magnetizing force is required to make steel molecules swing around so that their axes are parallel. Once in line, however, they are equally difficult to get out of line. Thus, hard steel has a high *retentivity*—that is, it retains its magnetism well.

Soft iron is easy to magnetize because its molecules easily swing into line. However, when the magnetizing force is removed, the molecules tend to lose their alignment. Thus, soft iron has low retentivity.

Permeability

Because the *permeability* of iron or steel is much higher than that of air, these materials are attracted to a magnet. Permeability, in this sense, is a measure of the ability of a material to conduct magnetism. When a piece of soft iron, for instance, is placed in the air near a magnet, the magnetic lines tend to pass through the iron rather than through surrounding air (fig. 23). This phenomenon occurs because it is easier for the magnetic lines to pass through iron than to pass through air. Magnetic lines always seek the path of least resistance as well as the shortest path. Put another way, iron has greater permeability than air. As the lines pass through the iron bar, they magnetize the bar so that its south pole is nearest the north pole of the magnet. The closer the bar is moved toward the magnet, the stronger it is magnetized, and the greater is the attraction between it and the magnet.

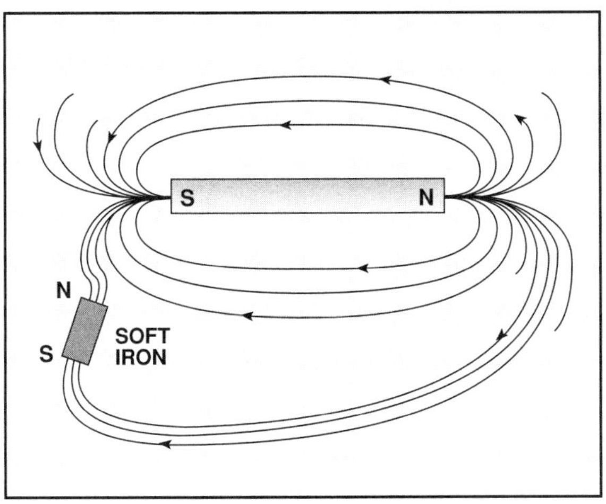

Figure 23 A piece of soft iron placed near a magnet

Soft iron or annealed steel allows lines of force to be easily distributed throughout the metal. Such material is a good conductor of magnetic lines and has high permeability. Air is assigned a permeability of 1.0. Iron and soft steel may have a permeability of 100 to 10,000 depending on the grade of iron or steel. This rating system indicates that magnetic lines can pass through soft iron or steel considerably easier than they can pass through air; indeed, 100 to 10,000 times easier.

Magnetic lines can also pass through a vacuum, but the difference in permeability between a vacuum and air is very small. Moreover, magnetic lines can pass through electrical insulators as well as conductors like copper or aluminum. The permeabilities of these materials are about the same as that of air. On the other hand, iron, cobalt, and nickel have high permeability.

Magnetic Flux

If a pane of glass is placed over a *bar magnet* and iron filings are sprinkled over the glass, the filings arrange themselves in an orderly fashion. (A bar magnet is a type of *permanent magnet* in that it is made of hard steel that tends to hold its magnetism for a long time.) The arrangement of the iron filings, which becomes more apparent if the glass is gently tapped, shows the lines of magnetic force—the *flux*—around the magnet. The concept of lines of force, or flux, is the best way to visualize the force around a magnet.

The lines of force are denser at the ends of the bar (see fig. 21). These points at the ends of the bar are the *magnetic poles*. The poles are the sites of the greatest attraction to other bits of iron. The space that magnetic lines occupy is a magnetic field, and the greater the number of lines in a given unit of space, the greater is the attraction to the iron in that vicinity.

Magnetic flux, or the lines of force, behaves similarly to several stretched rubber bands. Lines of force possess three distinct properties, and these properties explain magnetic attraction and repulsion.

1. Magnetic lines constantly tend to decrease their length and consistently seek the path of least resistance in much the same way as electrons that flow in the path of least resistance. Magnetic lines can be visualized as being in tension.
2. Magnetic lines never cross each other. Lines going in the same direction have a tendency to push away from each other.

The tendency to move away from other lines is similar to the repulsive force of two like charges.

3. Magnetic lines have direction and, as charged particles, are either positive or negative. They are assumed by most authorities to emerge from the magnet's north pole and enter its south pole.

Attraction and Repulsion

If a compass is held near the north pole of a magnet, the south end of the compass needle revolves so that the south pole of the compass points to the north pole of the magnet (fig. 24). This phenomenon demonstrates the first important rule of attraction: unlike poles attract. If the north poles of two magnets are held together, they repel each other. Thus, the second rule is that like poles repel. The third rule of magnetic attraction and repulsion is that the force of attraction or repulsion varies directly with the product of strength of the separate poles and inversely with the square of the distance separating the magnetic poles. This rule can be shown in a written-out equation as—

$$Force = \frac{product\ of\ strength\ of\ two\ poles}{(distance\ between\ poles)^2}. \quad \text{(Eq. 18)}$$

In figure 25A–B, two bar magnets are shown. In 25A, the unlike poles of the magnets are near each other. Attraction occurs between the magnets because the lines of force are attempting to shorten their path in the air gap between the magnets' north and south poles. In 25B, the like poles (the north poles in this case) of the magnets are near each other. Repulsion occurs between the magnets. That is, the lines of force repel each other. Notice that the air

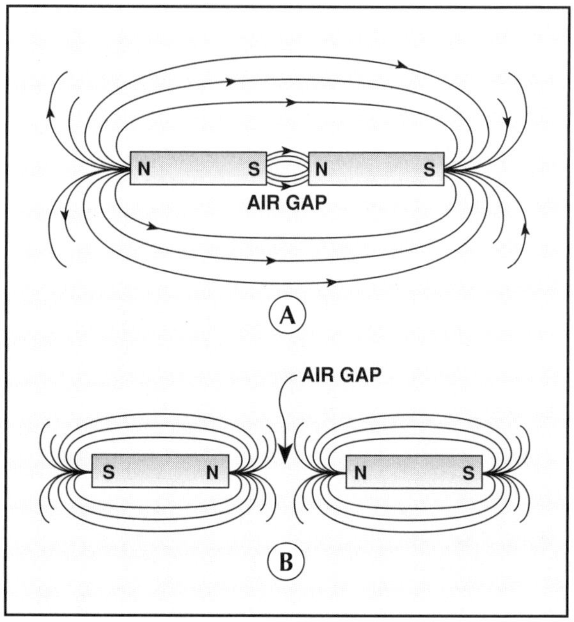

Figure 25 Reaction between two magnets

gap is large and that the lines of magnetic force do not flow directly from one pole to the other as in the case where unlike poles are placed near each other.

TYPES OF MAGNETS

Magnets may be classified according to how they were created, shaped, or formed. Included in this classification are *temporary magnets* and permanent magnets.

Temporary Magnets

It is possible to create *artificial magnets* by placing certain materials in a magnetic field. Artificial magnets may be classified as either temporary or

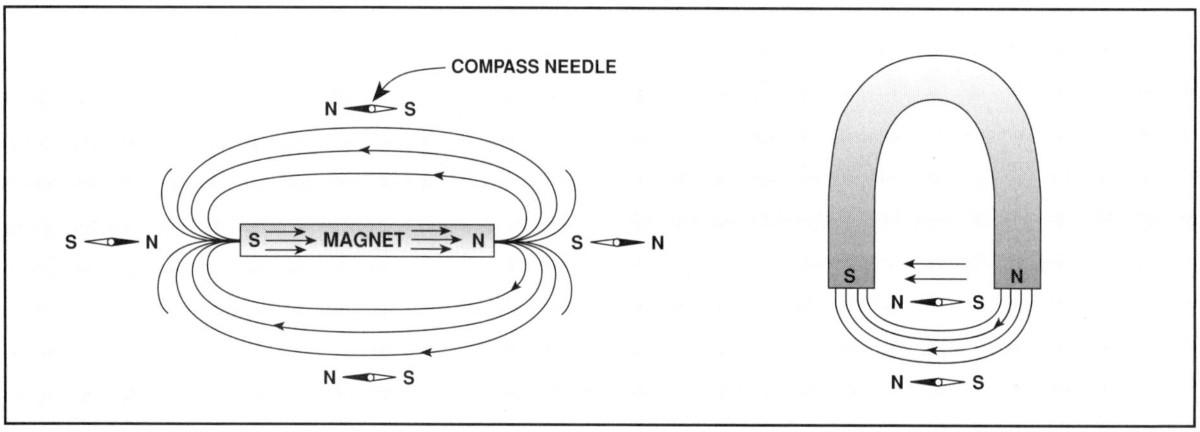

Figure 24 The position of a compass needle in relation to north and south poles of magnets

permanent magnets, depending on their ability to retain their magnetic strength (their retentivity) after the magnetizing force has been removed. For example, when soft iron or steel is placed in a magnetic field, it becomes magnetized temporarily. When the soft iron bar is removed from the field, it loses most of its magnetism. If the bar is placed in a strong field, it becomes strongly magnetized; if it is placed in a weak field, it becomes weakly magnetized. In either case, however, the soft iron or steel loses most of its magnetism when the magnetic field is removed. Thus, soft iron and steel are temporary magnets.

When a soft iron bar is removed from a magnetic field, it does not, however, lose all its magnetism. The small amount it retains is *residual magnetism*. The residual magnetism of iron objects affects the accuracy of a compass. Residual magnetism accounts for the warning that cross-country hikers should not take a bearing with a magnetic compass near a gun, a wire fence, or a pipeline. Residual magnetism in these objects attracts the compass needle and it gives erroneous readings.

As mentioned earlier, the ability of a material to retain its magnetism once it is removed from the magnetizing force is its retentivity. A bar of soft iron or steel has high permeability because it easily conducts magnetic lines. However, soft iron or steel has low retentivity because it does not remain strongly magnetized when removed from the magnetizing force. A magnet of this type is a temporary magnet. Temporary magnets are used in motors, generators, transformers, relays, and other electrical apparatus.

Permanent Magnets

A material that retains a large percentage of its magnetism once it is magnetized is a permanent magnet. Permanent magnets are made of very hard steel or steel with other metals alloyed with it. Although this type of material is relatively hard to magnetize and has low permeability, it has high retentivity. Through the years, efforts have been made to produce stronger permanent magnets—that is, magnets that have more lines of force per volume of material used. By alloying steel with other metals (some of which are not magnetic), powerful magnets have been produced. One example is the *alnico magnet*. It is made of an alloy of aluminum, nickel, cobalt, and other elements, such as copper and titanium. It is interesting to note that aluminum, by itself, is not magnetic.

Shapes

Permanent magnets are available in many shapes, but the most common are the horseshoe, ring, and bar (fig. 26). The closed ring magnet is interesting because virtually no magnetic effect is felt as long as the ring is complete. However, if a section of the ring is cut out to make an open ring magnet, magnetic poles are established, and the ring magnet behaves similarly to a horseshoe magnet.

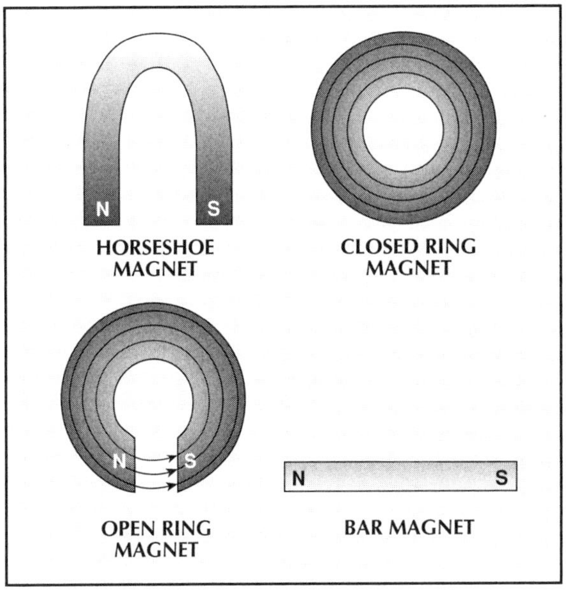

Figure 26 Various shapes of magnets

Manufacture

Pouring molten alloy into the proper form is the first step in making a permanent magnet. After the material has cooled, the pole faces are ground flat. Because the alloys are very hard, they are difficult to drill or cut. Therefore, most permanent magnets have a rough finish. Although the material used to make magnets is very hard, it is also quite brittle. Brittleness means that permanent magnets can be broken relatively easily. After the magnet has been properly shaped and finished, it is then magnetized.

Care

Permanent magnets lose their magnetism if they are heated to high temperatures. Therefore, they should not be subjected to even moderately high temperatures. Also, magnets should not be struck with a hammer or dropped. Either form of abuse upsets the alignment of the molecules in the magnet and causes the magnet to lose strength.

Storage

When magnets are not in use, a soft iron bar should be placed between the poles. Such a bar is a *keeper*. The keeper protects the magnet's poles if the magnet is dropped or accidentally struck by a hard object. Similarly-shaped magnets should be stacked in the attracting position—that is, the north pole of one magnet should be attached to the opposite, or south, pole of the other magnet (fig. 27).

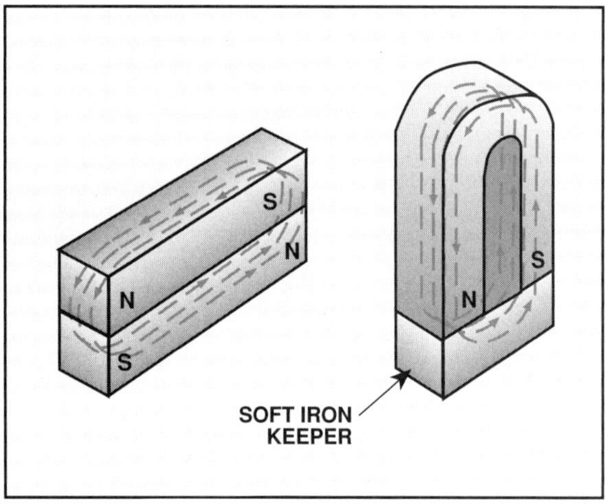

Figure 27 Proper storage of magnets

CURRENT AND MAGNETIC FIELDS

As mentioned earlier, a close relationship exists between electricity and magnetic fields. Current produces a magnetic field and a magnetic field can produce voltage.

Oersted's Discovery

In 1820, Hans Christian Oersted, a Danish scientist, discovered the relationship between electric current flow and magnetism. He accidentally held a compass near a current-carrying wire and noticed that current flowing through the wire deflected the compass needle from its normal position. He also noticed that the needle went back to its normal position when current flow was stopped. This deflection of a compass's needle showed that the compass was being influenced by a magnetic field other than that of the earth. It also showed that the magnetic field was evidently coming from the current flow through the conductor.

To duplicate Oersted's experiment, pass a conductor vertically through a piece of cardboard in a horizontal position, and connect the conductor to a battery for current flow (fig. 28A–B). Note that the magnetic lines of force flow in a circle around the conductor. In figure 28A, the lines flow in a clockwise direction. In reality, you cannot see the lines of force, but you can determine their direction of travel by observing the direction in which the compass needle points. It points in the direction the lines flow. Thus, in figure 28A, the needle's various positions around the conductor show that the lines of force flow in a clockwise direction. On the other hand, in figure 28B, the lines of force flow counterclockwise, which was achieved by reversing the polarity of the battery connections. Reversing the battery connections changes the direction of current flow through the conductor, which, in turn, changes the direction of the lines of force.

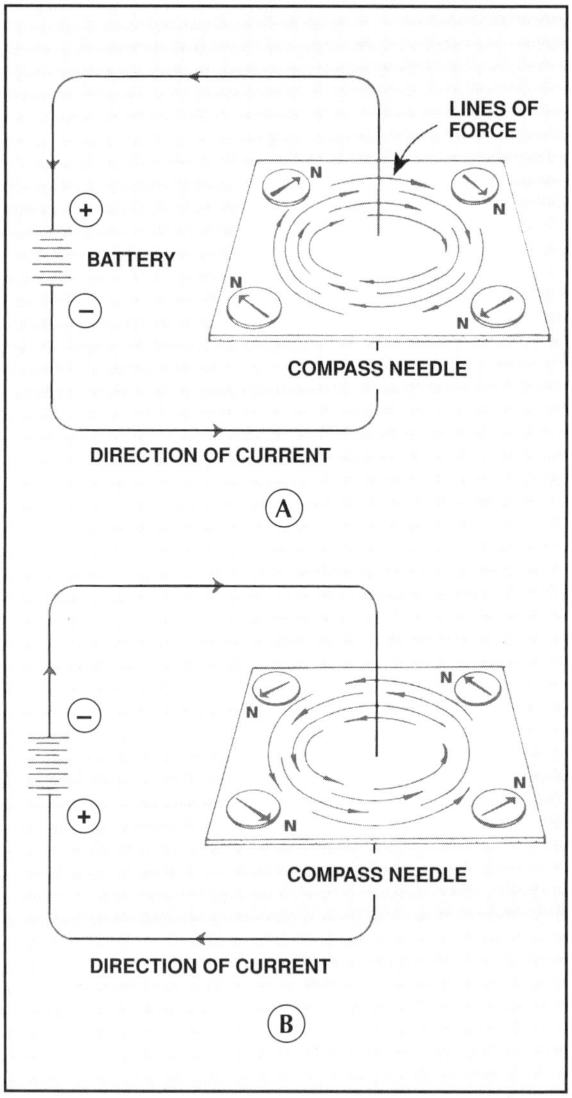

Figure 28 A and B. Effect of current flow on a compass

Current Strength and Magnetic Fields

The strength of the magnetic field varies with the distance from the current-carrying conductor. If the compass is placed close to the wire, the compass needle responds quickly, indicating a strong magnetic field. The magnetic field is densest at the axis of the wire and rapidly becomes less dense (weaker) as distance from the wire increases.

The strength of the magnetic field is also affected by the strength of the current flowing through the conductor. The stronger the current, the stronger the magnetic field. Moving charges (electrons in motion) set up the magnetic field. So, an increase in current through the conductor causes a greater number of electrons to move through the conductor and sets up a greater number of magnetic lines.

Right-Hand Rule for Conductors

The relation between the direction of the magnetic lines of force around a conductor and the direction of conventional current flow along the conductor may be determined by using the *right-hand rule for conductors*. If a conductor is grasped in the right hand with the thumb pointing in the direction of current flow—that is, away from the positive terminal of the battery—the fingers point in the direction of the magnetic lines. This direction is the same that the north pole of the compass would point if it were in the magnetic field.

Parallel Conductors and Magnetic Fields

When two parallel conductors carry currents in the same direction, an attraction occurs between the fields (fig. 29A). So, the conductors move toward each other. The magnetic fields attract in the same manner that unlike poles of two magnets attract. As shown at top right in figure 29A, the magnetic lines of force rotating around the wires travel in opposite directions. Where the two magnetic fields overlap, the lines of force merge into continuous lines that travel around both wires. Since magnetic lines of force follow the path of least resistance, the two fields tend to combine and form a single field that encircles both wires.

Where two parallel conductors carry currents in reverse directions (fig. 29B), the wires push apart. The magnetic fields around the conductors repel just as the like poles of magnets repel. As shown at bottom right in figure 29B, the lines of force circle the wires in the same direction. Consequently, they collide and

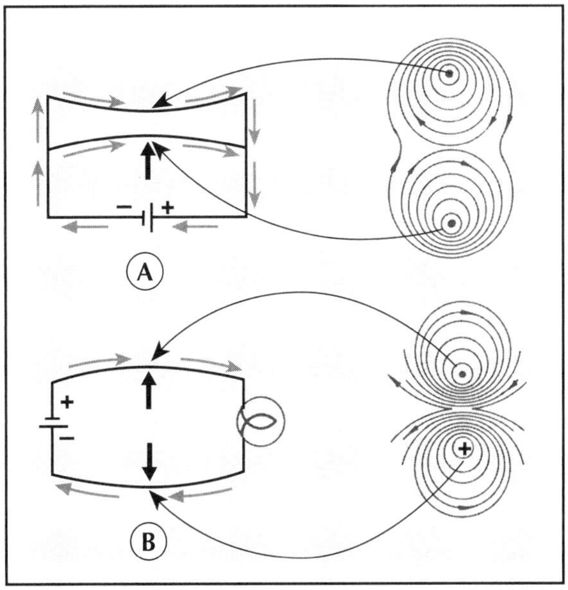

Figure 29 A. Two parallel conductors carrying current that flows in the same direction. B. Parallel conductors with current flowing in opposite directions.

cannot combine into one field. As the magnetic lines pass, they push away from each other, thereby pushing the wires apart.

Conductors at Right Angles

When conductors carrying currents are placed at right angles to each other, current flows around each leg of the conductor in opposite directions (fig. 30).

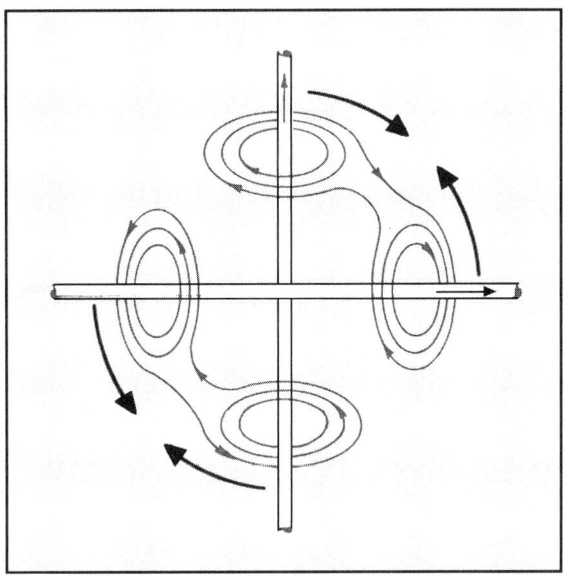

Figure 30 Two conductors at right angles to each other; current flow in each conductor tries to move the conductors into a parallel position.

In the upper right quadrant, current flow is counterclockwise. In the lower left quadrant, flow is clockwise. Because the flow in each quadrant is in the opposite direction, each leg of the conductor repels the other. Consequently, the conductors turn and become parallel to each other so that the currents flow in the same direction in each conductor. In general, current-carrying conductors tend to move away from a strong field to a weak field.

COILS

If a conductor is coiled into a loop and current is sent through it (fig. 31), the lines of force flow through the loop's center. Note, too, that lines of force circle the top of the loop. If a compass is placed beside the loop, the needle points toward the loop just as if the two sides of the loop were the poles of a permanent magnet. One side has a north pole while the other side has a south pole.

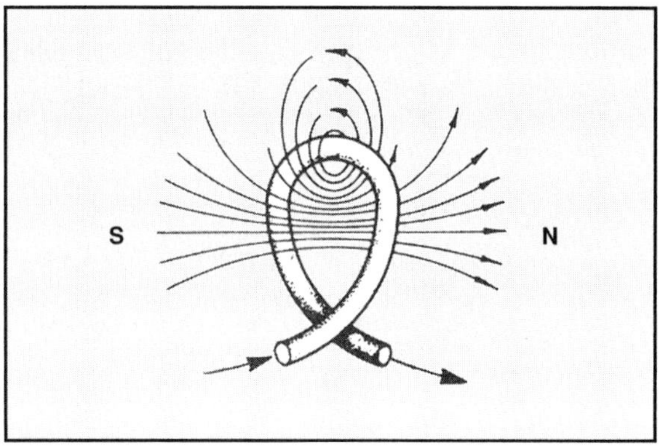

Figure 31 Current flow in a loop; note north and south poles

Current Flow in Coils

If the wire is formed into several loops, a *solenoid*, or *coil*, is created. If the coil is loosely wound (fig. 32), magnetic lines of force circle each loop of the coil and join to travel through the coil's center. If the coil is tightly wound (fig. 33), the lines circle the top and bottom of the whole coil instead of forming separate magnetic fields around the several loops that make up the coil. The loops are parallel conductors carrying current in the same direction (see fig. 29A).

In the loosely-wound coil, the lines between the turns go in opposite directions and try to cancel each other. To avoid canceling each other, the lines of force around the several loops combine into a single field that encircles all loops of the coil. The lines travel through the coil instead of around the entire coil because the two sides of each loop carry current in opposite directions (see figure 29B).

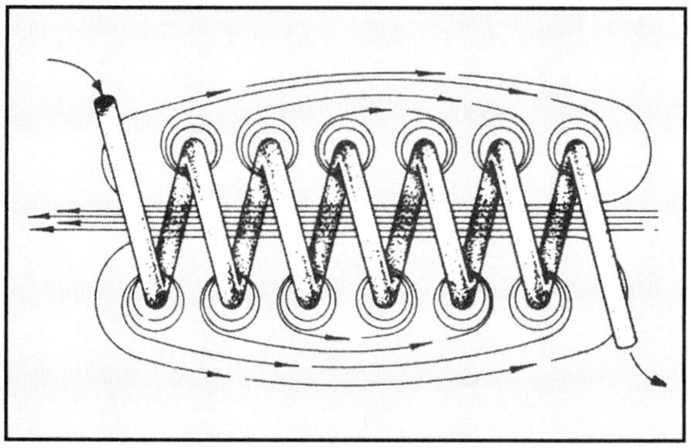

Figure 32 Magnetic field around a loosely-wound coil

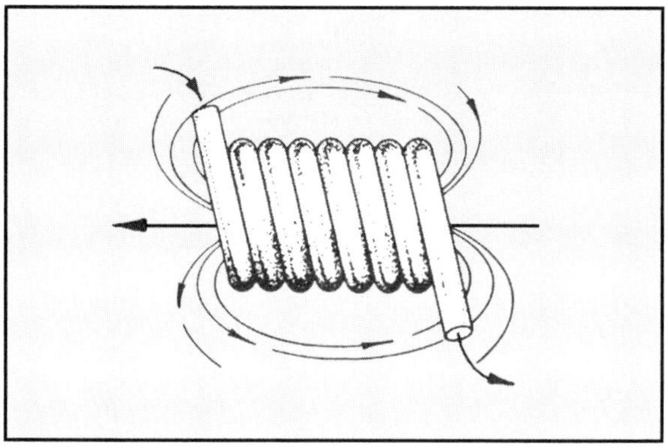

Figure 33 Magnetic field around a tightly-wound coil

Figure 33 also shows that the magnetic field around a tightly-wound, current-carrying coil is the same as the pattern of a permanent magnet. One end of the coil is a north pole and the other end is a south pole.

Right-Hand Rule for Coils

If the direction of conventional current flow through a coil is known, the coil's magnetic polarity can be determined using the *right-hand rule for coils* (fig. 34). If the coil is grasped in the right hand with the fingers wrapped around it in the direction of the current flow, the thumb points toward the coil's north pole.

Remember that the direction of current flow in the supply voltage is from positive to negative. Coils may be wound in a clockwise direction or in a counterclockwise direction. The rule holds for both winding directions. The poles of a coil may be reversed by reversing the battery connections to the

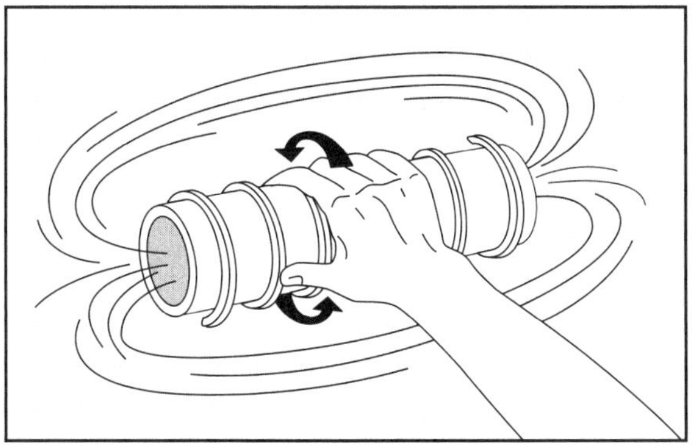

Figure 34 Right-hand rule for coils

coil, or by removing the turns of the coil and winding them in the reverse direction.

Electromagnets

A magnet's strength can be greatly increased if a soft iron bar is placed inside a coil. Iron provides a much better path for the magnetic lines than does air or other nonmagnetic materials. If a coil has iron inside of it, it is an *electromagnet*. Electromagnets are used in magnetic switches, relays, scrap-iron lifters, and other devices. Some electromagnets have a powdered-iron core. Powdered-iron cores are made of iron that is ground to a fine dust, formed into the proper shape, and held together by binding material.

Cores

The inside of the coil is its *core* whether a magnetic, or other, material is in the coil. Coils with iron inserted in them are *iron-core coils*. Coils without magnetic materials are *air-core coils*. Inserting an iron core into a current-carrying coil does not upset the location of north and south magnetic poles; instead, it greatly increases the number of lines set up because more lines encircle the entire coil. Incidentally, the right-hand rule for determining the magnetic polarity of a coil may be applied to electromagnets without any modifications.

Magnetic Field Strength

The strength of the coil's magnetic field depends on several factors. They include—

1. the number of turns of the coil,
2. the amount of current that flows through the coil,
3. the type of material in the core (the form that the coil is wound on), and
4. the ratio of the coil's length to its diameter.

MAGNETIC CIRCUITS

Because magnetism is a convenient way to transform electrical energy into mechanical energy, many electrical devices depend on magnetism for their operation. Among such devices are motors, generators, transformers, and magnetic switches. If an electrical current is to flow, the current must have a suitable circuit, or path, to flow in. Likewise, if magnetic lines of force are to be used to their full advantage, the lines must have suitable paths, or circuits.

Length, Cross-Sectional Area, and Permeability

In most electrical devices, iron cores provide the paths for the *magnetic circuits*. (A magnetic circuit is the path magnetic lines of force take in the electrical device in which a magnet is used.) When studying iron core magnetic circuits, several points should be considered.

- The length of a magnetic circuit is abbreviated *L*. As a magnetic circuit's length increases, it becomes more difficult to set up lines of force in it.
- The cross-sectional area of the magnetic circuit is abbreviated *A*. The larger the cross-sectional area of a magnetic circuit is, the greater the number of lines that can be set up in a given area of iron. Therefore, the larger the area, the greater are the number of lines that can be set up.
- The permeability of the iron used in the magnetic circuit is abbreviated μ, which is mu, the twelfth letter of the Greek alphabet. As mentioned previously, permeability is a measure of the relative ability of iron to conduct magnetic lines as compared to the ability of air to conduct lines. Permeability varies greatly, depending on the material. For example, each grade of iron or steel has different permeabilities. Permeability also varies with the degree to which the iron or steel is magnetized.

Similarity of Electric and Magnetic Circuits

The path that magnetic lines take, whether it is iron, steel, or air, is the magnetic circuit. *Rowland's laws*, which are laws that apply to magnetic circuits, are similar to those for electrical circuits. (Henry A. Rowland was an American physicist of the mid to late 1800s who studied magnetism.) One of

Rowland's laws is similar to Ohm's law. Because of this similarity, Rowland's law is sometimes called Ohm's law for magnetic circuits. As you recall, Ohm's law for electrical circuits is $I = E/R$, where I = current, E = voltage, and R = resistance. Rowland's law is—

$$\phi = \frac{F}{R} \qquad \text{(Eq. 18)}$$

where

ϕ = magnetic lines of force, or magnetic flux
F = magnetomotive force, ampere-turns
R = reluctance.

In an electric circuit, I stands for current, which, remember, is the movement of holes or electrons from atom to atom. In a magnetic circuit, ϕ, which is the Greek letter phi, stands for magnetic lines of force, or flux. In electric circuits, the current actually flows—that is, the holes or electrons move. In a magnetic circuit, however, the lines of force do not flow; instead, they simply come into existence and go out of existence in much the same way as a mechanical force. Figure 35 shows a schematic of a magnetic circuit with a schematic of an electrical circuit and compares Rowland's law in the magnetic circuit with Ohm's law in the electrical circuit. Note that a coil with current flowing through it creates the magnetic circuit, so Rowland's law, $\phi = F/R$, applies to it, while Ohm's law, $I = E/R$, applies to the electrical circuit. Thus, magnetic flux, ø, is equivalent to current, I; magnetomotive force, F, is equivalent to voltage, E; and reluctance, R, is equivalent to resistance, R.

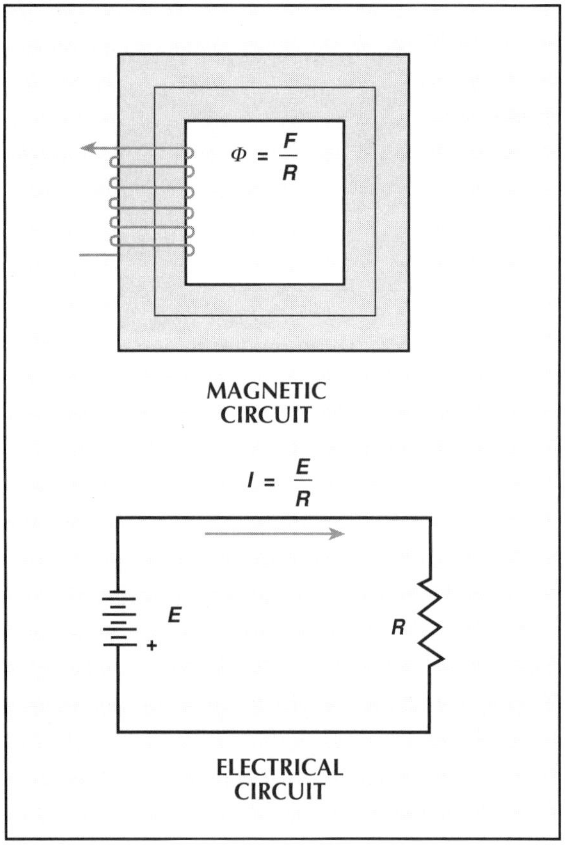

Figure 35 Series circuits; magnetic circuit (top) with Rowland's law and electrical circuit (bottom) with Ohm's law

Magnetomotive Force

In a magnetic circuit, F stands for the force that creates magnetic lines of force: the magnetomotive force (mmf). It is analogous to electromotive force, or voltage. Magnetomotive force is the product of two factors: the current that flows in the coil and the number of turns that the coil has. In equation form—

$$F = IN \qquad \text{(Eq. 19)}$$

where

F = magnetomotive force (mmf), ampere-turns
I = current, amperes
N = number of turns in coil.

Magnetomotive force is expressed in ampere-turns, just as electrical force is expressed in volts. To find the ampere-turns, the current is multiplied by the number of turns in the coil.

Reluctance

In electrical circuits, resistance is the opposition to current flow offered by the material of a conductor. In magnetic circuits, *reluctance* describes the opposition the core material offers to the establishment of magnetic lines within it. The letter R stands for reluctance, and electrical engineers and others have proposed a unit for reluctance they call the *rel*. However, rel has not been completely accepted; so, often no unit is used. Reluctance is usually expressed as so many units of reluctance, which is a number.

Just as length, cross-sectional area, and the nature of its material controls the resistance of a conductor, so does length, cross-sectional area, and the nature of its material (permeability) determine the reluctance of a magnetic circuit. In equation form—

$$R = \frac{L}{A\mu} \qquad \text{(Eq. 20)}$$

where

R = reluctance
L = length, in.

A = cross-sectional area, in.^{2}
μ = permeability.

In electrical circuits, the value of a conductor in ohms does not change with varying amounts of current flowing through it if the temperature is held constant. In magnetic circuits, however, the value of reluctance (R) of a magnetic circuit does vary with the number of lines in the magnetic material. The permeability of the core material changes with the degree of magnetization. And, because permeability changes, the reluctance also changes as shown by equation 20. Thus, calculating the performance of magnetic circuits is more difficult than similar calculations for electrical circuits.

Flux Density

The letter B stands for *flux density*, which is the number of lines of force in a square inch of a core's cross-sectional area. Flux density indicates the degree to which the core material is magnetized. For example, slightly magnetized steel has few lines per in.^{2}, while highly magnetized steel has thousands of lines per in.^{2}. To look at it another way, if 1 in.^{2} of core material is conducting thousands of flux lines, the core has high flux density.

If flux density is increased to a high level, a point is reached where an increase in magnetomotive force gives only a minimal increase in the number of lines in the core. At this point, the magnetic circuit, or core, is *saturated*. Figure 36 plots magnetizing force in ampere-turns per in. versus flux density in thousands of lines (*kilolines*) per in.^{2} for three materials: cast iron, cast steel, and silicon steel. The saturation point on the graph occurs where the curve ceases running vertically and curves in a horizontal direction. (This point is the *knee* of the curve.) A magnetic core of cast steel, for example, has a saturation point of 10 ampere-turns per inch and should not be operated above that point. Operation of magnetic circuits near or above the saturation point wastes electric power and is therefore avoided in well-designed equipment.

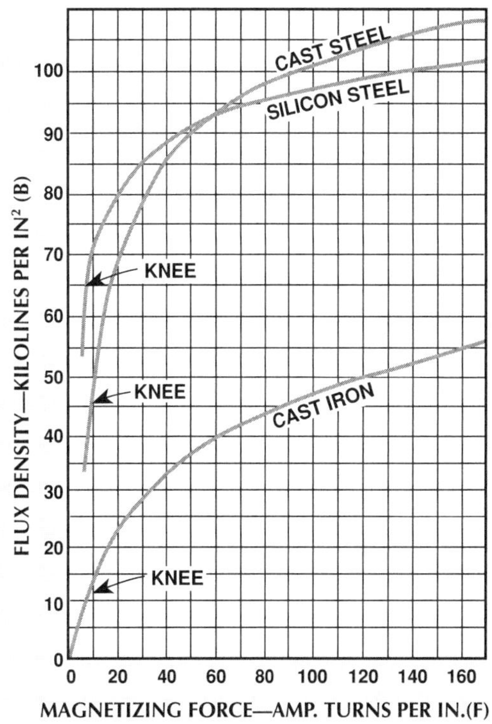

Figure 36 Magnetizing force vs flux density for three materials

Magnetizing Force

The letter H stands for the *magnetizing force*. Magnetizing force involves magnetomotive force in ampere-turns and the length of the magnetic circuit to be magnetized. Basically, magnetizing force is the ampere-turns per inch of the length of the core to be magnetized. Just as a voltage drop, or loss of voltage, occurs along the length of a conductor that is carrying current, a magnetomotive drop occurs along the length of the magnetic circuit. The longer the magnetic circuit, the greater is the magnetomotive force required to magnetize the core to the desired degree. Magnetizing force, or ampere-turns per inch, is expressed mathematically by the equation—

$$H = \frac{IN}{L} \qquad \text{(Eq. 21)}$$

where

H = magnetizing force
I = current, amperes
N = number of turns per in.
L = length of core, in.

Permeability

Permeability of a core material is a measure of the material's ability to conduct magnetic lines of force. Permeability is based on the conducting ability of air; air is assigned a permeability of 1. The permeability (μ) of a core material indicates how many times better it is than air for conducting magnetic lines of force. Permeability not only varies with the type of core material, but also with the degree of magnetization—that is, permeability depends on the number of flux lines per in.^{2} (B). The number of flux lines, in turn, depends on the number of ampere-turns per in. (H). As the number of ampere-turns increases, which causes an increase in the number of flux lines, permeability decreases. Decreased permeability means that more electrical power is required to produce a given increase in the level of magnetization of a core above the saturation point than is required to bring about the same increases below the saturation point. Permeability can be derived from the equation—

$$\mu = \frac{0.313B}{H} \qquad \text{(Eq. 22)}$$

where

μ = permeability
B = lines per in.^{2}
H = ampere-turns per in.
0.313 = a constant (adjusts equation to the units of measurement).

Parallel Magnetic Circuits

Figure 37 depicts an E-shaped core with two parallel magnetic paths and an equivalent parallel electric circuit. In the electric circuit to the right within the figure, note that current flow is counterclockwise in the left half of the circuit and clockwise in the right half. Flux line flow in the core is similar. That is, all flux lines are traveling in the center leg of the core. At the top of the center leg (the north pole) the lines divide; half of the lines flow in the core's left leg and the other half flow in the core's right leg. The lines then converge and enter the center leg of the core at the south pole. Many electromagnets and relays are constructed with E-shaped cores.

At the top of the drawing is a horizontal bar. If this bar is free to move—it could be hinged at one end—it is an *armature*. An armature is made of coils of wire placed around a metal core, in which electric current is induced. As it moves, the armature can produce electricity as it does in a generator, or it can interact with a magnetic field to produce torque in a motor.

The lower part of the figure shows a *yoke*. It is a yoke because it does not move—it is fixed. A yoke is a stationary part made of ferromagnetic material. A yoke does not have windings; rather, it simply connects the magnetic core. Yokes are commonly used in analog tape recorders where they form the recording head. The head, yoke, is the part on the recorder that magnetically affects the tape as it passes the head to record sounds on the tape.

Tractive Force

When a block of iron is placed near the poles of an electromagnet (fig. 38), it is attracted to the poles. However, in a typical electromagnet, an air space occurs between the poles of the magnet and the block of iron. The magnetic lines of force (shown by a single line in the figure) travel out of one of the poles of the electromagnet, through the air space, through the block of iron, through the other air space, through the other pole of the magnet, and through the armature. The lines of force cause the armature to move, or rotate, which, as mentioned previously, can generate electricity or produce torque in a motor.

Because the flux lines are shortened in the air space when the surfaces move closer together, a force of attraction exists in this area. This force of attraction

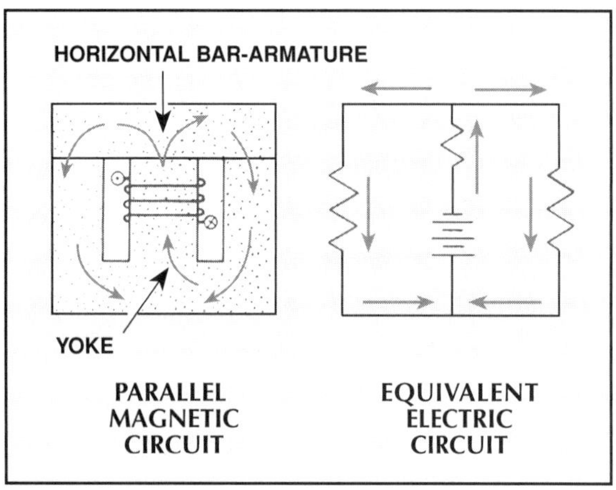

Figure 37 Magnetic parallel circuit and equivalent parallel electric circuit

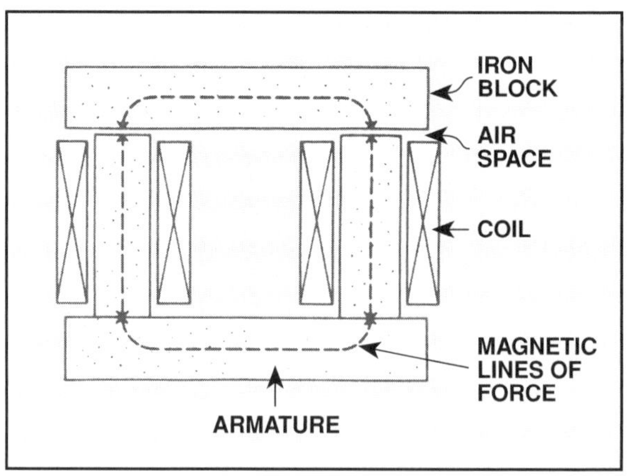

Figure 38 Electromagnet

is *tractive force*. The flux's density in the air space and the area of the air space determine the strength of the tractive force. An equation is available to find the tractive force of an electromagnet—

$$F = B^2 \times A \times 1.39 \times 10^{-8} \quad \text{(Eq. 23)}$$

where

F = force in pounds
B = lines per in.^{2}
A = area in in.^{2}
1.39×10^{-8} = a constant.

REVIEW QUESTIONS

1. Name some devices that depend on magnetism for their operation.
2. What is magnetite?
3. What is lodestone?
4. Define magnet.
5. Describe how to magnetize hard steel.
6. Draw a diagram of the field around a bar magnet.
7. What is a magnetic field?
8. At what part of a magnet are the lines of force at their densest?
9. What are magnetic poles and where are they located on a magnet?
10. What pole are the lines of force assumed to come out of?
11. What is permeability?
12. What is retentiveness?
13. Give the rules of attraction and repulsion in magnets.
14. Explain what a temporary magnet is.
15. What kind of material is a temporary magnet made of?
16. What type of device uses temporary magnets?
17. What is residual magnetism?
18. If you wished to permanently magnetize a metal, what kind of metal would you use?
19. What can cause a permanent magnet to lose its strength?
20. Describe how a magnet should be stored when it is not in use.

7 Induction Principles and AC Generators

ALTERNATING CURRENT

Alternating current (AC) is current that periodically reverses its direction of flow in a circuit and constantly changes its strength, or *amplitude*. One source of alternating current is an AC generator, which is an alternator. The current occasionally reverses its direction of flow in a circuit because the terminals of the alternator are constantly reversing their polarity. That is, a terminal of the AC generator is positive at one instant, negative the next instant, then positive, then negative, and so on. During the time that one terminal is positive, the other terminal is negative. When the terminals of the alternator reverse their polarity, the direction of current flow in the circuit also reverses, because conventional current always flows from positive to negative.

When current changes its direction of flow, the strength, or amplitude, of the current also changes. Current strength falls to zero at the instant the reversal in direction of flow occurs. This phenomenon is similar to a person's traveling due east and deciding to change direction. The person must come to a stop, at least momentarily, before he or she can travel due west. Similarly, the strength of alternating current changes from zero to a maximum in one direction, then falls to zero and rises to a maximum in the other direction, and returns to zero again. This action is repeated as long as current is flowing.

DIRECT CURRENT

Direct current (DC), on the other hand, flows in one direction; its strength, or amplitude, remains constant. When a circuit is connected to a battery, the current flows from the positive terminal of the battery through the circuit to the negative terminal of the battery in one direction. To reverse the current's direction through the circuit, the battery terminals must be reversed. Reversing the battery terminals is the only way to reverse the current's direction through the circuit. The positive terminal of a battery remains positively charged; it never becomes negative. Moreover, a battery's negative terminal never becomes positively charged. Likewise, the connecting wire that leads to the positive terminal of the battery is positive.

The two most important sources of DC current are batteries and DC generators. Each has a positive and a negative terminal, which are usually marked plus (+) for positive and minus (–) for negative.

ELECTRIC AND MAGNETIC FIELDS

Magnetic devices include such equipment as relays, motors, and transformers. Because they are magnetic devices, they are usually discussed in terms of their magnetic fields. On the other hand, conventional electrical devices, such as radio circuits and the like, are usually discussed in terms of their electric fields. Further, just as magnetic devices have magnetic fields and magnetic lines of force, conventional electrical devices have electric fields and electric lines of force. Electric lines of force exist whenever a difference in potential occurs, such as around the terminals of a battery, or around power lines.

Moreover, if an *atomic particle*, such as an electron or proton, is charged, electric lines of force are associated with it, too. If the particle is an electron, it is negatively charged. In this case, the lines go out of the particle (fig. 39A). If the particle is a proton, it is positively charged, and the lines go into the particle (fig. 39B).

Particles with unlike charges—that is, one has a positive charge and the other has a negative charge—are attracted to each other. This fact leads to the expression, "unlike charges attract." Electric lines of force, just as magnetic lines of force, always tend to travel the shortest route possible, and the shortest distance for the lines is between two unlike charges (fig. 39C). Thus, they attract.

On the other hand, when particles with like charges are brought together, they tend to repel—that is, "like charges repel." Like charges repel because the lines around the charges collide head on and therefore oppose each other (fig. 39D).

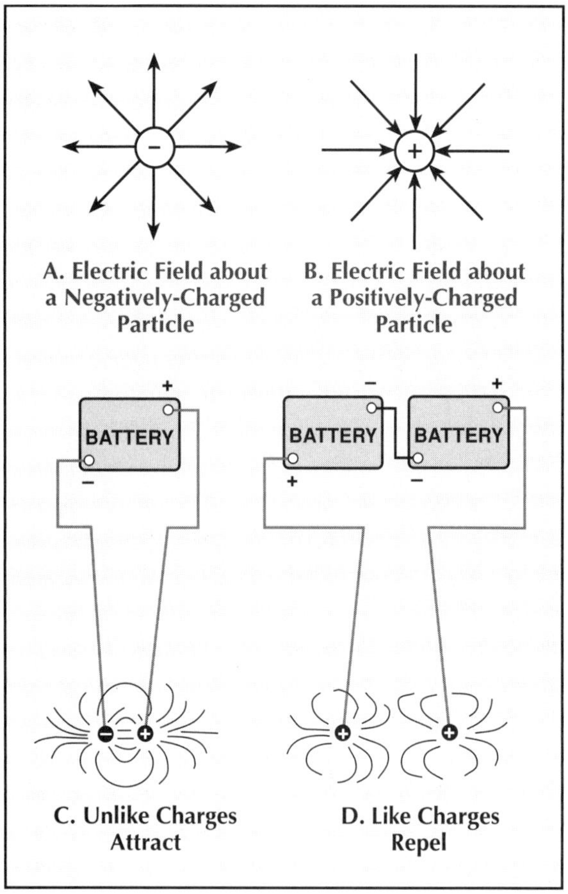

Figure 39 A-D. Electric fields and charges

ELECTROMAGNETIC INDUCTION

Up to now, this manual has mostly used a battery as an example of a voltage source. Indeed, batteries are an important source of voltage. However, in many cases, they are too expensive and impractical to use as a voltage source. For example, it is well beyond the capabilities of batteries to deliver commercial electrical energy on a large scale. Therefore, electric generators provide most commercial electricity.

A generator operates on the principle of *electromagnetic induction*. As you recall, induction occurs when a conductor is placed near an electrically-energized body. (Direct contact is not required.) Induction produces voltage in the conductor. Similarly, electromagnetic induction produces voltage when a conductor cuts across, or is cut by, a magnetic field. For example, voltage is induced in a coiled conductor when the coil is passed in and out of a stationary magnet's magnetic field. Or, moving the magnet past the coil can also induce voltage. Either the coil or the magnet can be moved, but one or the other must be moving to induce voltage in the conductor. Voltage is induced only when motion occurs.

Forming the conductor into a coil increases the induced voltage beyond that which is produced in a straight-wire conductor. Voltage is increased in the coil because voltage is induced in each loop, or turn, of the coil. Since the turns are in series—they follow one after the other—the voltages add up just as they add up in batteries that are in series. For example, suppose one turn of a coil has 0.1 volt induced across it. If ten turns make up the coil, then a total of 1 volt appears across the whole coil because $10 \times 0.1 = 1$.

The speed of the movement and the strength of the magnet also affect the value of the induced voltage in the coil. Increased speed and increased strength cause an increase in induced voltage. Put another way, the induced voltage is directly proportional to the number of lines that cut, or move across, the coil per second of time.

Induced Voltage

Figure 40 is a six-part drawing that illustrates voltage being induced in a conductor. In this case, the conductor is a straight wire. Figure 40A shows the conductor—the wire— moving downward through the magnetic field of a permanent horseshoe magnet. This movement induces voltage in the conductor.

Figure 40B represents the conductor and the magnetic lines of force as though an observer is at the magnet's north pole looking towards its south pole. Consequently, the magnetic lines of force are moving away from the viewer and are shown as dots. (The dots represent the tail end of arrows, which are used to show lines of force.) For simplicity, only one free electron is shown in the conductor. It is represented by a minus sign (–) inside a small oval. As indicated by the arrow, the free electrons are moving to the left in the conductor.

Figure 40C shows the free electron with magnetic lines around it from the same view as in figure 40B (only one line is shown for clarity). The magnetic lines appear around the free electron because the moving conductor moves the electrons and their electric fields. When an electric field is in motion, a magnetic field occurs.

Figure 40D shows the electron moving to the left. This leftward motion occurs because electrons move toward the weak side of the field. In this case,

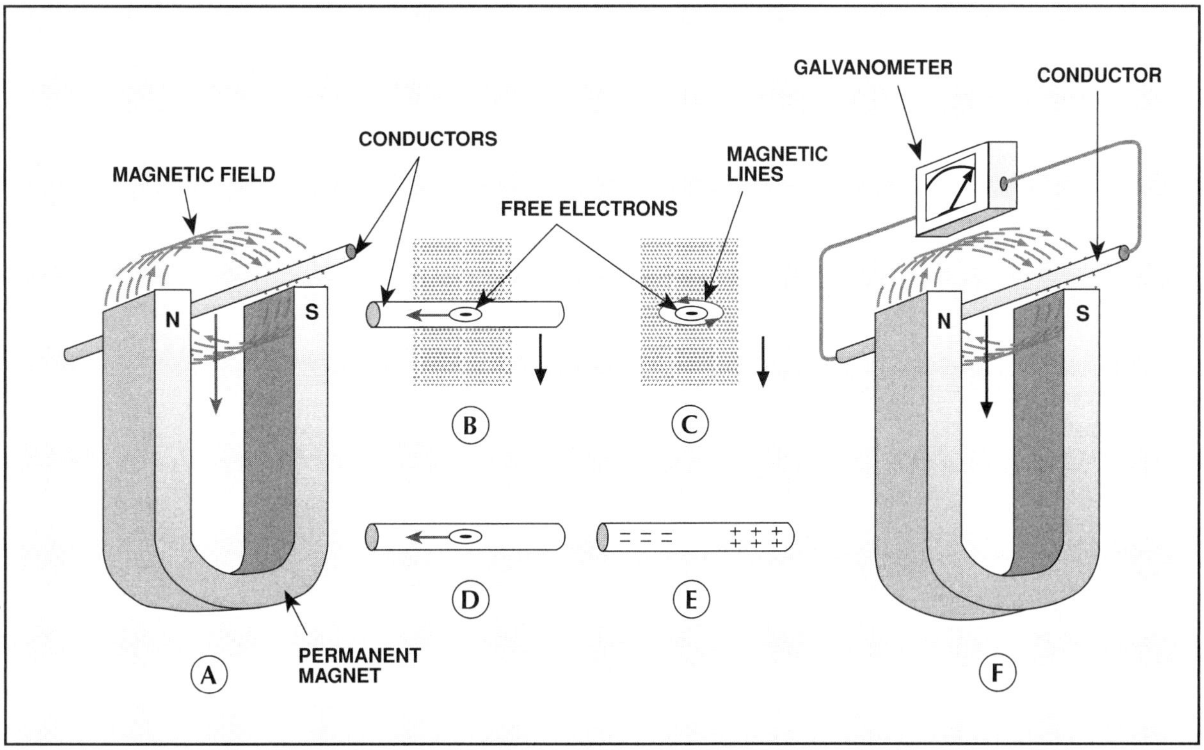

Figure 40 Inducing voltage by moving the conductor

the magnetic lines go around the electron in a counterclockwise direction (as indicated in figure 40C). Counterclockwise movement tends to strengthen the field to the right of the electron and to weaken the field to the left of the electron. So, the electron moves in the direction of the weaker field, which is to the left in figure 40D.

Figure 40E shows several free electrons (indicated by minus signs) that have moved to the left end of the wire. This collection of electrons represents a negative charge. Since these free electrons have moved from their normal positions in the wire, this leaves a deficiency of electrons (holes) at the right end of the wire, which constitutes a positive charge. The wire is now negatively charged at one end and positively at the other. The charge represents a difference in potential, or voltage. This voltage is *induced voltage* or *induced emf*. If a circuit were connected to each end of the wire, the voltage across the ends of the wire would drive current through the circuit much as a battery would. Such current is *induced current*.

Figure 40F shows a simple circuit connected to the ends of the conductor. A *galvanometer* is in the circuit. A galvanometer is an instrument that detects or measures a small electric current by movements of a magnetic needle or of a coil in a magnetic field. As the conductor and circuit are moved downward through the field, the galvanometer detects the current in the circuit, which is indicated by the galvanometer's arrow deflection to the right. Also, downward movement causes the charge on the left end of the wire to be negative and the charge on the right end to be positive. However, if the wire is moved upward, the charge on the left end of the conductor becomes positive, and the charge at the right end of the conductor becomes negative. Upward movement also causes the current flowing through the galvanometer to reverse direction. In this case, the galvanometer's arrow deflects to the left.

Induced Current

Figure 41A–B shows a stationary coil and a magnet being moved in and out of the coil. A galvanometer is connected to the coil to indicate current flow. As the magnet is plunged downward into the coil (fig. 41A), the moving magnetic field sets up electric fields in the turns of the coil. This electric field causes the free electrons to move down to the bottom end of the coil, placing a negative charge at the lower end of the coil. The upper end of the coil is positively charged because electrons have moved away from

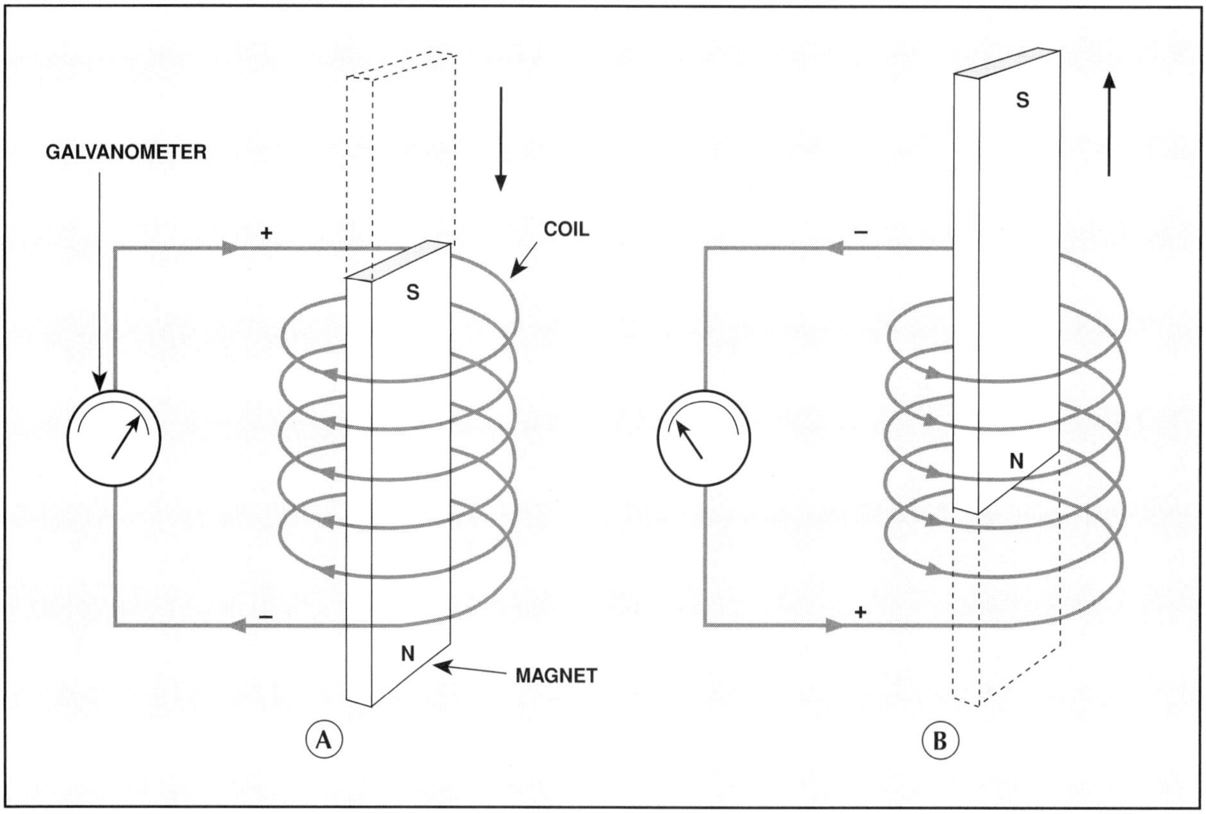

Figure 41 Inducing voltage by moving the magnet

this section of the coil, which leaves a lack of electrons to create a positive charge.

When the magnet is moved upward (fig. 41B), the electric lines assume the opposite direction and drive the free electrons to the upper end of the coil, making it negatively charged. The lower end is positively charged. The meter pointer kicks to the right when the magnet is plunged into the coil and kicks to the left when the magnet is pulled out of the coil. The reversal of the current direction through the meter causes this effect.

Conservation of Energy

According to the *law of conservation of energy*, energy can neither be created nor destroyed. Therefore, the voltage (electrical energy) produced by electromagnetic induction must be produced at the expense of some other form of energy. In this case, mechanical energy, which is the motion of the conductor or the magnet, is changed to electrical energy in the coil. The law of conservation of energy means that as induction produces more electrical energy, more mechanical energy is required to push the magnet in and out of the coil.

Lenz's Law

In 1834, Russian physicist Heinrich Lenz discovered the directional relationships between induced magnetic fields, voltage, and current when a conductor is passed within the lines of force of a magnetic field. Subsequently, he published *Lenz's law*, which states that an induced electromotive force (induced voltage) generates a current that induces a countermagnetic field that opposes the magnetic field generating the current. Put another way, Lenz discovered that the voltage created by electromagnetic induction must be of such a direction that the current flow and its resultant magnetic field oppose the action that produces them.

Simple Generators

A generator is a machine that employs the principles of electromagnetic induction to convert mechanical energy into electrical energy. Since it is not practical to employ a machine that plunges a magnet in and out of a coil, generators either have a coil that rotates in a stationary magnetic field or a magnet or an electromagnet that rotates within a stationary coil.

Rotating Coil

Figure 42 shows two views of a one-turn coil suspended in a magnetic field. The coil is mounted on a shaft so that it can be rotated. The rate at which the coil rotates through the flux lines and the density of the flux lines cutting (or crossing) the coil determines the coil's induced voltage.

When the coil is rotated in the magnetic field at a constant rate, the instantaneous voltage induced in the coil depends on the flux density of the magnetic field at each point of measurement. At the instant when the *flux linkage* is greatest, the induced voltage is the greatest. (Flux linkage is the product of the number of turns in the coil and the magnetic flux passing through the coil.)

In the case of a one-turn coil, the flux linkage is greatest when the coil is traveling through the point of maximum flux density (fig. 42A). At this instant, coil sides *ab* and *cd* are moving perpendicularly to the flux. At this point, the coil cuts the maximum number of flux lines. At the instant during the coil's revolution when the flux linkage is at a minimum—that is, when the coil is passing through the point of minimum flux density—the induced voltage is at a minimum, as shown in figure 42B. At this instant, the coil sides are moving parallel to the flux instead of cutting across it, and no voltage is induced in the coil.

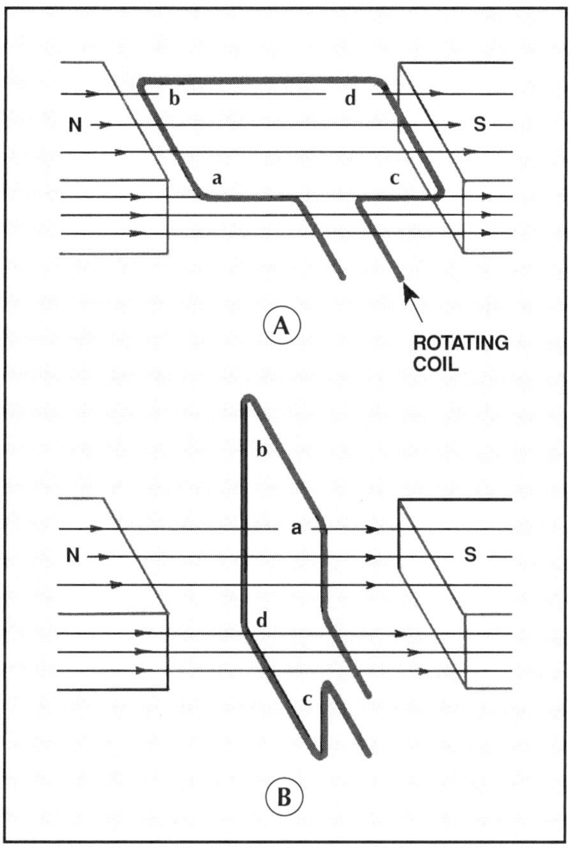

Figure 42 Inducing voltage in a rotating coil

The voltage induced in the coil while it is rotating in the magnetic field occurs because of the action of the conductor cutting the flux lines. Put another way, the voltage induced in a coil occurs because of the relative motion between the coil and the flux linking it. The only parts of the coil that cut the magnetic field when the coil is rotated are sides *ab* and *cd*.

To rotate the coil continuously, the ends of the coil must be connected to devices that do not twist off as the coil rotates. *Slip rings* and *brushes* are used for this purpose (fig. 43). Two copper rings, insulated from each other, are mounted on the shaft, and the ends of the coil are connected to the rings. Thus, the rings rotate as the coil rotates. The brushes are made of carbon and are mounted in a stationary position so that they contact the rings. Conductors attached to the brushes carry electricity to a circuit. In the figure, a *load* resistor is shown. It is placed in many generator circuits to absorb, or consume, electric power. Resistors of the required value are often installed in such circuits to ensure that the correct amount of power flows into the circuit.

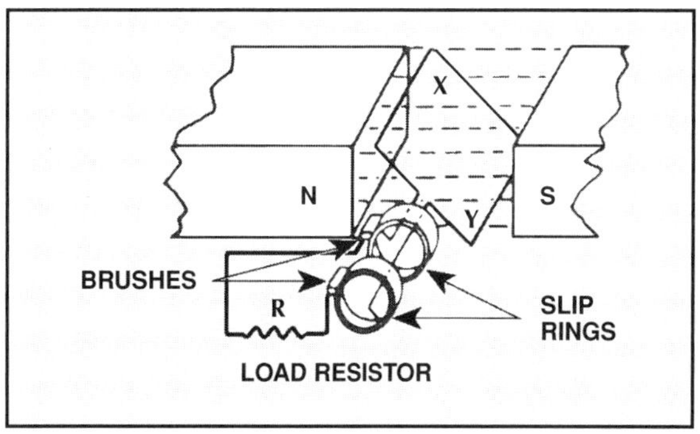

Figure 43 Coil connected to a generator

Flux Cutting Angles

Figure 44 shows a single coil with sides *X* and *Y* rotating 360 degrees in the magnetic flux. The coil's position at various points in its rotation are labeled *A* through *I*. To the right and below each position, a graph plots the induced voltage's *waveform* at that position. (The waveform is a pictorial representation of the voltage as the coil generates it.) The graphs show the instantaneous voltage of side *X* as it rotates.

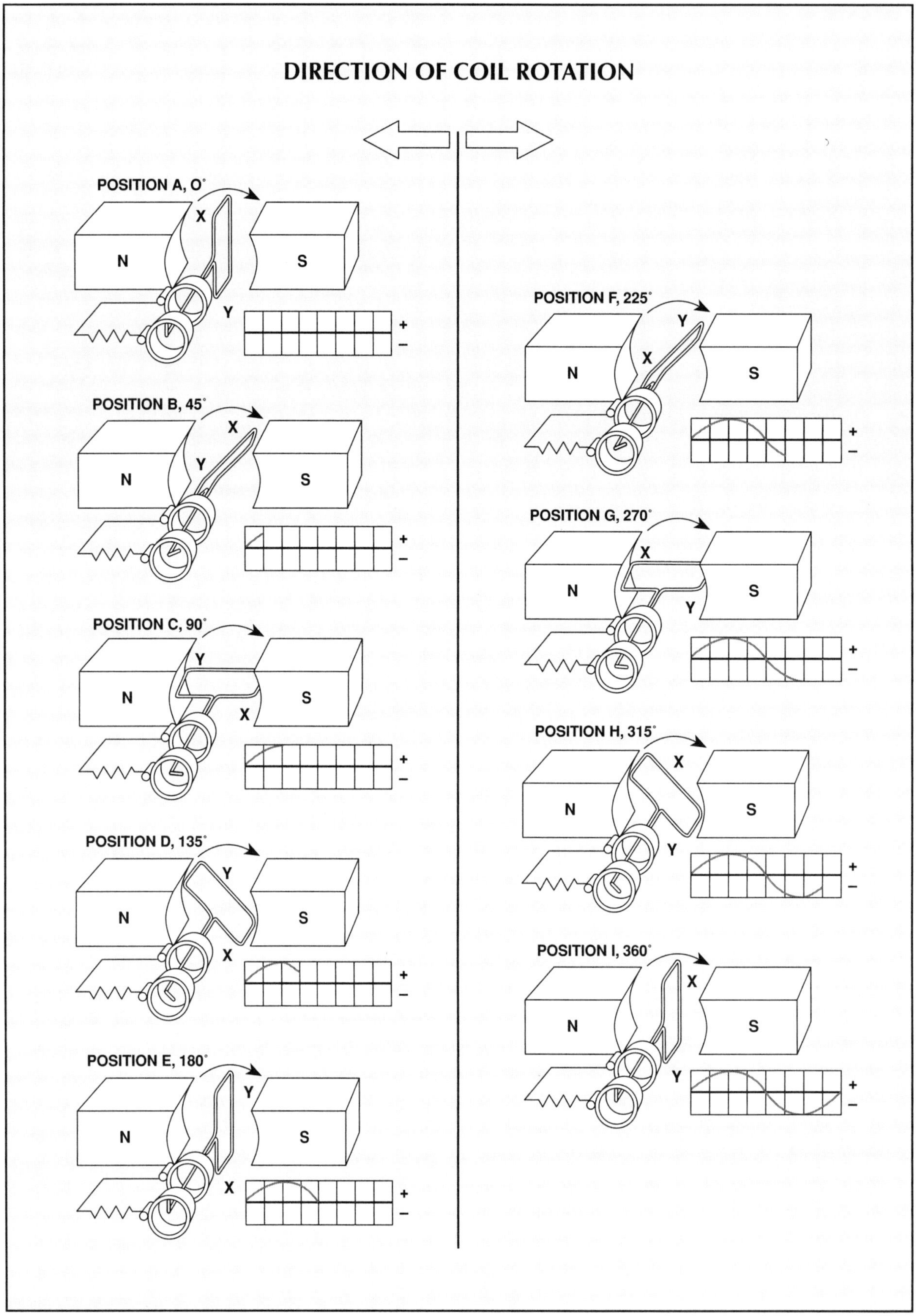

Figure 44 Voltage waveforms produced by a rotating coil

Also, as side X begins its rotation from its starting point (position A), it induces positive voltage. On the other hand, side Y is the opposite—that is, when side X is positive by a certain amount, then side Y is negative by the same amount. (For clarity, figure 44 only shows the waveform for side X.)

At position A, or 0 degrees, side X is parallel to the lines of force. So, it does not cut across the lines; rather it moves along with them. Consequently, the coil does not induce voltage during this period. As side X rotates to the right, it begins to cut downward through the lines, and it induces a small voltage. At position B, side X has rotated 45 degrees, and positive voltage has risen to 70 percent of the maximum positive voltage. Maximum positive voltage is reached at position C, where the coil has rotated to 90 degrees. Here, side X cuts the maximum number of lines per degree of rotation.

Side X continues to cut downward through the flux, but, as the angle increases, it cuts fewer flux lines per degree. At position D, or 135 degrees from the starting point, the voltage has fallen to 70 percent of the maximum positive voltage, which occurred at position C. At position E, or 180 degrees from the starting point, side X again cuts no lines of force because it moves with them. Here, the induced voltage is zero.

When side X begins to cut upward through the flux lines, the polarity changes to negative. At position F, or 225 degrees, side X cuts upward through the flux lines, and it induces 70 percent of the maximum negative voltage. At position G, or 270 degrees from the starting point, side X cuts the maximum number of lines per degree of rotation, and it induces the maximum negative voltage in the generator. At position H, side X still cuts upward, but it cuts a fewer number of lines, and negative voltage falls. At position I, side X has completed 360 degrees of rotation, or one revolution. Since it is moving with the flux lines at this position, no voltage is being induced. Side X is now ready to start another revolution, and the whole process is repeated as long as the coil is rotated through the flux.

ALTERNATING CURRENT PROPERTIES

Alternations and Cycles

A coil rotating through a magnetic flux illustrates alternating current. On the waveform graph, you saw how positive voltage rises and falls as the coil rotates from 0 to 180 degrees through the flux. This rising and falling of the voltage in the positive direction makes up one *alternation*. Similarly, the rising and falling of the induced voltage in the negative direction as the coil rotates from 180 to 360 degrees also constitutes one alternation. One positive alternation and one negative alternation is 1 *cycle*.

Thus, in the two-pole generator illustrated in figure 44, one complete revolution produces one cycle. If the coil is rotated at a speed of 60 revolutions per minute (rpm), the generator produces 60 cycles per minute or 1 cycle per second. If the coil is rotated at a speed of 3,600 rpm, it produces 60 *cycles per second*, or 60 *hertz*, because 3,600 ÷ 60 = 60. Named after a German physicist of the late nineteenth century, hertz means cycles per second. Its symbol is Hz.

Frequency

The number of cycles per second that alternating current undergoes is its *frequency*. The frequency of the alternating current produced by a generator depends on the coil's rotating speed. If the coil turns faster than 3,600 rpm, the frequency is higher than 60 hertz (Hz). Similarly, if the coil rotates slower than 3,600 rpm, the frequency is lower than 60 Hz.

For rotating machines that produce frequency based on the number of magnetic poles and rpm, the frequency can be calculated using the equation—

$$f = \frac{PN}{120} \qquad \text{(Eq. 24)}$$

where

f = frequency, Hz
P = number of magnetic poles
N = rotational speed, rpm
120 = constant.

For example, if a generator has two poles (P) and rotates at 3,600 rpm (N), then—

$$f = \frac{2 \times 3{,}600}{120}$$

$$f = \frac{7{,}200}{120}$$

$$f = 60 \text{ Hz.}$$

The standard frequency for commercial current in the United States is 60 Hz. Some countries use 50-Hz current, and others use frequencies as low as 25 Hz. In companies that generate their own power, generator operators must strive to maintain a constant speed of the engine that drives the generator. If engine speed varies, the frequency of the generated current also varies. While it may be difficult and costly to maintain a precise frequency in relatively small and isolated generator installations, changes in frequency create problems. For example, electric clocks in the U.S. run on 60-Hz alternating current. If the frequency is greater than 60 Hz, the clock will run fast; if the frequency is lower than 60 Hz, it will run slowly. Frequency changes also affect the speed of induction motors.

Mechanical generators can be designed to produce any frequency up to about 20,000 Hz. Above 20,000 Hz, special electronic generators called *oscillators* are employed. They can generate alternating currents with frequencies as high as several billion Hz.

Sine Waves

The graphs in figure 44 show how the induced voltage varies throughout the cycle. The produced voltages make a curve on the graph that takes the shape of a *sine wave.* (A sine wave is an undulating form—a wave—whose height, or amplitude, above and below a reference line varies as the *sine* of a linear function of time. A sine is a trigonometric function that has to do with right triangles. Specifically, the sine is a function of the ratio of the side opposite a given acute angle to the hypotenuse.) You do not need to fully understand what a sine wave is to understand a single-coil generator. It is enough to know that the shape of the waveform, or curve, on the graphs is a sine wave. It has an easily recognized characteristic shape.

Because the generator's output voltage and current waveform make a sine wave, the waveform is also *sinusoidal*, which simply means that it draws a sine wave on a graph. Voltage and current waveforms of generators are sinusoidal because the generator has a circular construction. Rotating, circular coils in generators generally produce waveforms that appear as sine waves.

Three-Phase AC

A generator with a single coil produces *single-phase voltage*. In this case, a phase is one complete revolution of the coil in the magnetic field. In figure 44 at position *I*, the sine wave on the graph illustrates single-phase voltage. As long as the coil turns, this single, complete wave is generated over and over. That is, a single sine wave occurs, which curves upward, indicating positive voltage, during the first half of the coil's revolution and curves downward, indicating negative voltage, during the last half of the coil's revolution.

Often, however, three single-phase sources, such as an AC generator, are connected together. They are connected so that the output from each generator is electrically displaced—that is, while voltage output is decreasing in one generator, it is increasing in another. Such displacement avoids generating electricity with gaps in its output, which, as you recall, occur twice in a cycle where the coil is parallel to the flux. Displacing the output in three single-phase generators eliminates gaps and thus increases efficiency. Three-phase power can be transmitted efficiently and is compatible with AC induction motors, transformers, and generators.

Normally, three single-phase sources are electrically displaced from each other by 120 degrees (fig. 45). That is, in the complete 360-degree rotation of the coil, each is timed so that at the 120-degree rotation point of the first coil, where voltage is decreasing toward zero, the second coil is going upward and voltage is increasing. The third coil is timed to begin rising at the point where the second coil is going down at the 120-degree point in its travel. Meanwhile, the first coil is producing increased negative voltage (shown by the sine wave's being below the zero line on the graph). Connecting three single-phase sources in this manner makes it *three-phase voltage*.

Characteristics of three-phase voltage include—

- the three voltages that basically have the same magnitude (voltage).
- the frequency of each phase is identical.

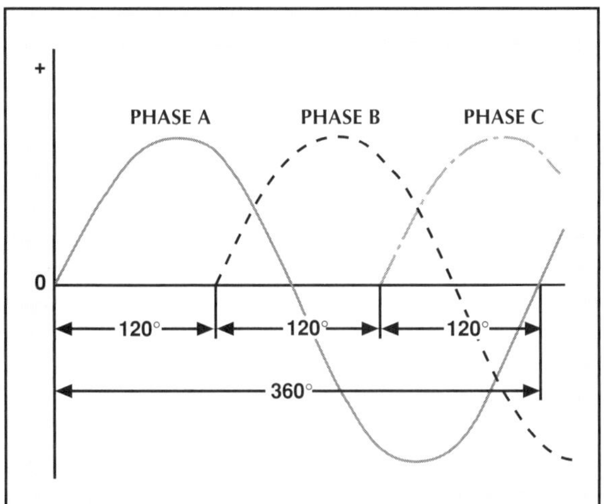

Figure 45 Three-phase voltage

- each voltage is phase-separated from the other by 120 electrical degrees.

REVIEW QUESTIONS

1. What sets up magnetic lines around a conductor?
2. At what point are the flux lines denser around a current-carrying wire?
3. What effect does reversing the direction of current flow have on the direction of magnetic lines?
4. If the current is increased through a conductor, how is the magnetic field affected?
5. If parallel wires are carrying currents in opposite directions, do they attract or repel?
6. Describe the right-hand rule for conductors.
7. Write Ohm's law for electric circuits and the law that is commonly called Ohm's law for magnetic circuits (Rowland's law).
8. Reluctance is analogous to what electrical term?
9. Magnetomotive force can be likened to what electrical term?
10. What electrical term can magnetic flux be likened to?
11. What three factors determine the reluctance of a magnetic path or circuit?
12. What is a core?
13. What is permeability? How does permeability vary?
14. What is magnetizing force?
15. What is flux density?
16. An alternator has three basic functional elements; name them.
17. What is the frequency of an eight-pole alternator rotating at 900 rpm?
18. What is the phase separation between the three-phase voltages?

Electrical Components, Impedance, and Power

RESISTORS

Although several materials resist the flow of electric current, most circuits require a specific resistance to operate properly. For example, electronic amplifiers, control circuits, instrumentation loops, and similar devices require resistors of a fixed value. When electricity flowing in a circuit encounters a resistor, the resistor reduces the voltage to a specific and required amount. Resistors provide the correct operating voltage to a component or to another part of the circuit. In short, resistors create voltage drops where they are wired into the circuit.

Resistors are made from (1) a special wire, (2) a mixture composed of conductive and nonconductive materials termed a *composition*, or (3) a thin membrane, or skin, called *film*. Figure 46 shows a wire-wound resistor, a composition resistor, and the symbol for a resistor, which is used in schematics (drawings) of circuits. Also shown is the Greek letter Ω (omega), which stands for ohms. The resistor in figure 46 is a 15-ohm, or 15-Ω, resistor.

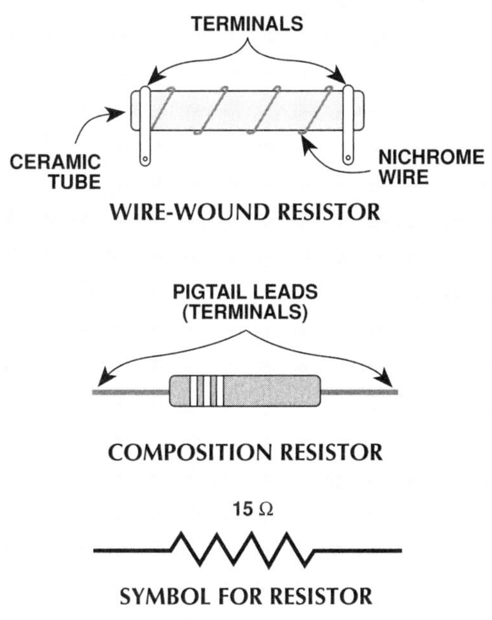

Figure 46 Two types of resistors and the symbol for a resistor

Wire-Wound Resistors

Wire-wound resistors consist of high-resistance wire wound on a ceramic spool or tube. Wire-wound resistors give off a large amount of heat; so, they are used in well-ventilated circuits. Manufacturers often employ Nichrome wire in wire-wound resistors because of its high resistance. A wire-wound resistor's value in ohms depends on the length of the wire wound on the tube, the diameter of the wire, and the grade of Nichrome wire. Resistance in wire-wound resistors is usually given as *ohms per mil-foot*. Thus, the longer the wire and the heavier, or thicker, its gauge, the higher is its resistance.

As mentioned earlier, wire-wound resistors give off heat and heat speeds up *oxidation*. Nichrome wire not only has high resistance, but also it resists oxidation. That is, Nichrome wire does not readily combine with oxygen in the air and oxidize. Just as rust damages or destroys an iron or steel object, oxidation can damage or destroy the wire on a wire-wound resistor. Therefore, iron wire is not satisfactory for resistors because it rapidly oxidizes.

Wire-wound resistors can handle large amounts of power and can be made to relatively close tolerances. They are rugged and reliable, they have low temperature coefficients—that is, temperature changes do not significantly affect their resistance—and they are stable under most operating conditions. Also, low-resistance wire-wound resistors with close tolerances are very economical to make.

On the other hand, wire-wound resistors are not available in as wide a range of resistance values as film and composition resistors. Also, for a given *power rating* and resistance value, they are significantly larger than the other two. (A resistor's power rating is a measure of how much power, in watts, the resistor can handle without failing or malfunctioning.) Although manufacturers can make high-resistance wire-wound resistors, they are very expensive.

Composition Resistors

Composition resistors are the most common. They are frequently used in electronic equipment. Composition resistors are made of conductive and nonconductive materials that are placed inside a ceramic tube, or shell. By varying the ratio of conductive and nonconductive material and the physical size of the materials used, manufacturers can produce resistors of practically any *ohmic* value. Indeed, composition resistors of thousands or millions of ohms can be economically constructed.

Composition resistors are small in size; consequently, they dissipate only small amounts of heat. Manufacturers carefully select the materials from which they make resistors to keep the temperature coefficient (the amount resistance increases as temperature increases) small.

Composition resistors are available in a wide range of values and can be produced economically in power ratings of up to 2 watts. That is, they can handle up to 2 watts of constant power without malfunctioning or failing. Composition resistors are rugged and reliable when used properly. However, some composition resistors suffer a loss of stability if operated in conditions of high-moisture. Also, their resistance may vary an unacceptable amount if operating temperatures change rapidly and drastically.

Because of its relatively small size, color bands, rather than printing, on the body of a composition resistor indicate its ohmic value and its *resistance tolerance* (fig. 47). The tolerance of a resistor refers to the accuracy of its resistance value—that is, how close the resistor's indicated resistance is to its actual value. For example, if the resistor's tolerance is 20 percent, the actual value of the resistance may vary as much as plus or minus 20 percent from the value indicated by the color code.

Figure 47 shows that twelve color bands, plus a lack of color, indicate resistance and tolerance. Bands A, B, and C indicate resistance, while band D indicates tolerance. For example, suppose a resistor with *axial leads* has bands of orange, white, red, and gold. (Resistors with axial leads have wires, or leads, that come out from the center of each end of the small cylinder that makes up the resistor's body.)

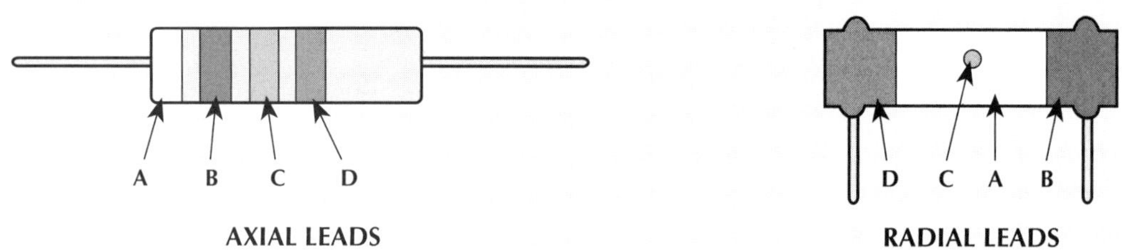

Band or Body A		Band B		Band or Dot C		Band D	
Color	*Digit*	*Color*	*Digit*	*Color*	*Multiplier*	*Color*	*Tolerance*
Black	0	Black	0	Black	1	Silver	10%
Brown	1	Brown	1	Brown	10	Gold	5%
Red	2	Red	2	Red	100	No band	20%
Orange	3	Orange	3	Orange	1,000		
Yellow	4	Yellow	4	Yellow	10,000		
Green	5	Green	5	Green	100,000		
Blue	6	Blue	6	Blue	1,000,000		
Violet	7	Violet	7	Silver	0.01		
Gray	8	Gray	8	Gold	0.1		
White	9	White	9				

Figure 47 Resistor color code

From figure 47, you can find that band A, which is orange, represents the first digit of 3, band B, which is white, represents the second digit of 9, and band C, which is red, indicates a multiplier of 100. So, in this example, 39 × 100 = 3,900, or 3,900 ohms (Ω). Since band D is gold, the resistor's tolerance is 5 percent. Five percent tolerance means that the resistor's actual resistance is between 3,705 and 4,095 ohms. In other words, 3,900 × 0.05 (5 percent) = 195; 3,900 – 195 = 3,705 ohms; and 3,900 + 195 = 4,095 ohms. If a resistor's tolerance is 20 percent, band D is not present—that is, no fourth color band exists.

As another example, consider a 3.9-ohm resistor with a tolerance of 10 percent. It has four bands: orange, white, gold, and silver. Orange represents 3, white 9, gold a multiplier of 0.1 (thus, 39 × 0.1 = 3.9 ohms), and silver a tolerance of 10 percent.

A resistor with *radial leads* (see fig. 47) has essentially the same color-coding, except that a dot, rather than a band, represents the multiplier. Also, in the case of precision resistors, whose tolerance is below 5 percent—1 or 2 percent, for example—the resistance is usually printed on the resistor's body.

Film Resistors

Film resistors are similar to composition resistors because they are about the same size for given values of resistance and power rating and they have similar stability characteristics, temperature behaviors, and manufacturing tolerances. They can be obtained in precision grades—that is, the tolerances are tighter than 5 percent—at a cost less than that of other types. The resistance element in a film resistor may be a carbon film, a metal film, or a metallic oxide film.

Preferred Resistance Values

A set of *preferred resistance values* exists for any electronic circuit's *capacitors* and resistors. (A capacitor is a component that stores an electrical charge in a circuit and then returns the charge to the circuit when certain electrical conditions occur. Capacitors are discussed later in this section.) Preferred values are standard values that manufacturers use to make resistors of various tolerances. Most circuits can tolerate a 10 to 20 percent deviation from a resistor's stated value without a significant loss in performance. Because such deviation does not affect performance, manufacturers can make large quantities of resistors at low prices that meet standards.

Publishers print many versions of preferred values in tabular form. The table in this book (Table 12) assigns the letter E followed by a one- or two-digit number to designate preferred values in resistors—for example, E3 and E12. Six series of preferred values are available: E3, E6, E12, E24, E48, and E96.

The numbers after E, which are 3, 6, 12, 24, 48, and 96, indicate the number of resistance values in each series. For example, the E12 series has 12 resistance values and the E24 series has 24 resistance values. Each series has a tolerance specification—that is, resistors in a particular series are manufactured to a certain tolerance. E12 resistors, for example, have a 10 percent tolerance and E96 resistors have a 1 percent tolerance. So, a 47-ohm, E12 resistor could vary as much as plus or minus (±) 4.7 ohms and still fall within tolerance (47 × 0.1 or 10 percent = 4.7). On the other hand, E96 resistors have only 1 percent tolerance. So, a 280-ohm E96 resistor could vary only ±2.8 ohms.

The most popular series of resistors is the E12 series with 10-percent tolerance values. As you can see from table 12, the preferred resistance values in the E12 series are 10, 12, 15, 18, 22, 27, 33, 39, 47, 56, 68, and 82. Each number represents a set of ten—a *decade* of—resistor values in ohms. For example, consider an E12 value of 47. This number means that E12 resistors are available with a resistance value of (1) 4.7 ohms, (2) 47 ohms, (3) 470 ohms, (4) 4,700 ohms, (5) 47,000 ohms, (6) 470,000 ohms, (7) 4,700,000 ohms, (8) 47,000,000 ohms, (9) 470,000,000 ohms, and (10) 4,700,000,000 ohms. Any of the resistors in this decade—this group of ten—may vary ±10 percent from its designated resistance.

Incidentally, some tables of preferred resistance values place a decimal point after the first number in two-digit values and after the second number in three-digit values. For example, such a table of E12 resistor values would show values of 1.0, 1.2, 1.5, and so on, instead of 10, 12, 15, and so on. Similarly, a table of E48 values would show values of 10.0, 10.5, 11.0, and so on, instead of 100, 105, 110, and so on. The values are the same whether or not a decimal point appears. Some table publishers simply prefer to show the first and lowest resistance value in the decade of values.

TABLE 12
Preferred Values of Resistors

10	22	47

E3 Preferred Values (10 percent tolerance)

10	12	15
33	47	68

E6 Preferred Values (20 percent tolerance)

10	12	15
18	22	27
33	39	47
56	68	82

E12 Preferred Values (10 percent tolerance)

10	11	12	13	15	16
18	20	22	24	27	30
33	36	39	43	47	51
56	62	68	76	82	91

E24 Preferred Values (5 percent tolerance)

100	105	110	115	121	127	133	140
147	154	162	169	178	187	196	205
215	226	237	249	261	274	287	301
319	332	348	365	383	402	422	442
464	487	511	536	562	590	619	649
681	715	750	787	825	866	909	953

E48 Preferred Values (2 percent tolerance)

100	102	105	107	110	113	115	118
121	124	127	130	133	137	140	143
147	150	154	158	162	165	169	174
178	182	187	191	196	200	205	210
215	221	226	232	237	243	249	255
261	267	274	280	287	294	301	309
316	324	332	340	348	357	365	374
383	392	402	412	422	432	442	453
464	474	487	499	511	523	536	549
562	576	590	604	619	634	649	665
681	689	715	732	750	768	787	806
825	845	866	887	909	931	953	976

E96 Preferred Values (1 percent tolerance)

Notice that gaps in the decade of values for resistors exist. For example, E12 resistors are not available in the decade of values between 39, 47, and 56. Thus, you can obtain, let's say, a 470-ohm, E12 resistor, but not a 420-ohm, E12 resistor. The reason is because of the tolerance to which E12 resistors are made. For example, a 470-ohm, E12 resistor with 10 percent tolerance could have a resistance as low as 423 ohms or as high as 517 ohms (470 – 47 = 423 and 470 + 47 = 517). Therefore, no need exists for manufacturing E12 resistors closer to 470 than 390 ohms on the low side or 560 ohms on the high side because E12 resistor tolerances are greater than the spacing between values. For example, a 390-ohm E12 resistor could have a resistance as high as 468 ohms, which is very close to 470 ohms (390 + 39 = 468). Moreover, a 560-ohm resistor could have a resistance as low as 504 ohms (560 – 56 = 504). This resistance is very close to 514 ohms and is the value a 470-ohm, E12 resistor would have if its tolerance were 10 percent high.

While most electronic equipment uses resistors with the preferred standard values shown in table 12, special devices may need resistors with tolerances tighter than 1 percent. Such resistors cost more to produce, of course, but precision equipment may require them. One piece of equipment that requires resistors with narrow tolerances is a programmable logic controller (PLC). A 250-ohm resistor with a tolerance of only ±0.1 percent is used in certain PLC circuits to provide compatibility with other components.

High Resistance Values

Technicians and engineers often write high resistance values, such as 47,000 ohms and higher, with prefixes that indicate thousands, millions, and billions. For example, 47,000 ohms can also be written as 47 *kilohms*, which is abbreviated 47 kohms, or 47 kΩ. This 10 × 10 = 1,000 and 47 × 1,000 = 47,000. Similarly, 47,000,000 (47 million) ohms is easier to write as 47 *megohms*, which is abbreviated 47 Mohms, or 47 MΩ. (This figure can also be written as 47×10^6.) And 4,700,000,000 (4 billion, 700 million) ohms is 4.7 *gigohms*, which is abbreviated 4.7 Gohms, or 4.7 GΩ. (This figure can also be written as 4.7×10^9.) In short, kilo stands for 1,000 (thousand); mega stands for 1,000,000 (million); and giga stands for 1,000,000,000 (billion). Note that when these prefixes are used with "ohms," the vowel "a" is dropped to avoid confusion. Thus, it is "kilohms" instead of kilo-ohms, "megohms" instead of mega-ohms, and "gigohms" instead of giga-ohms. Table 13 provides a list of prefixes used with various units.

Power Ratings of Resistors

When current passes through a resistor, heat is produced. The resistor must be able to dissipate this heat into the surrounding air. Generally, the larger the surface area of the resistor, the greater the amount of heat it can give off to the surrounding air. A resistor's ability to dissipate heat is expressed in terms of watts rather than in heat units such as *British thermal units* (Btus). Watts is an electronic term that is

TABLE 13
Prefixes for SI Units

Prefix	Symbol	Power	Example	Prefix	Symbol	Power	Example
yotta	Y	10^{24}		deci	d	10^{-1}	decilitre (dL)
zetta	Z	10^{21}		centi	c	10^{-2}	centimetre (cm)
exa	E	10^{18}		milli	m	10^{-3}	milligram (mg)
peta	P	10^{15}		micro	*m*	10^{-6}	microgram (*m*g)
tera	T	10^{12}	terawatt (TW)	nano	n	10^{-9}	nanosecond (ns)
giga	G	10^{9}	gigawatt (GW)	pico	p	10^{-15}	picofarad (pF)
mega	M	10^{6}	megahertz (MHz)	femto	f	10^{-15}	femtosecond (fs)
kilo	k	10^{3}	kilometre (km)	atto	a	10^{-18}	
hecto	h	10^{-2}		zepto	z	10^{-21}	
deka	da	10^{-1}	dekalitre (daL)	yacto	y	10^{-24}	

appropriate for resistors, whereas a Btu is appropriate for objects that have nothing to do with electronics. In any case, the higher a resistor's watts rating, the better is its ability to dissipate heat.

Figure 48 shows the relative sizes of three composition and two wire-wound resistors. Often, the larger the resistor, the higher is its power rating in watts. Usually, manufacturers do not print the wattage rating on resistors. What is more, resistors may have the same wattage rating but be of different sizes. Fortunately, the packaging or containers for the resistors usually specifies the wattage ratings. If not, the manufacturer should be able to provide it.

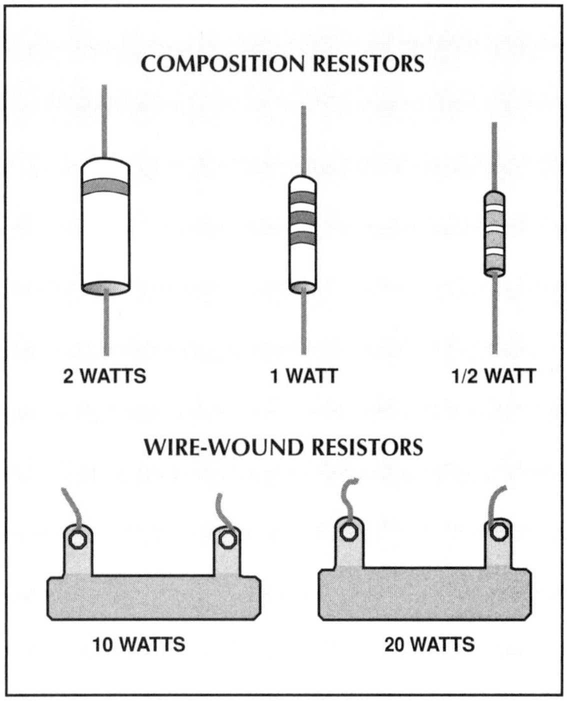

Figure 48 Relative resistor sizes in various power ratings

The circuit in which a resistor is used determines the required wattage rating. The wattage dissipated by a resistor in a circuit can be determined by one of three power law equations—

$$P = VI \quad \text{(Eq. 25)}$$

$$P = I^2R \quad \text{(Eq. 26)}$$

$$P = \frac{V^2}{R} \quad \text{(Eq. 27)}$$

where

P = power dissipated in resistor, watts
I = current through resistor, amperes
R = resistance, ohms
V = voltage across resistor, volts.

As an example of using the power law equations, suppose a circuit requires a 3-ohm resistor that has a 6-volt drop across it and current flow is 2 amperes. What wattage rating should the resistor have? Using equation 25—

$$P = VI$$
$$P = 6 \times 2$$
$$P = 12.$$

Thus, a resistor with a 12-watt power rating is needed. If, however, a supplier only stocks resistors with power ratings of 10 watts and 20 watts, the resistor with the next higher rating should be chosen—in this case, the 20-watt resistor. The 20-watt resistor operates cooler and lasts longer.

Composition resistors are commonly manufactured with 0.1-, 0.25-, 0.5-, 1-, and 2-watt ratings. Wire-wound resistors are generally employed when a circuit requires wattage ratings greater than 2 watts.

Power Ratings of Series and Parallel Circuits

Wiring resistors in series or in parallel affects the total amount of resistance created by the resistors. For example, wiring two 5,000-ohm resistors in series creates 10,000 ohms of resistance. On the other hand, wiring two 5,000-ohm resistors in parallel creates 2,500 ohms of resistance. Similarly, if 10,000 ohms of resistance is required in a parallel circuit, it can be achieved by wiring two 20,000-ohm resistors in parallel. In short, wiring resistors in series adds their resistance while wiring resistors in parallel halves their resistance.

Resistor power ratings, on the other hand, add together, whether the resistors are in series or parallel. Therefore, the equivalent of a 10,000-ohm, 10-watt resistor may be made by placing two 5,000-ohm, 5-watt resistors in series or by placing two 20,000-ohm, 5-watt resistors in parallel (fig. 49). In both cases, the power rating in watts adds up, but the resistance is either doubled or halved depending on whether the resistors are in series or in parallel.

Resistors in AC Circuits

When resistors are used in AC circuits, they react in much the same way as in DC circuits. Ohm's law applies to the various electrical components when *root-mean-square (RMS) values* are used to calculate

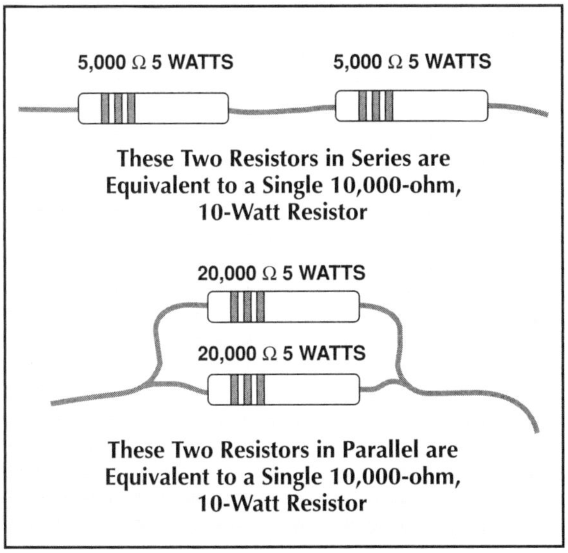

Figure 49 Series and parallel resistors

current, voltage drops, and power. An RMS value is the square root of the time average of the square of a quantity. RMS values are used to characterize the frequency response of certain devices.

AC voltage applied to a circuit that alternates polarity in cycles does not affect the resistance of most resistors. Put another way, most resistors are insensitive to the frequency of the AC voltage. However, AC voltage frequency can affect some wire-wound resistors. For example, supply frequencies considerably higher than 50 or 60 Hz affect a wire-wound resistor's resistance. Also, *inductance* in a wire-wound resistor may affect the resistor's value. (Inductance is the property of an electric circuit or of two neighboring circuits wherein an electromotive force is generated in one circuit by a change of current in it or in the other circuit. Because a wire-wound resistor is coiled wire, it generates inductance.)

A significant relationship also exists between resistance, current, and voltage in an AC circuit. If current and voltage are flowing in a circuit in which only resistors or resistive components exist, and the current and voltage are displayed as waveforms, the waveforms show complete harmony (fig. 50). Or, in electronic terms, the current and voltage waveforms do not show any *phase shift*. (Phase shift occurs when the waveforms are not in harmony. The waveforms move out of phase. For example, part of the current waveform may be on the positive side of the zero line while part of the voltage waveform may be on the negative side or vice versa.)

POTENTIOMETERS

A *potentiometer* is a variable resistor. Potentiometer is formed from the words "potential" (voltage) and "meter," which means to payout, monitor, or change. Thus, a potentiometer is used to change, or vary, the potential—the voltage—in the circuit in which it is installed. For example, a potentiometer is often used to increase or decrease the volume (loudness) on a radio because it can increase or decrease the resistance to the voltage flowing in the radio's volume circuit. Potentiometers also control and adjust electronic and control equipment. They can regulate the set points of instruments and they can control alarms.

A potentiometer has three terminals: two are fixed and one is mechanically connected to a shaft. Voltage enters the potentiometer through one of the

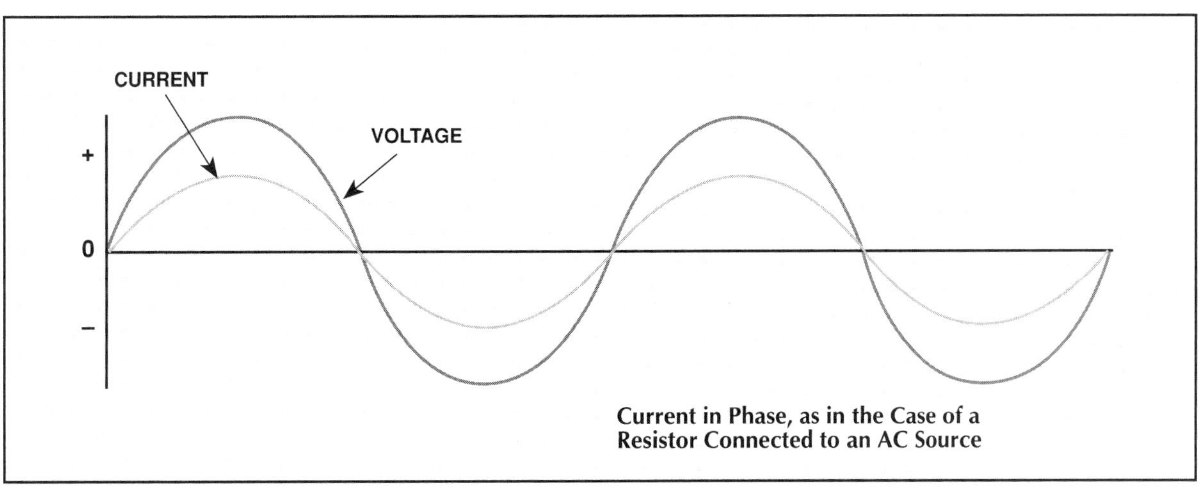

Figure 50 AC voltage and current relationship in resistive circuits

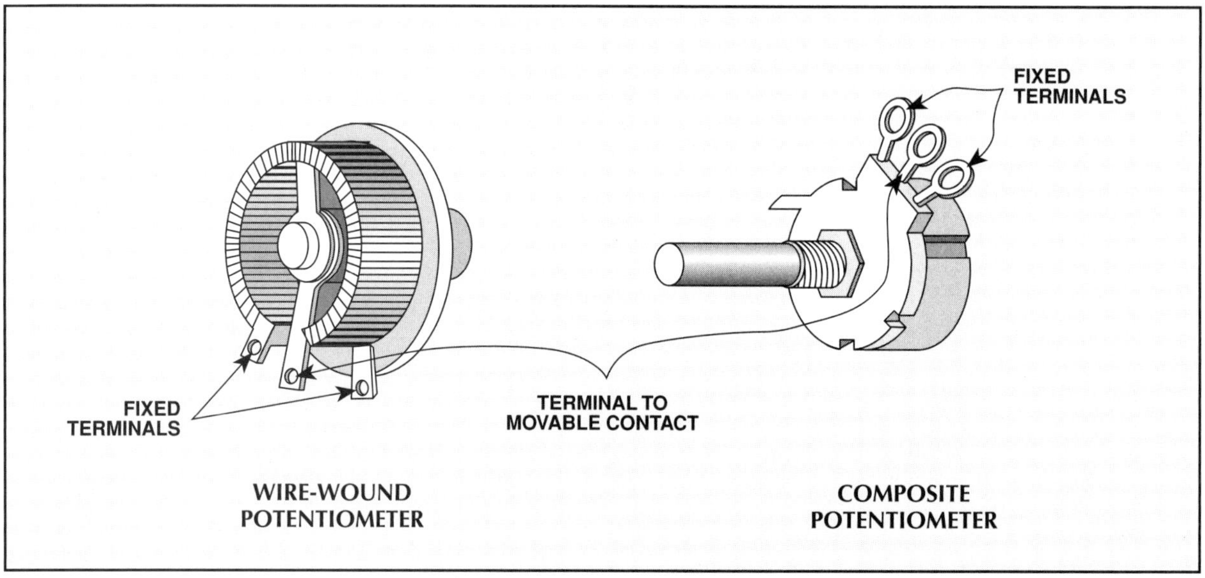

Figure 51 Types of potentiometers

fixed terminals, goes to the terminal connected to the shaft, and travels to the rest of the circuit by exiting the other fixed terminal. When rotated, the shaft moves the terminal's contact to any point along the conducting material of the potentiometer. So, by rotating a potentiometer's shaft, which is connected to the middle, adjustable terminal, fixed voltage is varied and sent to additional components in the circuit.

Two kinds of potentiometer are available: wire-wound and composite (fig. 51). The material used in the potentiometer is a conductor such as composition material, Nichrome wire, or resistance films. Also, note the fixed terminals and terminal to the movable contact.

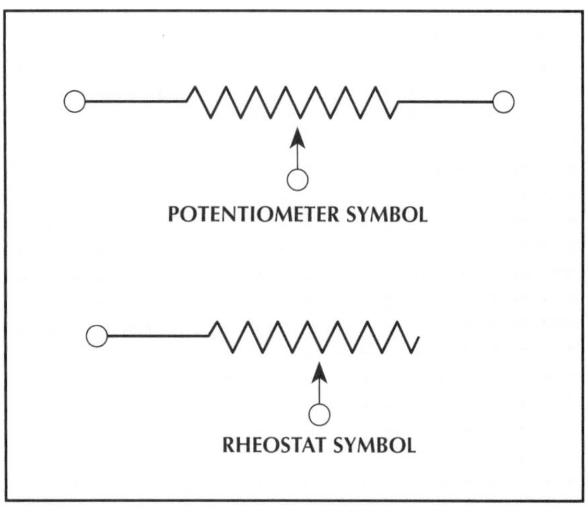

Figure 52 Schematic symbols of a potentiometer and a rheostat

Potentiometers can also provide a variable resistance function. When the variable terminal is used with only one fixed terminal, the resistance between the two changes as the shaft is rotated. When used in this manner, the device is a *rheostat*. Figure 52 shows the schematic symbols for a potentiometer and a rheostat. Note that the rheostat has only one fixed terminal, while the potentiometer has two.

Potentiometer Resistance Ratings

Potentiometers are rated in terms of their resistance and maximum power dissipation in watts. The resistance value of a potentiometer is measured from one fixed terminal to the other fixed terminal. The two outside terminals are connected to the fixed connections of the resistive element. Then, an ohmmeter is used to make the measurement. The ohmmeter should always be connected to the outside, fixed terminals of the potentiometer.

No color code indicates the value in ohms of the potentiometer. Instead, the ohmic value is stamped on the cover of the unit. Use caution when reading the value because the number may have a multiplier. For example, a 2,000-ohm potentiometer may be stamped 2K or 2kΩ. The latter is a better designation because it agrees with the accepted symbols and prefixes of the *International System of Units (SI)*, which is a worldwide system of measurement units based on the metric system. For either designation, however, k or K is the symbol for kilo, which is a multiplier of 1,000, or 10^3. The resistance of a 2k or 2kΩ potentiometer is 2,000 ohms or 2 kilohms (kΩ).

The letter M may also follow a number stamped on an electronic component. M stands for mega, which multiplies the value by 1 million (1,000,000, or 10^6). On a potentiometer, it designates 1 megohm (MΩ) when it is associated with a number stamped on a resistance component. Again, the proper designation is MΩ, because these symbols are in agreement with accepted SI measurement standards.

Bear in mind that the k or M that follows a resistance can indicate values less than 1,000 or 1 million. For example, if a potentiometer is marked 0.8 k (0.8 kΩ), its resistance is 0.8 kilohms, or 800 ohms. In the same way, a potentiometer marked 0.2 M (or 0.2 MΩ) has a resistance of 0.2 megohms, or 200,000 ohms (0.2 × 2,000,000 = 200,000). Potentiometers are manufactured with resistance values ranging from a few ohms to 10 megohms (MΩ) and higher.

Power Dissipation Ratings

Potentiometers with composition elements usually have a power dissipation rating of 1 or 2 watts. These ratings are ample for most electronic equipment. If, however, a potentiometer must dissipate a substantial amount of wattage, then wire-wound potentiometers are used. Electrical control panels use potentiometers rated at hundreds of watts, which are much larger than those on electronic gear. The wattage rating is stamped on the control near the resistance rating of the unit.

Multiple Potentiometer Units

Two or more potentiometers may be coupled together as one unit (fig. 53). One shaft telescoped inside the other allows each potentiometer to be adjusted independently. Further, an on-off switch may be mounted on the back of the control and operated by the shaft that controls one of the potentiometers. Such an arrangement provides an on-off range control for the potentiometer.

Resistance Tapers

Some potentiometers are wound with two or more sizes of Nichrome wire. Multiple sizes of wire are employed in circuits that require a greater or a lesser change in resistance per degree of rotation of the shaft. Usually, the dissimilar resistance wire is connected to one of the fixed terminals. If the potentiometer's control is uniformly wound, it has a *linear taper*. Linear taper gives the potentiometer proportionate control over the resistance. For example, in a potentiometer with a linear taper, turning the control knob one-quarter turn varies the resistance by one quarter, either up or down, depending on the direction the control is rotated.

Other tapers are also available—for example, some controls may be wound with a *logarithmic taper*. (A logarithm is the exponent to which a number is raised to produce a given number. For example, the logarithm of 100 to the base 10 is 2 because 100 = 10^2.) A potentiometer with a logarithmic taper provides nonproportional resistance control. That is, turning the control knob one-quarter turn does not

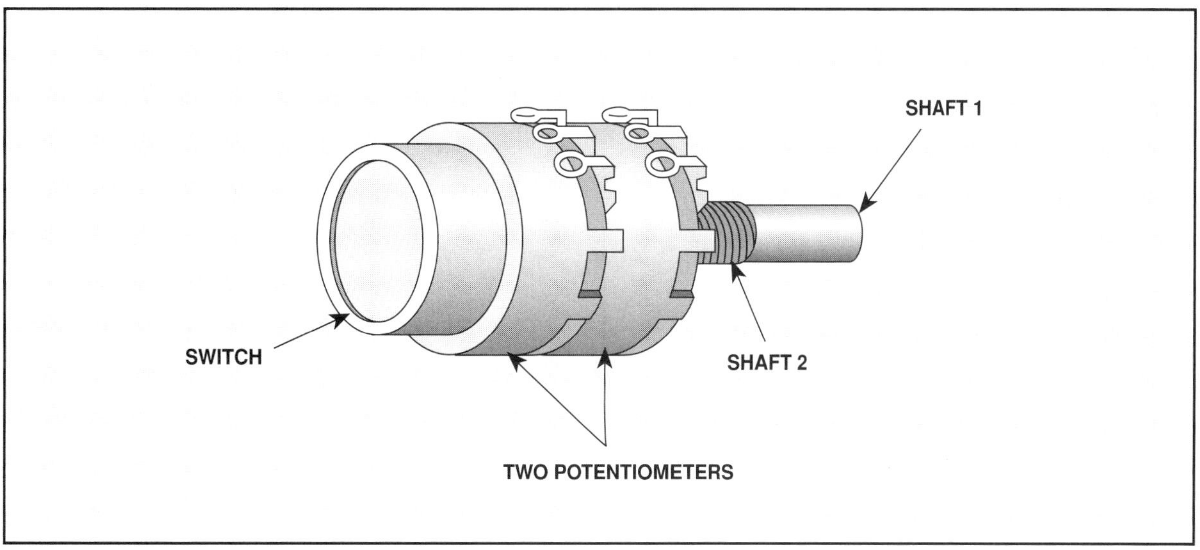

Figure 53 Multiple-stacked potentiometers

vary the resistance by one quarter; rather, it may increase or decrease the resistance two- or four-fold, depending on the control's taper.

INDUCTORS AND INDUCTANCE

Certain electric circuits have inductance, particularly those made up of coils of wire. Inductance not only occurs in a single circuit, but also in nearby, or neighboring, circuits—circuits close to each other. Inductance occurs when changes in current flow in a circuit induces voltage in itself or in a neighboring circuit. In the case of a single circuit, the induced voltage is *countervoltage*—that is, the induced voltage's polarity is opposite the voltage that is applied to the circuit. When variations in current flow occur in a primary circuit, the variations can induce voltage in a neighboring circuit.

Figure 54 illustrates inductance and induced voltages in a circuit energized by a battery. Besides a battery, the circuit has a switch, two voltmeters (V_s and V_p) and two windings, or coils (S and P). When the switch is closed (fig. 54A), battery voltage is applied to the primary winding (P) of the setup. At the moment the voltage is applied, a voltage is also induced into the secondary winding (S). Voltmeter V_s's needle tips quickly to the right or left and then returns to zero, which indicates induced voltage. Voltmeter V_p continues to read battery voltage until the switch is opened. At the moment the switch is opened (fig. 54B), the needle in voltmeter V_s moves to indicate a voltage opposite in polarity to that observed when the switch was first closed. The needle in voltmeter V_p also moves to indicate an opposite polarity. Then both meters quickly return to zero.

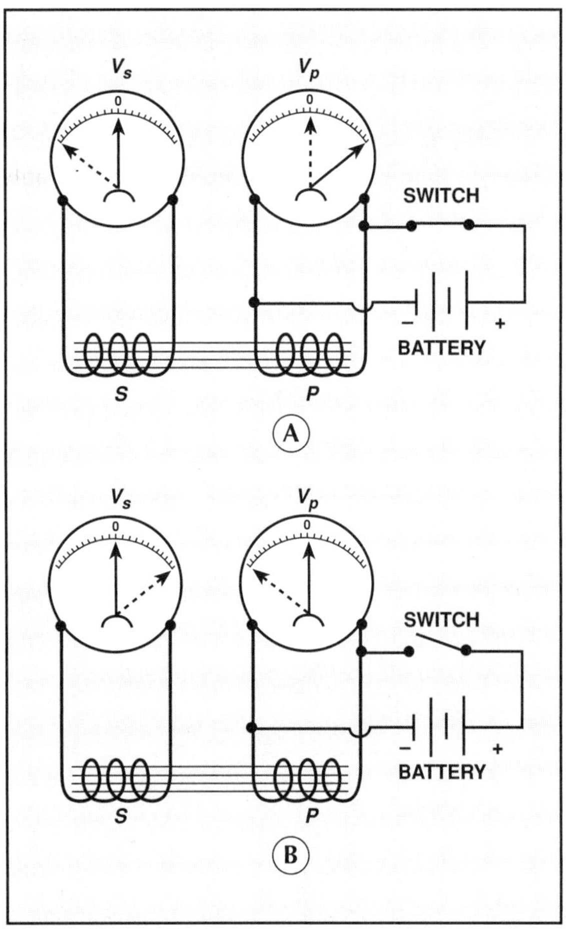

Figure 54 Inducing voltages in inductive circuits

An inductive device opposes a change in current flow. Put another way, when current flow through an inductive device rises or falls, it induces a voltage that tends to oppose the change in current. The induced voltage tries to maintain a steady value of current flow.

The unit of inductance is the *henry*, named for Joseph Henry, a nineteenth century American scientist who pioneered induction research. Traditionally, the symbol for inductance in henries is the capital letter L, although the SI system prefers the capital letter H to avoid confusion with the abbreviation for litre, which is also L. (This manual uses H.)

If current flowing through an inductance coil changes at the rate of 1 ampere per second and a countervoltage of 1 volt is generated, then the coil has an inductance of 1 henry (1 H). An equation for determining inductance in a circuit is—

$$H = \frac{4\pi \times 10^{-9} \times \mu \times A \times N^2}{l} \quad \text{(Eq. 29)}$$

where

H = inductance, henries
$4\pi \times 10^{-9}$ = a constant (arises from the mathematical derivation of the equation and helps adjust for the use of particular units of measurement)
μ = permeability of the core material
A = cross-sectional area of core, in.^{2}
N = number of turns in the coil
l = length of the core, in.

Unfortunately, the equation does not always hold true because the permeability of an iron core varies with flux density. Also, manufacturers of inductance circuits rate their products for a specific amount of current flow. As the amount of current flow changes, so does the inductance. Small inductors used in radios are usually rated in millihenries. A *millihenry* (mH) is one-thousandth of a henry, or 10^{-3} henries.

High-frequency coils may have an air core. The magnetic permeability of air is 1, or, in mathematical terms, is *unity*, which means that it has virtually no permeability. Therefore, the coils retain a reasonably stable value of inductance over wide ranges of current flow because flux density does not change appreciably in coils with an air core.

Inductive Reactance

Inductive reactance is the retarding effect on the passage of alternating current through a circuit because of inductance. Put another way, inductive reactance occurs in a circuit when inductance in the circuit impedes the flow of AC electricity. Inductive reactance is important to electronics because it affects the operation of motors and *transformers*. The retarding effect of inductance is different from normal resistance. If the *inductor* (a coil of wire) has high inductance and low resistance, figures 55 and 56 show what happens in an inductive circuit in the presence of varying current and voltage.

Figure 55 depicts a circuit with a battery, a switch, a voltmeter, an *ammeter*, and an inductor with an iron core. (An ammeter measures current flow in the circuit in amperes.) At the moment the switch is closed, the voltmeter immediately indicates the battery voltage applied across the inductor, which is shown in figure 55 by the dotted arrow in the voltmeter. On the other hand, the ammeter indicates a significant delay before the current reaches its maximum value. In figure 55, the dotted arrow indicates maximum value. It takes some time for the arrow to deflect all the way.

When inductors are placed in AC circuits, the voltage's frequency has a direct effect on the inductors. As stated earlier, one effect is that the inductance retards, or impedes, the passage of AC current. Although this resistance to current flow is similar to that caused by resistors, engineers and technicians call inductive resistance *reactance*. Like resistance, reactance is measured in ohms.

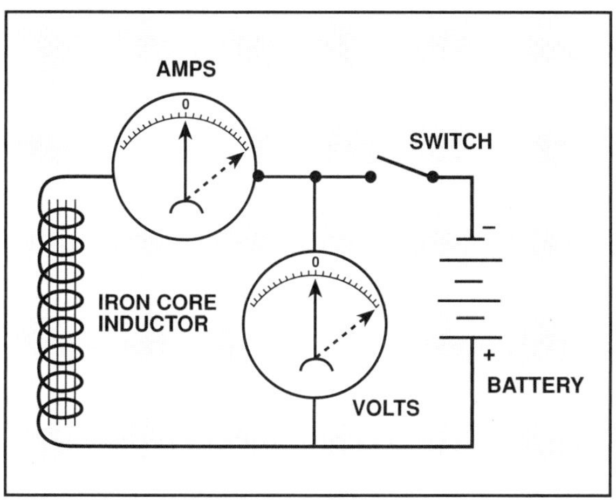

Figure 55 ***The retarding effect of inductance in a circuit***

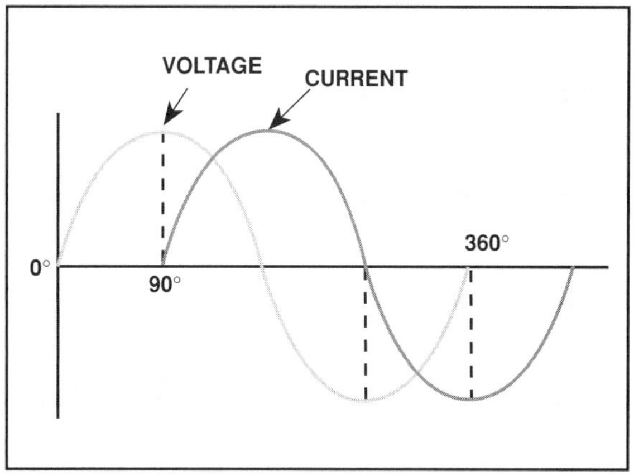

Figure 56 ***Waveforms of current values and voltage values in an inductive AC circuit***

In a circuit with an inductive device, values of current lag behind values of voltage. That is, if the circuit's current and voltage are displayed as waveforms, the current's maximum value occurs after the voltage's maximum value. Figure 56 shows the waveforms of current and voltage in a circuit that only has inductance and no ohmic resistance. In this case, the current lags behind the voltage by 90 degrees. That is, the voltage peaks 90 degrees earlier than the current, both positively and negatively. Note that when the voltage is at its maximum value, the current value is zero, and vice versa.

The unit of inductive reactance is the ohm, and its symbol is X_H. The X represents reactance, and the subscript H ($_H$) identifies it as inductive reactance. Inductive reactance is an important aspect of AC circuits because its retarding effect is far more pronounced than ohmic resistance in such devices as motors and transformers. The factors that affect inductive reactance are given in the equation—

$$X_H = 2\pi f H \qquad \text{(Eq. 30)}$$

where

X_H = inductive reactance, ohms
2π = a constant (arises from derivation of the equation representing the number of radians in a hertz; 2π radians = 360°)
f = frequency, Hz
H = inductance, henries.

CAPACITORS AND CAPACITANCE

Capacitors are electrical devices capable of storing an electrical charge. (Capacitors are sometimes called condensers.) They behave differently in DC and AC circuits. If a source of DC voltage is impressed across the terminals of a capacitor, current flows in the circuit until the capacitor takes on a charge that makes it capable of delivering voltage equal to that of the DC source. When used in AC circuits, the capacitor charges alternately with positive and negative charges and produces current flow in external circuits. The unit of capacitance is the *farad*, which is derived from Michael Faraday's last name. Faraday was an English scientist who studied electricity in the 1800s.

Capacitor Construction

Manufacturers make capacitors by separating two or more metallic plates with a thin sheet of insulating material, such as plastic, paper, or mica. This insulator is called a *dielectric*. (A dielectric is a material that is a nonconductor of electricity; it is an insulator.) In many cases, the plates are made of thin metal foil to reduce the size, weight, and cost of the capacitor (fig. 57). Figure 58 illustrates another method of capacitor construction. In this case, sheets of mica separate thin copper plates.

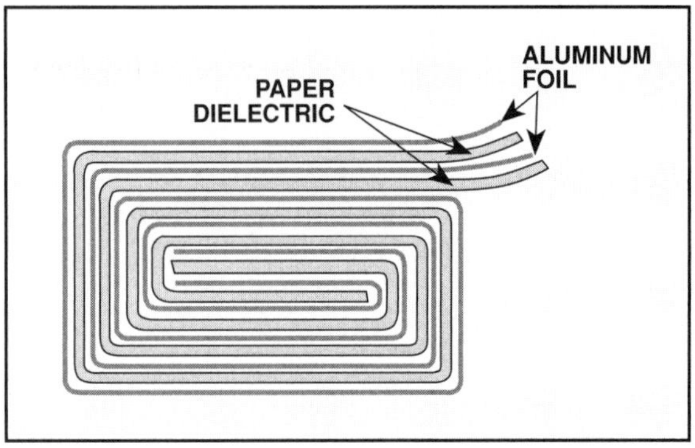

Figure 57 Coil-wrapped capacitor

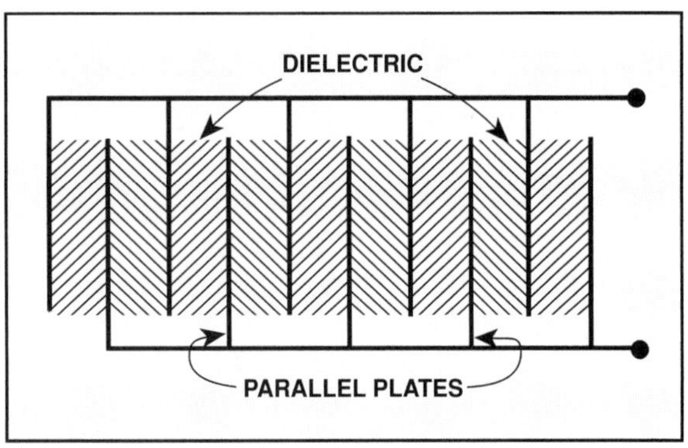

Figure 58 Rectangular capacitor plates

Figure 59 shows a *variable capacitor* with movable and fixed, or stationary, plates. Because the movable plates are attached to a shaft, turning the shaft rotates the plates into or out of the stationary plates. (Fig. 59 shows the movable plates rotated into a position that places them mostly outside of the stationary plates—that is, less than about one-fourth of the movable plate's area meshes inside the stationary plates.) The movable and stationary plates mesh with each other but do not touch. As the movable plate is turned, it changes the amount of area of the capacitor plates. In this case, rotating the shaft clockwise moves more of the movable plate's area in between the stationary plates, which increases the capacity of the capacitor. The dielectric or insulator in this case is air between the plates.

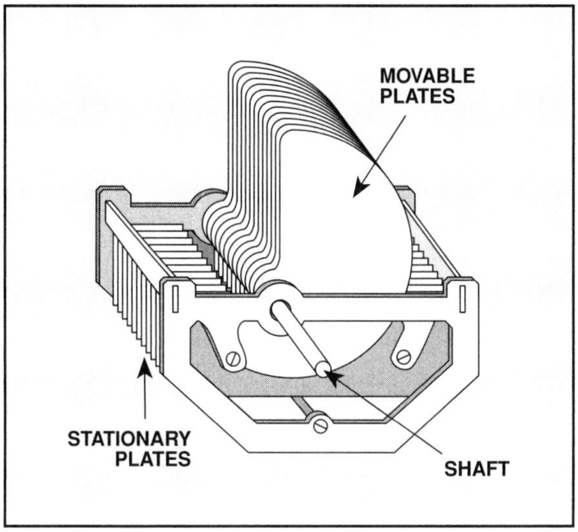

Figure 59 Variable capacitor

Capacitor Insulators

As stated before, the insulating material between the plates of a capacitor is a dielectric. The dielectric can be made from a variety of materials, which includes paper, mica, air, gases, oils, and chemicals. Capacitors are classified according to their dielectric material and to the material that surrounds the plates and the dielectric. For example, a capacitor with paper dielectric is a *paper capacitor* and a capacitor with mica dielectric is a *mica capacitor*. However, one paper type is housed in a metal container, filled with oil, and sealed. Although its dielectric is paper, the capacitor is an *oil-filled capacitor*.

A perfect insulator or dielectric does not exist. Therefore, a small amount of current usually flows between the plates of a capacitor when voltage is applied. This current is *leakage current* and varies depending upon the quality of insulation used. Likewise, conductors used for plates are not perfect either. So, they have a small amount of resistance, which gives a capacitor some internal resistance.

Capacitor Charging and Discharging

Figure 60 shows a circuit with a 10-volt battery as an electric energy source, two switches, and a capacitor made up of two metal plates separated by an insulator, or dielectric. The plates are labeled A and B. Before battery voltage is applied (fig. 60A), each plate is *electrically neutral*, or uncharged.

When the switch is closed (fig. 60B), electrons flow to plate A from the battery's negative terminal. At the same time, electrons flow from plate B to the positive terminal of the battery. In other words, plate A is negatively charged and plate B is positively charged. The voltage, or potential difference, across the plates of the capacitor instantly rises from 0 to full battery voltage. When the switch is opened (fig. 60C), the charge is trapped and remains on the capacitor. The charge remains for some time, but slowly drains because of leakage in the dielectric.

Keep in mind that, before the voltage drains, a large capacitor can charge to a high voltage level—high enough to injure or kill a person. Therefore, always take care when handling capacitors, even if the circuit is not energized (unplugged) or even if the capacitor has been completely removed from the circuit. Safety standards require that capacitors be properly discharged, shorted, and grounded before any maintenance work is performed on or around them.

Another important consideration is the behavior of capacitors in AC circuits. In AC circuits, voltage not only reverses direction, but also changes in value each time it changes direction. Moreover, the voltage changes from positive and negative each time it changes direction, or goes through one cycle. Thus, if 240 volts, for example, is flowing in an AC circuit, this voltage momentarily increases, or peaks, every time it reverses direction. (The peak may be either positive or negative volts.) This momentary voltage increase is termed *peak voltage*. In a 240-volt AC (VAC) circuit, the peak is about 340 volts. So, if a capacitor is operating in a 240-VAC circuit, and its disconnect switch is opened at the moment the voltage across the capacitor reaches its peak, the charge remaining on the capacitor is the voltage's peak value, or about 340 volts. This peak value may be either positive or negative.

After the voltage peaks, it gradually decreases to zero, then climbs until it peaks again. So, if the capacitor's circuit switch is opened at a point other than at peak voltage, the capacitor's charge will be whatever the value is when the switch is opened. Put another way, the voltage charge on a capacitor in an AC circuit can be any value between zero and its positive or negative peak value.

Figure 60D shows a charged capacitor with the battery switch open. A shorting switch is provided in the circuit to discharge the capacitor. At the instant the shorting switch is closed, electrons rush out of the negatively-charged plate, through the completed short circuit, and on to the positively-charged plate. This current flows until the positive and negative charges on both plates are balanced. The capacitor is then discharged and remains discharged until the battery is connected to it again.

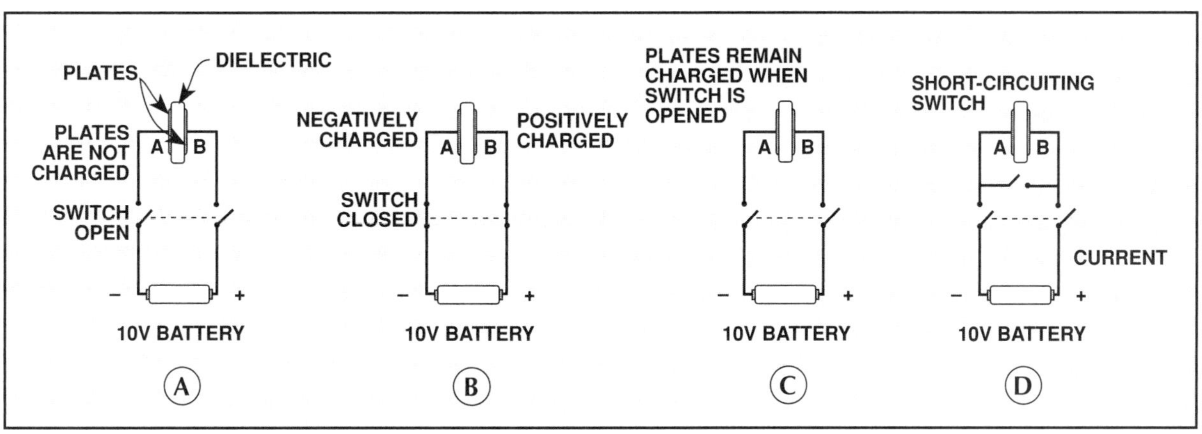

Figure 60 Charging and discharging a capacitor

Discharging a capacitor through a switch only illustrates the events that take place during discharge. In reality, never discharge capacitors with a switch, screwdriver, or any low-resistance conductor because high currents may result. Discharged high currents can act as an electric welding torch, creating molten metal that can spray onto a person and cause injury. In addition, equipment can be damaged.

The preferred way to discharge a charged capacitor is to connect a resistor across the capacitor's terminals. Discharging through a resistor releases the capacitor's energy in a controlled manner. Be sure to employ a resistor with adequate resistance and sufficient watts to safely discharge the capacitor. To determine the value of the resistor and its power rating, use Ohm's law and the power law. Also, you must consider the time required to discharge the capacitor. A convenient way to determine the time required, the resistance needed, and the power rating required to discharge a capacitor is to use an equation. One equation gives the time required to discharge a capacitor to its 63-percent level (meaning 37 percent of the charge remains in the capacitor). This equation is—

$$t = RC \qquad \text{(Eq. 31)}$$

where

t = discharge time, seconds
R = discharge resistance, ohms
C = capacitor value, farads.

To discharge a capacitor to its 90-percent level, where only 10 percent of the charge remains in the capacitor, R and C are doubled. The equation is—

$$t = 2RC. \qquad \text{(Eq. 32)}$$

As an example of determining discharge time, assume a capacitor has a value of 100 microfarads (1 millionth of a farad) and its voltage charge level is 200 volts DC (VDC). To safely discharge the capacitor, assume that the maximum discharge current is 0.1 ampere at the moment discharge begins. The resistance value is determined with the Ohm's law equation, $R = E/I$. In this example, voltage, E = 200 volts and current, I = 0.1 amps. Thus, $R = 200/0.1$, which works out to resistance R being 2,000 ohms.

The peak power at the initial discharge point is determined with the power law equation, which is $P = EI$. In this case, E = 200 volts and I = 0.1 amps. So, $P = 200 \times 0.1$, which works out to 20 watts (P = 20 watts). With this information, the time to discharge the capacitor to its 90-percent level is determined by plugging the values into equation 32. Thus, $t = 2 \times 2{,}000 \times 100 \times 10^{-6}$, which solves to $4{,}000 \times 100 \times 10^{-6}$ and $400{,}000 \times 10^{-6}$. The answer is 0.4 seconds. Therefore, the time required to discharge the capacitor to its 90-percent level is 0.4 seconds. This discharge time is quite fast.

To extend the discharge time, a resistor with higher resistance and lower wattage can be used. However, equation 32 establishes the basic requirement, which can be altered as needed—for example, a 10-watt, 4,000-ohm resistor could be used to lengthen the discharge time. Also, bear in mind that safe operations require a wait of at least 5 minutes after the capacitor is discharged before touching it to remove it from the circuit.

Capacity

Just as resistors are rated in terms of their ability to oppose the flow of current, capacitors are rated according to the amount of charge that results from a given voltage. Several factors affect the capacity of a capacitor.

Area of Plates

The capacity of a capacitor varies directly with the area of the plates used. A low-capacity capacitor has plates with a small area. A high-capacity capacitor has plates with a large area. In a capacitor with only two plates, area, and thus capacity, is determined only where the plates face each other. To achieve high capacity with a large plate area, many capacitors are constructed with thin metal foil, which is rolled between a paper dielectric (see fig. 57). Rolling it in this manner makes for a small capacitor with high capacity.

Distance between Plates

As the distance between plates of a capacitor increases, the capacity decreases. The closer the plates are, the higher is the capacitor's capacity. The thickness of the dielectric between the plates determines the distance between the plates. With a given plate area, a thinner dielectric increases capacity because it decreases the distance between the plates. However, the amount of voltage applied to the capacitor limits the dielectric's thinness. If the dielectric is too thin for a given voltage, the voltage ruptures the dielectric and shorts out the capacitor. Once a hole is burned in the dielectric, the capacitor is no longer useful.

Dielectric Materials

The kind of dielectric used in a capacitor also affects the capacity of a capacitor. Capacitors with the highest capacity use air for the dielectric; such capacitors are usually immersed in a nonconductive liquid such as oil. Air is a standard dielectric—that is, all dielectric materials are rated in reference to air.

Air has a *dielectric constant*, K, of 1; mica, for example, has a dielectric constant of 6. Thus, a capacitor using mica as a dielectric has six times the capacity of a capacitor using air as the dielectric, assuming that the plate areas and spacing between the plates is the same in both. Table 14 lists the dielectric constants of several materials.

TABLE 14
Dielectric Constants

Material	Dielectric Constant
Air	1.0
Glass	9.0
Mica	6.0
Paraffin Paper	3.5
Ethyl Alcohol	28.4
Oil (average)	3.0
Distilled Water	81.0

Besides constants, dielectric materials also have strengths. The *dielectric strength* of a material is based on its thickness and the amount of voltage that breaks down the material. Put another way, dielectric strength is the voltage per *mil* of thickness required to break down the material. (A mil is 0.001 in.) Dielectric strength is a measure of the material's insulating ability. On the other hand, dielectric constant is a measure of the material's capacitive effect. Table 15 lists the dielectric strength of several materials.

TABLE 15
Dielectric Strengths

Material	Dielectric Strength, volts/mil
Air	76
Bakelite	150 to 500
Glass	760 to 3,800
Paper	250 to 650
Mica	760 to 5,000
Oil	100 to 500
Distilled Water	380

Determining Capacitance

Capacitors are rated in terms of their capacitance—that is, their ability to store a charge. As previously mentioned, the unit of capacitance is the farad; however, a capacitor of 1 farad is too large for practical use. Therefore, small units of capacitance are used, including the *microfarad*, *picofarad*, and, rarely, the *nanofarad*.

High-capacity capacitors are rated in microfarads, which is abbreviated mF or μF. (This manual uses μF. Also, be aware that manufacturers sometimes abbreviate microfarad as MFD and imprint it on the body of the capacitor.) The Greek letter mu (μ) stands for the prefix "micro." Micro before any fundamental unit (farads in this case) means to multiply the fundamental unit by one-millionth, which is 0.000001, or 10^{-6}.

Manufacturers usually rate low-capacity capacitors in picofarads, which is abbreviated pF, or PFD. The prefix "pico" means that the fundamental unit is to be multiplied by one-trillionth, which is 0.000000000001, or 10^{-12}. (Incidentally, while capacitors are not usually rated in nanofarads, the prefix nano means that the fundamental unit is to be multiplied by one-billionth, which is 0.000000001, or 10^{-9}.)

To determine the capacity of a capacitor, use the following equation—

$$C = \frac{2.25 \times 10^{-7} \times K \times A}{d} \qquad \text{(Eq. 33)}$$

where

C = capacity, μF
K = dielectric constant of dielectric
A = area of one of two equal-sized plates forming the capacitor, in.^{2}
d = distance between plates, in.
2.25×10^{-7} = a constant.

As an example of using equation 33, determine the capacity of a capacitor made of two metal plates, each of which is 10 in. by 10 in., and that are separated by a 0.001-in. thick paraffin paper with a dielectric constant of 3.

The solution—

$$C = \frac{2.25 \times 10^{-7} \times 3 \times 10 \times 10}{0.001}$$

$$= \frac{2.25 \times 10^{-7} \times 3 \times 10 \times 10}{0.001}$$

$$= \frac{2.25 \times 10^{-7} \times 3 \times 100}{0.001}$$

$$= \frac{2.25 \times 10^{-7} \times 300}{0.001}$$

$$= \frac{0.0000675}{0.001}$$

$$C = 0.0675\ \mu\text{F, or } 67{,}500 \text{ farads.}$$

Capacitor Voltage Ratings

Capacitors are rated not only in terms of their capacity, but also in terms of the maximum voltage that may be safely applied to them over long periods. This rating is the *maximum working voltage rating*. Capacitors can be tested for short periods at higher voltages than the working voltage. High-voltage application for a short period is *test voltage*. Capacitors undergo such tests to ensure that they will perform well at working voltages. While working voltages for capacitors is specified, test voltages are usually not.

A capacitor's voltage rating is determined by the strength of the dielectric used. If, for example, the voltage rating of a capacitor is low, the dielectric may be thin and the plate area small to provide the required capacitance.

To construct a capacitor of higher voltage rating, the thickness of the dielectric must be increased. To maintain a given capacity, the area of the plates must also be increased. Thus, a 0.01 μF capacitor rated at 200 volts is about half the size of a capacitor of 0.01 μF capacity rated at 600 volts. The 600-volt capacitor has a thicker dielectric and a greater plate area. Figure 61 shows three capacitors of different sizes and voltage ratings. As is the case with most electrical components, the cost of a capacitor goes up with increased voltage rating.

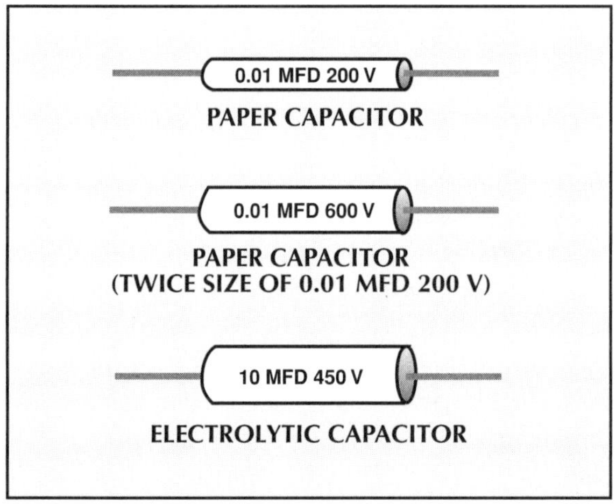

Figure 61 Relative size and voltage ratings of capacitors

Notice that one of the capacitors in figure 61 is an *electrolytic capacitor*. Electrolytic capacitors use an *oxide film* as the dielectric. (An oxide film is a thin layer of material—a film—that is coated with metallic oxides.) Electrolytic capacitors are so called because an electrolyte (a chemical compound, which, when molten or dissolved in certain solvents, usually water, conducts electricity) separates their *electrodes*. (An electrode is simply a conductor through which electric current enters or leaves the capacitor.) Oxide films not only have excellent voltage breakdown characteristics, but also they are thin. This thinness makes it possible to manufacture capacitors of relatively great capacity and of small size. They are also reasonable in cost, considering their high capacity.

Figure 62 shows two types of electrolytic capacitors. One is a *can type*, which is installed in the circuit in a vertical position on top of a device's *chassis*. (An electronic chassis is a metal frame that houses the electrical components of a particular device.) The capacitor's negative terminals connect, or ground, to the chassis. In the other electrolytic capacitor in figure 62, the dielectric film and plates are wound inside a small cylinder and two leads for wiring the capacitor into a circuit exit are provided. Advantages of electrolytic capacitors include low cost, small size, and desirable voltage ratings.

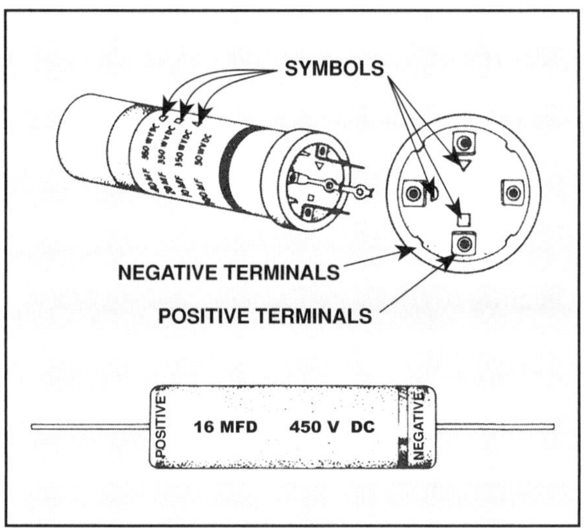

Figure 62 Two types of electrolytic capacitors

Polarized Capacitors

Polarized capacitors are electrolytic capacitors that are used only in DC circuits. Capacitors that use mica or paper dielectrics—that is, those that do not have electrolytic dielectrics—are nonpolarized and may be employed in either AC or DC circuits. Polarized electrolytic capacitors have terminals that are marked plus (+) and minus (–). The positive terminal must be connected to the positive polarity in a circuit or component and the negative terminal connected to the negative polarity in a circuit or component.

If the positive terminal of a polarized electrolytic capacitor is connected to the negative terminal of a DC supply or circuit, leakage current becomes so large that the capacitor appears to be shorted. If it is connected properly, the capacitor functions normally.

Capacitors in Parallel and Series Circuits

When capacitors are connected in parallel in a circuit, the resulting capacity is the sum of the capacitors. In equation form, the statement is—

$$C_{parallel} = C_1 + C_2 + C_3 + C_4 + \text{etc.} \quad \text{(Eq. 34)}$$

where

$C_{parallel}$ = total capacitance in a parallel circuit, μF

C_1, C_2, etc. = capacitance of each capacitor in parallel, μF.

For example, if two 50-μF capacitors are connected in parallel in a 120-volt DC circuit, the resulting capacitance is 50 + 50 = 100 μF. Assume, in this case, that the voltage rating of each capacitor is the same and is a minimum of 120 VDC.

When capacitors are connected together in series in a circuit, the resulting, or total, capacitance decreases. The relationship of series-connected capacitors is shown by the equation—

$$C_{series} = \frac{1}{\frac{1}{C_1} + \frac{1}{C_2} + \frac{1}{C_3} + \text{etc.}} \quad \text{(Eq. 35)}$$

where

C_{series} = total capacitance in a series circuit, μF

C_1, C_2, etc. = capacitance of each capacitor in series, μF.

For example, if two 50-μF capacitors are connected in series and their combination is applied across a 120-VDC supply, what is the resulting capacitance and voltage ratings?

$$C_{series} = \frac{1}{\frac{1}{50} + \frac{1}{50}} = \frac{1}{0.02 + 0.02} = \frac{1}{0.04}$$

$$C_{series} = 25\ \mu\text{F}.$$

The voltage across each capacitor divides in accordance with its voltage rating because the dielectric thickness of each capacitor is proportional to its voltage rating. However, capacitors with similar voltage ratings usually have slightly different leakage currents. With different leakage rates, the voltage does not divide in proportion to each capacitor's rating. Therefore, it is advisable to use capacitors whose voltage ratings are equal to the supply voltage. Doing so avoids problems created by unequal leakage currents.

Capacitor Color Codes

If the capacitor's container is big enough, manufacturers print capacitor ratings on it. (See the bottom drawing in figure 62, which shows a 16-μF capacitor with a voltage rating of 450 volts DC.) On the other hand, color codes are employed on small capacitors. They closely resemble resistor color codes (fig. 63). Three codes may be encountered: *military standards* (*MIL*), *Electronic Industries Alliance* (EIA) standards, and an outdated *American War Standards* (*AWS*) specification. Another outdated code is the *Radio Manufacturers Association* (*RMA*) code, which evolved into the EIA code. As you can see in figure 63, these codes come into play with disk, standoff, button head, and mica capacitors.

Three colored dots or bars indicate the capacity of a color-coded capacitor. The first two colors indicate the first two digits of the capacity and the last color indicates the multiplier used to obtain the total capacity. Location of the color dots denotes the type of capacitor. The value is in picofarads (pF) and the decimal point must be moved six places to the left to get a reading in μF. Usually, capacitors with capacities greater than 1,000 pF are expressed in μF.

As an example of reading a capacitor's color code, assume the first color is red, the second color is orange, and the multiplier color is red. As you can determine by reading the chart in figure 63, the capacity of the capacitor is 2,300 pF. Red indicates 2, orange 3, and the red multiplier indicates 10^2, or 100. Thus, $23 \times 100 = 2{,}300$ pF. To change pF to

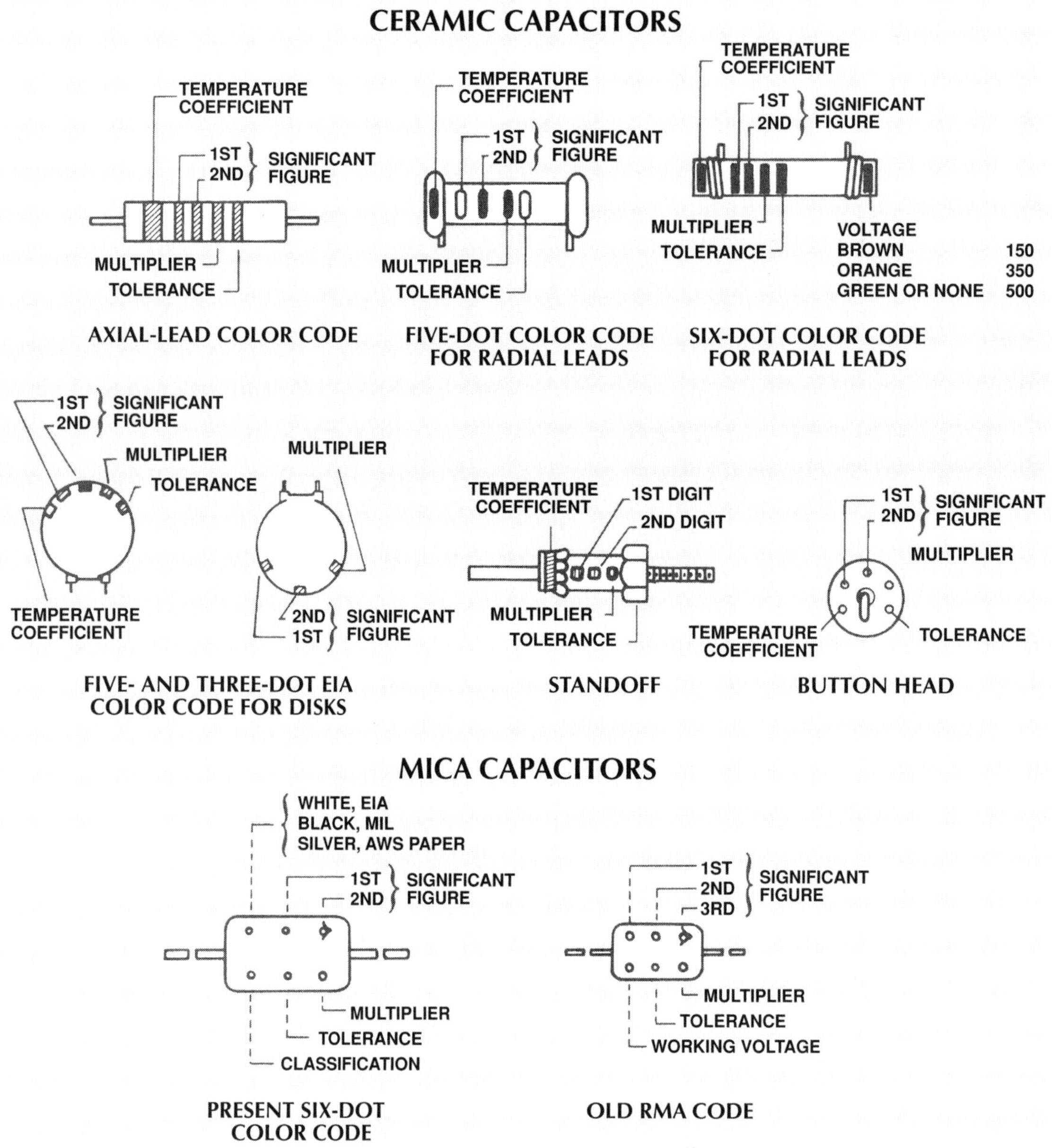

Color	Significant Figures	Decimal Multipliers	Tolerance (%)	Voltage Rating	Coefficient (PPM/°C)
Black	0	1	20		0
Brown	1	10	1	100	–30
Red	2	10^2	2	200	–80
Orange	3	10^3	3	300	–150
Yellow	4	10^4	4	400	–220
Green	5	10^5	5	500	–330
Blue	6	10^6	6	600	–470
Violet	7	10^7	7	700	–750
Gray	8	10^8	8	800	30
White	9	10^9	9	900	500
Gold		0.1	5	1,000	
Silver		0.01	10	2,000	
No color			20	500	

Figure 63 Color code chart for capacitors

μF, move the decimal point to the left six places, which yields 0.000023 μF.

Always read the colors from left to right. Some molded mica capacitors have an arrow between the two rows of dots. To read this type, orient the capacitor so that the arrow points to the right. When no arrow exists, but a brand name appears between the row of color dots, orient the capacitor so that the brand name can be read properly. Then, read the color dots from left to right.

Molded tubular paper and molded ceramic capacitors are the most common capacitors used in electronic equipment. The capacity and voltage ratings are usually printed on them; however, many manufacturers use a color code for these capacitors. Consequently, paper capacitors both with and without color codes exist.

The fourth color band on a molded paper capacitor indicates the voltage. For example, if the color is blue, which stands for 6, add two zeroes to obtain the voltage rating of 600 volts. Voltage ratings are in steps of 100 if the voltage is over 100 volts. The voltage ratings in electronic equipment range from about 10 volts to 1,000 volts. If the voltage rating is over 1,000 volts, the voltage is stamped or printed on the capacitor. Reading capacitor values by the color code with confidence requires practice. Frequent reference to the code chart helps.

Capacitor Time Delay in DC Circuits

A resistor in a circuit with a capacitor delays the time it takes for the capacitor to achieve a full charge. The time required to charge the capacitor depends on the value of the resistor in ohms and the value of the capacitor in μF. Figure 64 is a schematic of part of a DC circuit. It has a 10-volt battery, a switch, a resistor, and a capacitor. Closing the switch causes the capacitor to start charging up at a rate determined by the current supplied through the resistor. An equation is available to determine the time it takes for the capacitor to be charged to 63 percent of the total voltage it can store. (The 63 percent figure is derived from an advanced mathematical equation beyond the scope of this book. Suffice it to say that a 63 percent charge is representative of the behavior of capacitors and the time required for them to achieve a charge.) In the example shown in figure 64, total voltage is 10 volts because the battery supplies 10 volts to the circuit. Therefore, 63 percent of its total capacity is 6.3 volts. The equation is—

$$T = RC \qquad \text{(Eq. 36)}$$

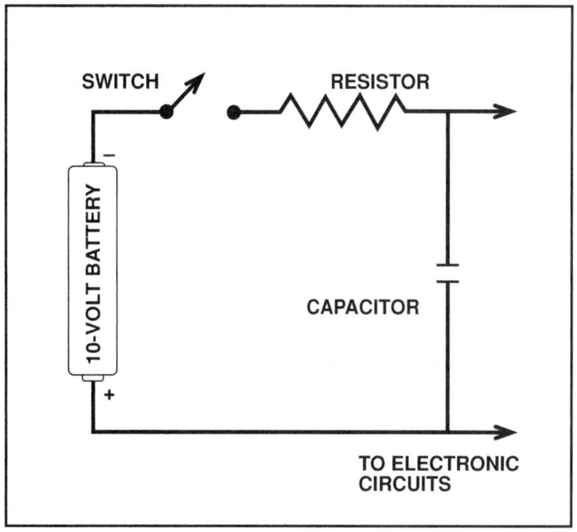

Figure 64 Time delay using capacitors

where

T = time to reach 63 percent of final value, microseconds (1 microsecond = 1-millionth of a second)
R = resistance, ohms
C = capacitance, μF.

For example, suppose a source of 10 volts is applied to a circuit that has a 1-million-ohm resistor in series with a 1 μF capacitor. How long does it take the capacitor to charge to 6.3 volts (63 percent of 10 volts)? The solution—

$$T = 1{,}000{,}000 \times 1$$
$$= 1{,}000{,}000 \text{ microseconds}$$
$$T = 1 \text{ second.}$$

Circuits and devices such as control circuits, threshold-detecting circuits, and *operational amplifiers*, use capacitor time delay in their operation.

Capacitors in AC Circuits

When a capacitor is in an AC circuit, current flows to the capacitor, but does not flow through it because of the insulator, or dielectric, between the plates. Nevertheless, it appears that current flows through it. Actually, that which appears to be current is the charging and discharging that occurs in the capacitor during each half cycle of the AC voltage.

The capacitor's size—that is, its ability to store small or large charges—influences the current's magnitude. Therefore, the higher the capacitor's capacity—the higher its microfarad or farad rating—the greater is the amount of AC charging current. The frequency of the applied AC voltage also influences the magnitude of current. If the frequency of the alternating current being applied to the capacitor is increased, the number of charges and discharges per second is increased. Consequently, more current flows in the circuit.

An ohmic value for a capacitor can be obtained by applying Ohm's law, whereby applied voltage to a capacitor is divided by the current. Keep in mind, however, that this ohmic value is not resistive; instead, it is reactive.

Capacitive Reactance

Capacitive reactance is the opposition to AC current flow a capacitor causes in a circuit. This impedance, or reactance, is not the same as ohmic resistance or the retarding effect of inductive reactance. Instead, capacitive reactance causes current flow to precede voltage flow through the circuit. If the current and voltage flowing in an AC circuit with a capacitor are shown as waveforms (fig. 65), the current leads the voltage by 90 degrees. This phenomenon is a phase shift because the capacitor changes, or shifts, the phase of current and voltage as they flow in a circuit. (If they were in phase, current and voltage would achieve maximum values at the same time.) Note that when voltage is at a maximum, current is zero and that when current is at a maximum, voltage is zero.

The unit of capacitive reactance is the ohm, and its symbol is X_C. X represents reactance, and subscript C ($_C$) identifies it as capacitive reactance. Capacitive reactance is an important aspect of AC circuits because it causes current flow to precede voltage flow in the circuit. Capacitive reactance is more pronounced than ohmic resistance, especially in electronic filters and motor-starting capacitors. Capacitor reactance can be determined with the equation—

$$X_C = \frac{1}{2\pi fC} \qquad \text{(Eq. 37)}$$

where

X_C = inductive reactance, ohms
2π = a constant
f = frequency, Hz
C = capacitance, farads.

As an example, determine the reactance of a 100-μF capacitor if 60 Hz AC is applied to it. (Note that 100 μF must be expressed in farads. To convert farads to μFs, simply move the decimal point six places to the left. Or simply multiply 100 by 10^{-6}, which works out as $100 \times 106 = 10^{-4}$; $10^{-4} = 0.0001$ farad. Also note that $\pi = 3.14$.)

$$X_C = \frac{1}{2 \times 3.14 \times 60 \times 0.0001}$$

$$= \frac{1}{6.28 \times 60 \times 0.0001}$$

$$= \frac{1}{0.0377}$$

$$X_C = 26.5 \text{ ohms.}$$

AC Capacitor Applications

A *single-phase motor* uses the phase shifting characteristics of a capacitor. (A single-phase motor is simply a motor that has only one source of alternating current to make it run.) Figure 66 is a schematic of a

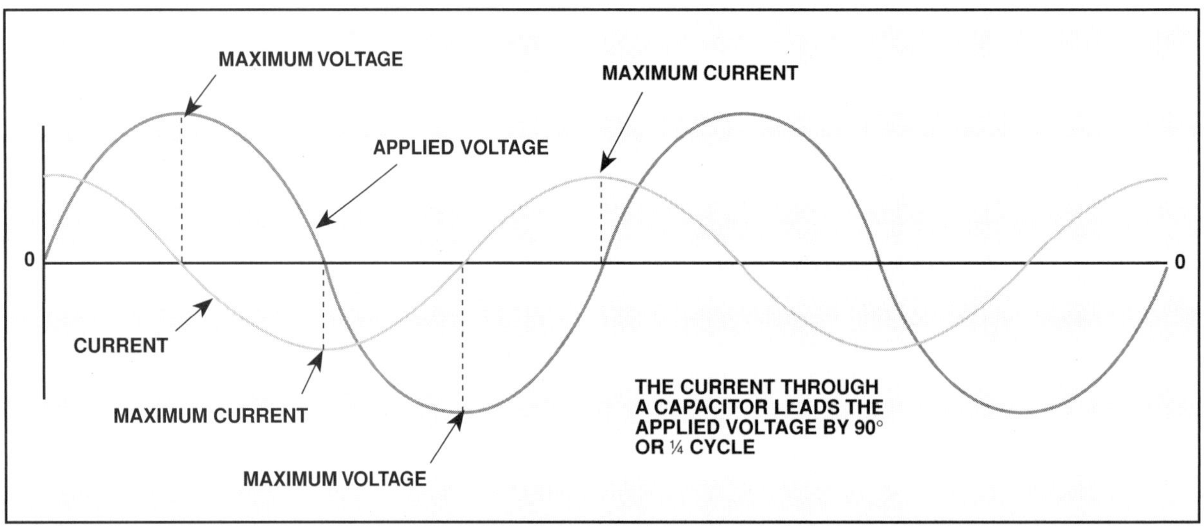

Figure 65 Graph of current values and voltage values in a capacitive alternating current circuit

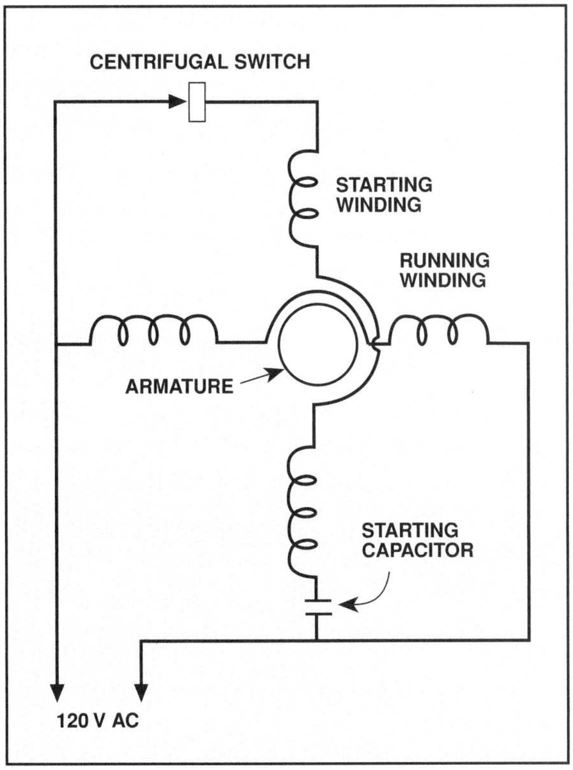

Figure 66 Single-phase AC induction motor

single-phase AC induction motor. A circle represents the motor's armature. The figure also depicts four windings (or coils), a centrifugal switch, and a starting capacitor. A 120-volt AC circuit provides electrical energy for the motor. To develop starting torque in the motor, a rotating magnetic field must be developed. To achieve this magnetic field, two out-of-phase currents are applied to separate windings—one to the starting winding and one to the winding connected to the capacitor. The capacitor creates the out-of-phase current. The magnetic fields created in each winding oppose each other and produce a rotating magnetic field because they are out of phase. The capacitor connected to one of the windings shifts the current phase to achieve the magnetic fields. The magnetic fields initiate starting torque. Once the motor achieves full speed, centrifugal force created by the rotating motor causes the centrifugal switch to disconnect the starting circuit.

IMPEDANCE

As stated earlier, impedance is the total retarding effect on the current in a circuit. It is a combination of resistive opposition and reactive opposition. Impedance is measured in ohms, and its symbol is Z. The impedance of a circuit cannot be determined simply by adding the ohmic resistance and the reactance. The current flowing through the ohmic resistance is *in phase* with the voltage. (Being in phase means that voltage and current values reach maximum and minimum values at the same instant.) As mentioned before, a waveform of current and voltage in inductive reactance shows current values lagging behind voltage values by 90 degrees (see fig. 65). Stated another way, in capacitive reactance, current values lead voltage values by 90 degrees.

To determine the impedance of an electrical circuit consisting of resistance (R), inductance (H), and capacitance (C) components connected in series, a mathematical procedure called *vector addition* is used. A *vector* is the representation of a quantity that gives the quantity both magnitude and direction. In the case of impedance, the vector describes the magnitude and the angle of the phase shift. The resistive component (R) is in phase with the current. The inductive component (H) lags behind the resistive component by 90 degrees. The capacitive component (C) leads the resistive component by 90 degrees.

Figure 67 shows R, H, and C components as a vector diagram from which impedance can be determined. The R component is resistance, the H component is inductive reactance (labeled X_H on the diagram), and the C component is capacitive reactance (labeled X_C on the diagram.)

The vector components can be added if they are in phase, which is the case with inductive reactance (X_H) and capacitive reactance (X_C). However, to add the resistive component (R) to the reactive components, the *Pythagorean theorem* is used.

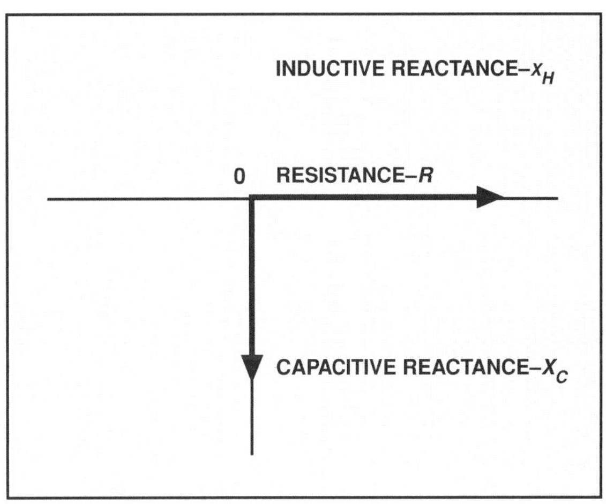

Figure 67 Vector diagram of an R, H, and C circuit to determine impedance

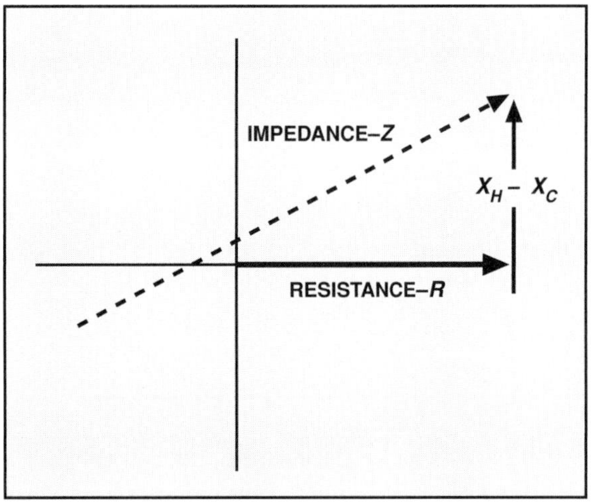

Figure 68 Combining R, H, C vectors into impedance

Pythagoras was an ancient Greek mathematician who discovered that in a right triangle, the square of the length of the hypotenuse is equal to the sum of the squares of the lengths of the other two sides. Put another way, the hypotenuse is the square root of the sum of the squares of the other two sides. Figure 68 shows R, H, and C vectors combined to form a right triangle from which impedance, Z, can be determined. Resistance, R, forms one leg of the triangle, the H and C vectors, expressed as $X_H - X_C$, form another leg, and impedance, Z, forms the hypotenuse. Using the Pythagorean theorem, total impedance, Z, (the hypotenuse) in an R, H, and C circuit is—

$$Z = \sqrt{R^2 + (X_L - X_C)^2} \quad \text{(Eq. 38)}$$

where

Z = total circuit impedance, ohms
R = resistance of circuit, ohms
X_H = inductive reactance, ohms
X_C = capacitive reactance, ohms.

POWER

Up to now, power has been discussed in conjunction with a DC circuit where the current and voltage do not vary with time and whose frequency is zero. In this case, power (P) is in watts and is the product of voltage (V) times current (I), or $P = VI$.

When alternating current is applied to a circuit containing resistors, inductors, and capacitors, power takes on a broader meaning. When storage components such as capacitors and inductors are in a circuit, voltage and current are not in harmony with each other—that is, they are out of phase. When the voltage reaches a peak, the current is at zero and when the voltage is at zero, the current is at its peak (see fig. 65). Voltage and current do not work together as they do in a circuit only with resistors where the current and voltage work together in phase.

When an AC generator or transformer delivers voltage and current to a load, the voltage and current can be multiplied to determine power in the form of volt-amperes. This form of power is referred to as apparent power and for single-phase circuits, the equation is:

$$P = VI \quad \text{(Eq. 39)}$$

where

P = power, volt-amperes
V = voltage
I = current.

The only difference between equation 39 and power equation 11 (see page 14) is that equation 39 gives power in terms of volt-amperes instead of watts. Put another way, when current flows through a resistor, the product of voltage and current are in phase and produces power in watts. The apparent power in a resistor is the same as watts, so therefore, in a resistor, P = volt-amperes = watts.

When current flows through an inductor, an out-of-phase condition exists between current and voltage. Therefore, the product of voltage and current represents *reactive power*, which is expressed in terms of *volt-ampere reactive*, or *VAR*, for short. No real power is consumed in VAR, which is why reactive power is sometimes called *wattless power*. VAR in an inductive circuit is abbreviated VAR_H or VAR_L. (This manual uses H as the symbol for induction, instead of L, to avoid confusion with the SI symbol for litre.)

When current flows into a capacitor, the product of voltage and current represents reactive power because an out-of-phase condition exists between current and voltage, much like that in an inductor. Thus, VAR also exists in capacitor circuits and is abbreviated as VAR_C.

Apparent power represents the *total power* in a circuit and is the vector sum of the true power in watts plus the reactive power. While true power produces heat in a component, VAR does not. VAR circulates between the supply and the components or between reactive components.

A measure of the amount of real power—the power in watts—being delivered to a circuit as compared to its total, or apparent, power is referred to as the *power factor*. Power factor is a ratio with values between 0 and 1.0, and the equation is—

$$P_f = \frac{W}{VA} \quad \text{(Eq. 40)}$$

where

P_f = power factor
W = watts
VA = volt-amperes.

A way to visualize power is to use *power triangles*. These triangles are derived from vectors. The vectors display the types of power on a graph. For example, figure 69A shows a vector diagram of VAR and watts. The vectors are on a graph with an x and y axis. Inductive VAR (VAR_H) and capacitive VAR (VAR_C) are on the y axis, each going in opposite directions. Real power in watts is on the x axis. Figure 69B is a power triangle derived by combining the vectors. It is a right triangle where real power in watts forms one leg of the triangle, VAR_H minus VAR_C forms another leg, and apparent power (VA) forms the hypotenuse.

From the vector diagram and power triangle, several relationships between the types of power can be described.

1. Total reactive power is the difference between inductive VAR (VAR_H) and capacitive VAR (VAR_C) since their vectors are in phase, but in opposite directions. Stated mathematically, total reactive power = $VAR_H - VAR_C$.
2. Total real power, in watts, is shown on the x axis of the vector diagram and as one leg of the power triangle. Also, from equation 40, it can be determined that total real power is the product of volt-amperes (VA) times the power factor (P_f); or, total real power = $VA \times P_f$. Incidentally, real power can be measured with a wattmeter.
3. Total, or apparent, power in volt-amperes (VA) is the vector sum of reactive power (VAR_H and VAR_C) and real power. Total power (VA) is represented by the hypotenuse of the power triangle. The equation for determining total VA power uses the Pythagorean theorem and is—

$$VA = \sqrt{W^2 + (VAR_H - VAR_C)^2} \quad \text{(Eq. 41)}$$

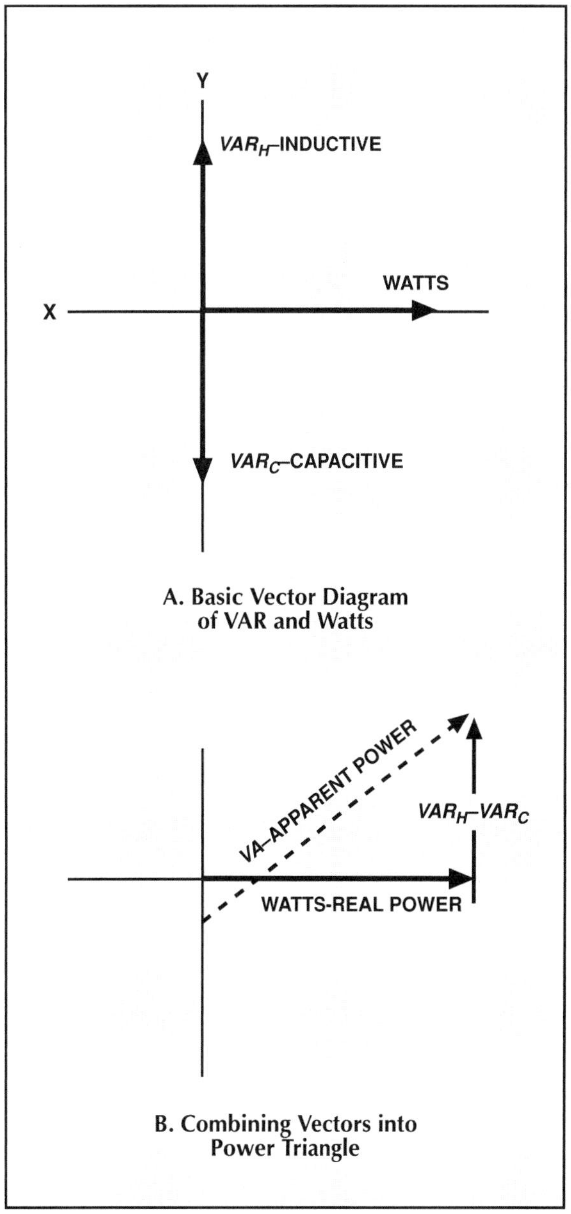

Figure 69 Vector representation of Watts and VAR

where

VA = total power, volt-amperes
W = real power, watts
VAR_H = inductive VAR
VAR_C = capacitive VAR.

In systems where a great deal of power is involved, power units are often expressed in units of 1,000 times the base unit and are thus prefixed with kilo. In these cases, the power units are kilovars (kVAR), which equals 1,000 VARs; kilowatts (kW), which equals 1,000 watts; and kilovolt-amperes (kVA), which equals 1,000 volt-amperes.

REVIEW QUESTIONS

1. What are two important sources of direct-current electricity?
2. What is the law of attraction and repulsion for electric charges?
3. What kind of lines of force exist around the terminals of a battery if the battery is not connected to a circuit?
4. How can voltage be induced in a wire?
5. Does current lead or lag the voltage in an inductive circuit?
6. Does current lead or lag the voltage in a capacitive circuit?
7. Is impedance determined by algebraic addition of resistance, inductive reactance, and capacitive reactance? Explain.
8. A transformer is delivering 1,000 kVA to a load whose power factor is 0.8. What is the real power component of this load in kW?

9 Electromagnetic Equipment

TRANSFORMERS

When a transformer is installed in a circuit, electricity of a given voltage and current flows into the transformer. The transformer then converts, or transforms, the voltage and current to a higher or lower value. The higher or lower values then flow to the remainder of the circuit in which the transformer is installed. For example, if 110 volts enters a transformer, the transformer can change it to, let's say, 1,110 volts before the voltage flows elsewhere in the circuit. In this case, input to the transformer is 110 volts, while output is 1,110 volts. On the other hand, 110 volts may flow into a transformer and the transformer can change it to 11 volts of output. Similarly, a transformer can change an input current of 1 ampere to 5 amperes and vice versa.

Transformers are essential in *power distribution systems*. Power distribution systems deliver electric power from a source of generation to a place of utilization. Without transformers, electric power cannot easily be transmitted over great distances. The electrical generation and transmission system of Hoover Dam in the western United States is a good example of how transformers are used in a power distribution system. The generators at the foot of the dam produce alternating current at 13,000 volts. Transformers step up this voltage to 287,500 volts for transmission through wires to Los Angeles, California, which is 200 miles (320 kilometres) away. In Los Angeles, additional transformers all over the city step it down to 440, 220, and 110 volts for use in industry, homes, and offices.

The reason for stepping voltage up and down to transmit it long distances is explained by the behavior of electricity when it travels through a conductor. When current passes through a conductor, such as a wire, some electrical energy is converted to heat, which is a loss of electrical energy. The amount of energy lost is proportional to the square of the number of amperes of current. That is, the higher the amperage of the current, the greater is the energy loss. The amount lost also depends on the conductor's size, or gauge. The larger the conductor, the less is its resistance and the lower is the heat loss through it. Another factor concerns the power law, which is $P = VI$, where power in watts (P) is equal to voltage (V) in volts times current (I) in amperes.

Let's say, for example, that 880 watts of power is to be transmitted. From the power law, note that any combination of volts and amperes that produces a product of 880 watts is possible. For example—

1. 110 volts × 8 amperes = 880 watts.
2. 220 volts × 4 amperes = 880 watts.
3. 440 volts × 2 amperes = 880 watts.
4. 880 volts × 1 ampere = 880 watts.

In this case, the most desirable condition for transmitting the power is 880 volts and 1 ampere, because low amperage causes less heat loss in the conductor. The most efficient way to transmit alternating current is at high voltage and low amperage.

Figure 70 illustrates the use of transformers between a generating station at a dam and the electric motor on an oilwell pumping unit. The power leaves the dam at high voltage and low amperage

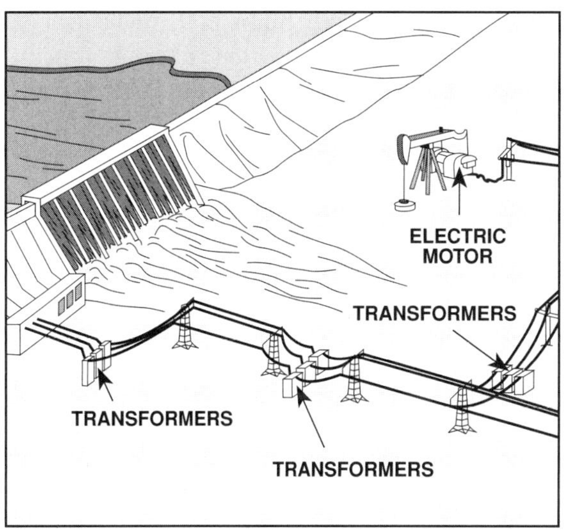

Figure 70 Use of transformers in a distribution system

and efficiently flows through the power lines. Step-down transformers near the oilwell's location lower the voltage to the amount the pump motor requires. Similarly, in virtually every city, step-down transformers are placed at many points to provide the required voltage to homes and businesses. In rural areas, almost every farm has a step-down transformer on a power line pole near the farmhouse.

Stage Transformation

Stepping up voltage to high values and then stepping it down near the point of use may require several stages of transformation. Electric transmission companies use transformers extensively when transmitting alternating current because *line loss*—the loss of electrical energy that occurs when electricity flows through a line, or conductor—is unacceptably high at relatively low voltage and high amperage. Yet, homes and businesses require relatively low voltages. Stepping up the voltage with transformers permit power companies to use smaller conductors in transmission lines and smaller components of every kind. Then, they use transformers to step down the voltage where required. The resulting savings in construction costs more than offsets the cost of transformers.

However, the number of volts that may be transmitted is limited. At very high voltages, transmission lines develop a *corona effect*. That is, the air surrounding a high-voltage conductor becomes *ionized*—loses electrons—and begins to conduct current. (The corona effect can be seen at night as a bluish light given off by the conductor.) Also, high voltages cause transformers to heat up, which causes energy loss. Because transformer heating causes an efficiency loss, manufacturers constantly conduct research aimed at producing more efficient transformers. This research pays off because voltages as high as 750,000 volts (750 kilovolts) are currently being transmitted. Soon, voltages of 1,000,000 volts (1 megavolt) and higher will be transmitted.

Principle of Transformer Operation

The key word in transformer operation is induction. When current flows through a conductor, a magnetic field occurs around the conductor. If a second conductor is placed near the magnetic field of the first and if the magnetic field changes direction, induction occurs. Induction produces voltage in the second conductor. Three elements in induction are (1) a conductor, (2) a magnetic field, and (3) a relative change in motion between the conductor and the magnetic field.

When a wire carries alternating current, the magnetic field changes because of the pulsating or alternating nature of the current. In a circuit carrying 60-Hz alternating current, current rises from zero to a peak and back to zero 120 times per second. Consequently, the magnetic lines of force surrounding the conductor also change 120 times per second. If another conductor is placed near the alternating magnetic lines of force surrounding the conductor, the magnetic lines induce voltage in the second conductor.

Figure 71 shows a primary circuit through which alternating current is flowing. AC flow creates a magnetic field in the primary circuit, or the *primary*, for short. A secondary circuit—a *secondary*—is placed near the primary's magnetic field to induce alternating current in the secondary. The two conductors are insulated from each other. The induced current has the same frequency—for example, 60 Hz—as the primary current.

In figure 72, an iron ring provides a path for the magnetic lines of force, which improves induction.

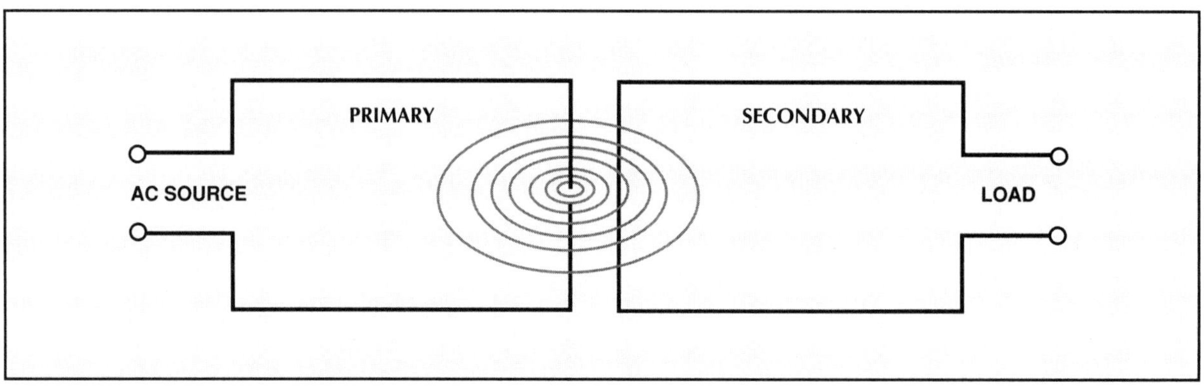

Figure 71 Induction of current in a secondary circuit

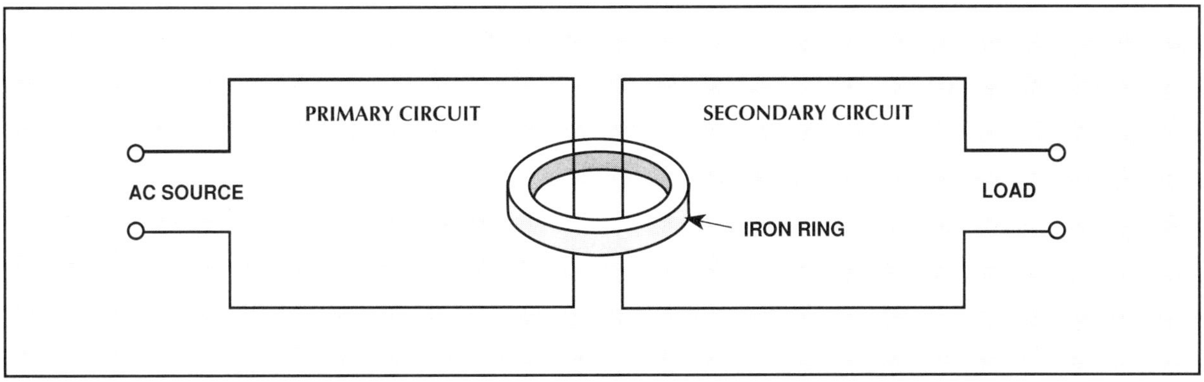

Figure 72 An iron ring provides a path for magnetic lines.

Forming the conductors into coils achieves further improvement (fig. 73). The coils intensify the magnetic field. Thus, coils, or windings, are employed in transformers.

Eddy Currents

The current flowing in the primary coil induces magnetic lines of force in the coil's core. These induced lines of force, in turn, induce currents that follow circular paths around the lines of force in the conductors as well as in the core. These currents are *eddy currents*. If the core is solid metal, eddy currents occur in a large part of the metal, and the resistance of the metal causes energy to be lost as the metal heats up. If, however, the core is made up of thin plates of metal with insulating material between them, the eddy currents cannot cross the insulation. The insulation confines the eddy currents to the thin plate and minimizes core heating.

Figure 74 shows a transformer core made of several thin sheets of *silicon steel* separated by insulators. (Silicon steel is steel that contains not only iron and carbon, but also from 0.5 to 4.5 percent silicon. The silicon in the steel makes it malleable, which allows manufacturers to make the thin, flat plates that are ideal for transformers.) The laminated construction of a core reduces eddy currents. Figure 75 shows a core with the primary and secondary windings in place. The dotted lines with arrowheads indicate the induced voltage.

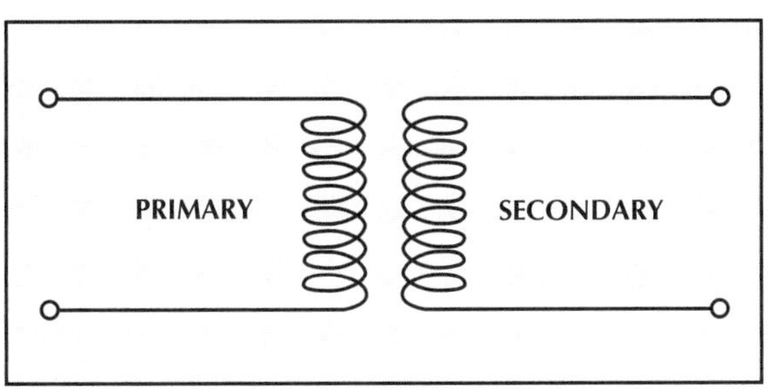

Figure 73 Using coils to intensify the magnetic field

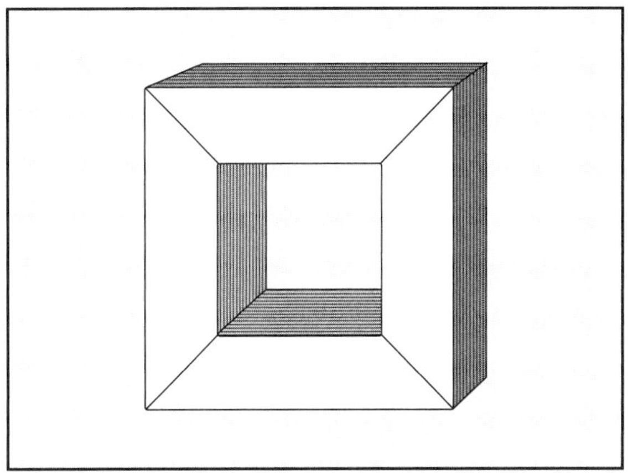

Figure 74 Laminated transformer core

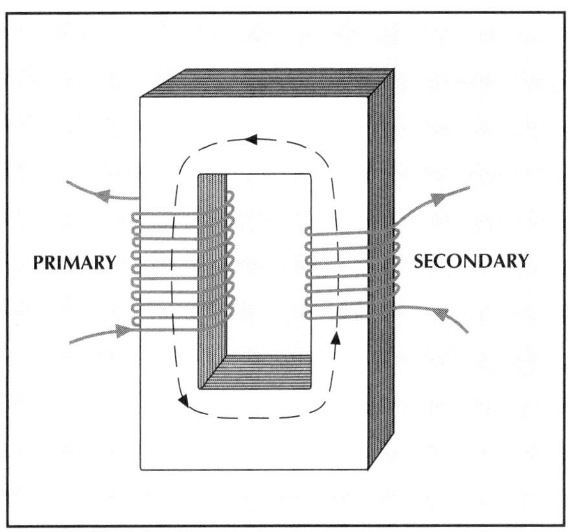

Figure 75 Core with coils in place

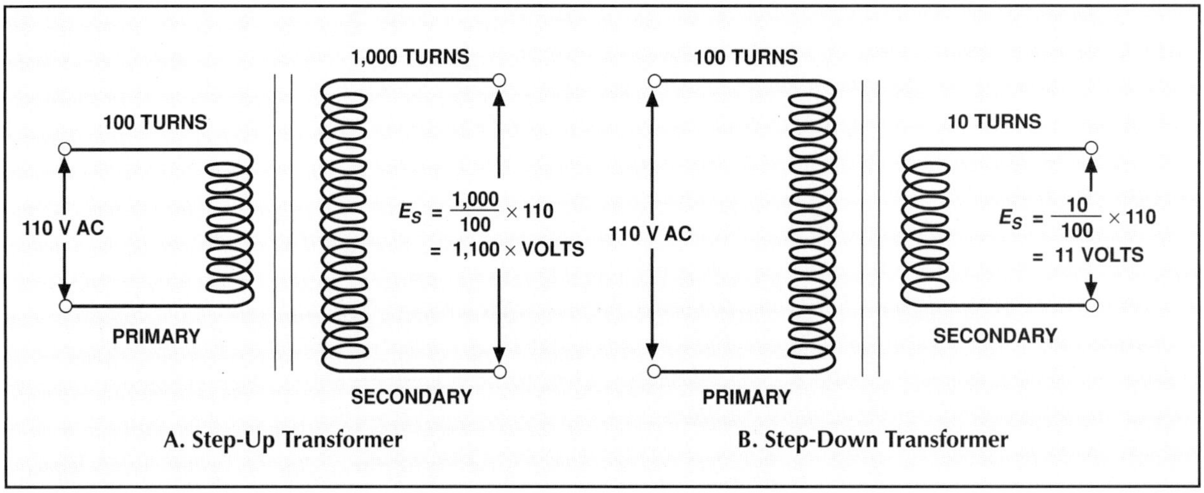

Figure 76 Examples of transformer windings

Step-Up and Step-Down Transformers

Figure 76 is a schematic of a *step-up transformer* and a *step-down transformer*. Notice that the primary and secondary windings differ from each other. That is, the number of turns in the primary coils is different from the number of turns in the secondary coils. Figure 76A shows a step-up transformer. The primary winding has 100 turns and the secondary has 1,000 turns. (The schematic does not show the actual number of turns; instead, it merely represents the actual number.) In the step-up transformer shown in figure 76A, the voltage induced in the secondary is ten times greater than the voltage applied to the primary. Consequently, the secondary voltage is 1,100 volts. So, this transformer is a step-up transformer because it increases, or steps up, the voltage from 110 to 1,100 volts. A transformer with more turns in the secondary coil than in the primary coil is a step-up transformer.

Figure 76A also shows a solved equation for this relationship. The unsolved equation is—

$$E_s = \frac{P_s}{P_p} \times V_s \qquad \text{(Eq. 41)}$$

where

E_s = step-up or step-down voltage, volts
P_s = number of windings in the secondary coil
P_p = number of windings in the primary coil
V_s = supply AC voltage, volts.

Plugging in the values shown in figure 76A works out to—

$$E_s = \frac{1,000}{100} \times 110$$
$$= 10 \times 110$$
$$E_s = 1,100 \text{ volts.}$$

Figure 76B shows a primary with 100 turns and a secondary with 10 turns. (Again, the schematic only represents the number of turns and not the actual number.) In this case, the voltage induced in the secondary is one-tenth of the voltage applied to the primary. Applying equation 41 shows that—

$$E_s = \frac{10}{100} \times 110$$
$$= 1.0 \times 110$$
$$E_s = 11 \text{ volts.}$$

A transformer is a step-down transformer when fewer turns exist in the secondary than in the primary.

The ratio of the number of turns in the primary coil to the number of turns in the secondary coil—that is, the *turns ratio* of the transformer—determines the voltage output of the secondary in relation to the primary supply voltage.

Variable Voltage Ratios

Transformers are available that allow an operator to change the step-up or step-down voltage ratio in a transformer by changing the turns ratio in transformers. Taps are placed at various points on the windings. An operator can manipulate the taps to increase or decrease the number of turns. Figure 77 is a schematic of a variable voltage transformer. Black dots on the schematic represent the taps. An operator can change from one tap to another by changing the line connection to the taps or by bringing the tap connections to a multiple-contact switch. Figure 77 also shows that a transformer's secondary windings may be in two or more sections. (The secondary winding in figure 77 is in two sections.) A secondary with

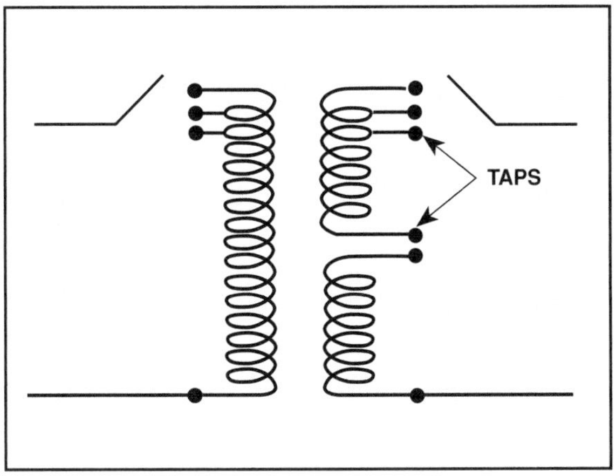

Figure 77 Variable voltage transformers

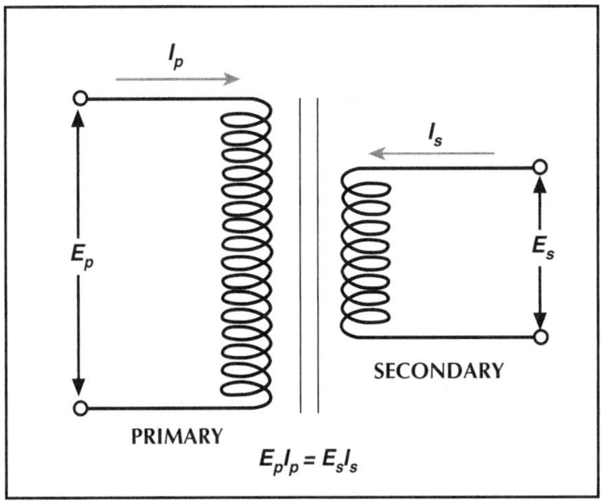

Figure 78 Current and voltage relationship

sections allows the voltage from one or more sections to be hooked up in series or in parallel as needed.

Current in Primary and Secondary

Current flows in the secondary when the secondary circuit is connected to a load. For a step-up transformer, current in the secondary is lower than current in the primary—the opposite of how a step-up transformer affects voltage. Current in the secondary is lower than current in the primary by the inverse ratio of the voltage. For example, in a step-up transformer with a five-to-one ratio, if current flowing in the primary winding is 5 amperes, then the current in the secondary is 1 ampere. On the other hand, voltage in the secondary is five times the primary voltage.

Bear in mind that transformer efficiency is not 100 percent—that is, the voltage and amperage in a five-to-one step-up transformer, for example, will not be an exact five-to-one ratio. Actual voltage and amperage will be a little lower. However, for purposes of explanation, consider a transformer to be 100-percent efficient.

An important relationship to remember about step-up and step-down transformers concerns power. Although transformers change the values of voltage and current, they do not change power. The product of the current and voltage in a transformer's primary circuit is equal to the product of the current and voltage in the secondary circuit. Put another way, the apparent power (in volt-amperes, or VA) is the same in the primary and secondary. The step-down transformer in figure 78 labels voltage in the primary as E_p and current as I_p. Similarly, it labels secondary voltage as E_s and secondary current as I_s. Figure 78 also shows the equation—

$$E_p \times I_p = E_s \times I_s \qquad \text{(Eq. 42)}$$

where

E_p = primary voltage, volts
I_p = primary current, amperes
E_s = secondary voltage, volts
I_s = secondary current, amperes.

Equation 42 holds true for both step-up and step-down transformers. (Bear in mind, too, that the equation disregards transformer inefficiency.)

Counter EMF in the Primary

The ohmic resistance of a transformer's winding is usually comparatively small in relation to voltage applied to the windings. Because the primary is connected directly across the line, a short circuit would occur if certain factors did not limit primary current. One limiting factor is the voltage induced in the primary winding. This voltage in the primary opposes and therefore limits the incoming line voltage. Consequently, the induced voltage in the primary is called *back*, or, *counter emf.* Counter emf limits the current that flows in the primary winding. Indeed, counter emf is almost equal to the line voltage. Current from the primary source sets up the magnetic flux that produces the induced voltage. As a result, enough magnetic flux exists to induce a counter emf equal to the line voltage, less the loss caused by the resistance in the primary circuit's *leakage reactance*. (Leakage reactance is the impeding of current flow caused by leakage current from the primary winding.)

In summary, the counter emf induced in the primary windings of the transformer is opposite in direction and almost equal in magnitude to the line voltage impressed on the primary. This counter emf prevents the incoming line current from overloading the primary and burning it out.

In a transformer, enough current flow always exists to set up the counter emf. When the transformer is not loaded—when it is in a no-load condition—counter emf is the majority of the current flowing and is termed *exciting current.* The exciting current may be as much as 40 percent of the full-load current in small transformers, such as those used in a doorbell ringer. In distribution transformers, it may range from 5 to 7 percent, and in large power transformers it usually is not more than 1 or 2 percent.

Exciting current is made up of two parts. One part is in phase with *impressed voltage* and supplies the losses in the core. (Impressed voltage is simply the voltage that is applied to the circuit in which the transformer is installed.) The other part of the exciting current is 90 degrees out of phase with the impressed voltage and supplies the *magnetizing current.* The magnetizing current is the part of the current that actually produces the magnetic lines of force (the flux) in the primary.

Keep in mind that a transformer does not generate electrical power; it simply transfers electrical power from one coil to another by magnetic induction. A transformer is, therefore, a device that transfers power from its primary circuit to its secondary circuit. Transformers are classified as step-up or step-down only in relation to their effect on voltage.

Transformer Power Losses

When current flows in electrical equipment, some electrical energy is always lost. The losses in large transformers are usually small and are expressed as a percentage of the *full-load* rating. Losses may be classified as *load losses* and *no-load losses.* A transformer is loaded when electric power is drawn from it. Thus, a load loss is the loss in power that occurs as power is drawn from the transformer. No-load losses occur in the primary, secondary, and core of the transformer and do not depend on the transformer's being loaded. No-load losses are classified as *primary losses, secondary losses,* and *core losses.*

A no-load core loss is made up of *eddy-current loss* and *hysteresis loss.* Eddy current losses occur in the iron core of the transformer. Recall that an eddy current is the electric current induced in the core by the same magnetic flux that induces voltage in the metal conductors of the windings. Eddy currents in the core are undesirable because they take energy from the circuit and produce no usable current. These eddy currents cause heat in the core where they flow, and increase the load in the primary and secondary windings. Hysteresis losses are losses in power caused by the lagging of changes in the core's magnetic field as the transformer's magnetic field varies.

Transformer Rating and Terminal Identification

Transformers used in electrical circuits are rated according to the *primary maximum capacity*—that is, the product of the maximum volts and maximum amperes in the primary. Volts times amperes is apparent power expressed as volt-amperes (VA). Small transformers are usually rated in terms of VA. Large transformers, on the other hand, are often rated in terms of *kilovolt-amperes,* or *kVA.* Kilovolt-amperes is an expression of volts times amperes divided by 1,000. For example, 5,000 volt-amperes (VA) are also 5 kilovolt-amperes (5kVA).

Figure 79 shows the schematics of two transformers. The schematics are electrical versions—that is, the manner in which they are drawn indicates that the transformers are used in an electrical system rather than in an electronic device such as an amplifier. In any case, both transformers have one primary and two secondary windings. The letters H_1 and H_2 identify the primary winding in each transformer. The letters X_1, X_2, X_3, and X_4 identify the secondaries. Each transformer's terminal is also labeled. Figure 79A shows the transformer's secondaries connected in series. Figure 79B shows the transformer's secondaries connected in parallel.

Electronic Equipment Transformers

Transformers that supply power in electronic equipment generally have capacities ranging from a few watts or VA to several hundred watts or VA. These transformers usually have several secondary windings. One of the secondary windings is a step-up winding, which usually transforms 120 volts to about 500 volts. The other secondary windings are step-down windings. They step down the voltage to values such as 5, 6.3, and 12.6 volts. Transformers used in electronic equipment are not rated by maximum

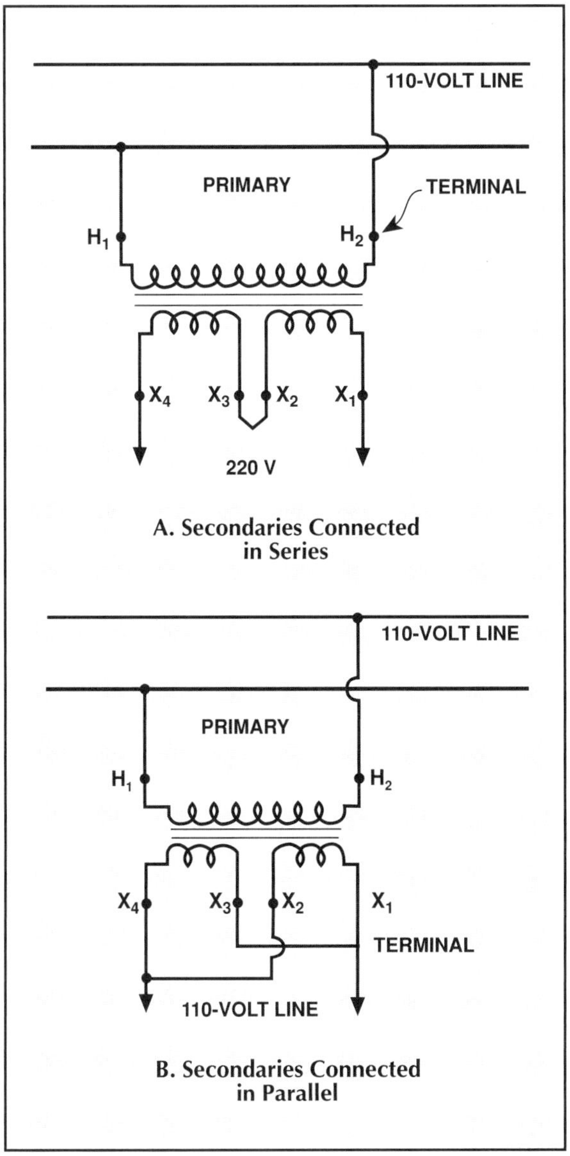

Figure 79 Identification of transformer terminals (electrical)

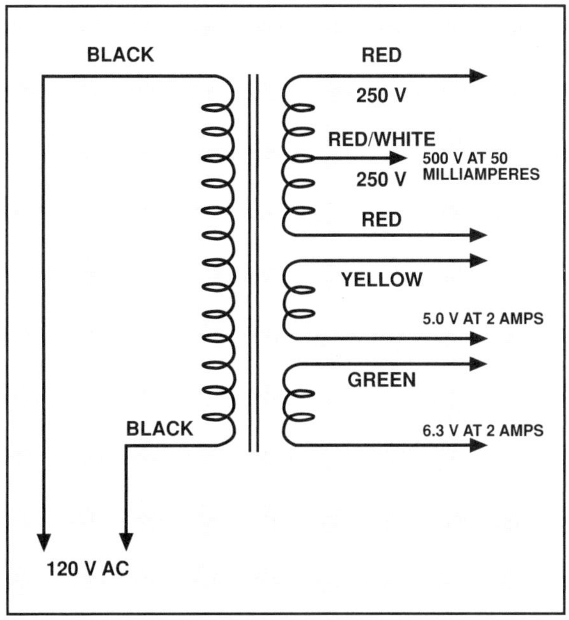

Figure 80 Identification of transformer terminals (electronic)

primary VA; instead, they are rated by the maximum voltage and current ratings of the secondary windings.

For instance, the small transformer shown in figure 80 is used in electronic equipment and its schematic is drawn accordingly. The transformer has three secondaries. One is a high-voltage secondary and is rated 500 volts at 50 milliamperes—that is, its output is 500 volts at 50 milliamperes. Another secondary has low-voltage output and is rated 5 volts at 2 amperes. The final secondary also has low voltage output and is rated 6.3 volts at 2 amperes. The transformer leads are covered with insulation of different colors to identify them.

Power Transformers

Most electronic devices use 120-volt alternating current for their source of power in the U.S. Many circuits also use *transistors*. A transistor is a small semiconducting device that is used as an *amplifier*, *detector*, or a switch. An amplifier increases the size (magnitude) of an electrical quantity such as a signal. A detector is the equipment in a *receiver* that recovers, or *demodulates*, a signal. (A receiver accepts radio waves and translates, or demodulates, them into intelligible forms such as sounds and pictures.)

Transistors and other electronic devices also require direct current for operation, so devices called *rectifiers* convert alternating current to direct current. Transistors also require that only certain parts, or segments, of the current be sent to them. So, *filters* in the circuit allow the desired segments of the electric current to flow and, at the same time, reduce, or attenuate, undesirable segments.

In any case, a transformer is used to obtain the proper voltage for operation. Such a transformer is a *power transformer*. The power transformer, capacitor, and filtering elements are grouped together and called the *power supply*. The power transformer in electronic equipment can often be recognized by its size, because it is usually larger than other transformers that may be found in the equipment.

Power transformers may have several secondary windings, with each winding rated in terms of its voltage and current output. Power transformers for transistor and other solid-state circuits typically have one or two secondary windings of relatively low-voltage and low-current capability ranging from about 100 milliamperes to several amperes. Transistor circuits use voltages that range from 4 to 24 volts and currents from a few milliamperes to several amperes.

Transformer Leads

If a power transformer uses flexible wire for its connecting leads, the insulating cover on the leads is usually of different colors. Figure 81 shows a typical power transformer and a schematic diagram of the transformer. The drawing of the actual transformer at top has a high-voltage secondary that provides 250 volts (V) at 50 milliamperes (mA). Notice that several leads come out of the bottom. The two on the right are the input leads that are connected to a 120-volt AC supply. The leads on the left come out of the transformer's secondary and are covered by insulation of different colors. The schematic at the bottom of figure 81 shows the secondaries, the voltage and ampere ratings of each secondary, and the color of the leads.

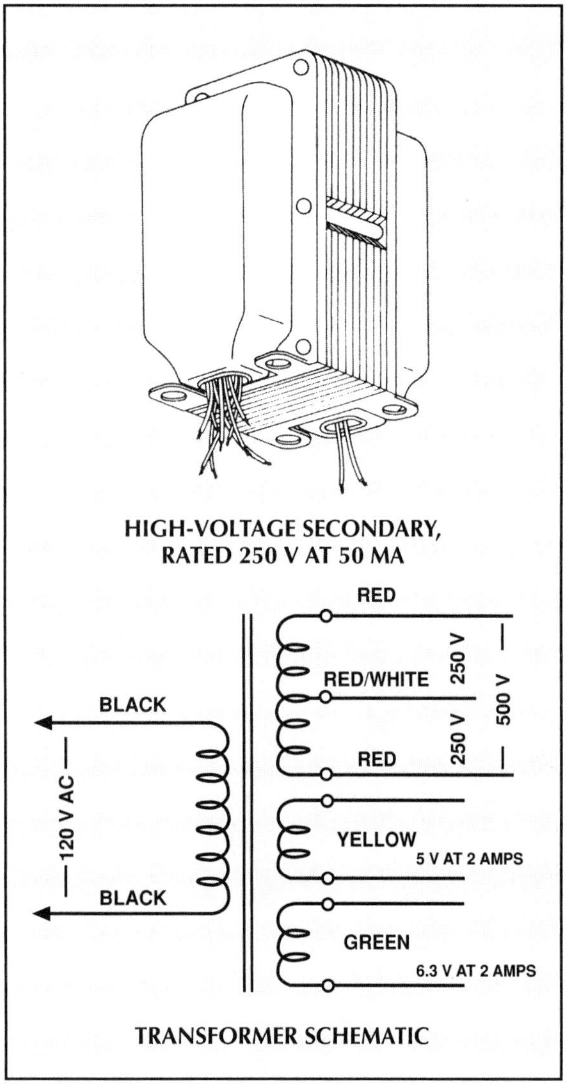

Figure 81 Power transformer

If a transformer has rigid metallic terminals, instead of color-coded wire leads, the manufacturer usually stamps the voltage rating on the insulation near the terminals. Sometimes the terminals are simply numbered. In this case, use the manufacturer's instruction sheet that comes with the new transformer to determine the ratings.

Regardless of the type of transformer, keep all the data and instruction sheets furnished by the manufacturer. This material provides information on the maximum voltage, the insulation between windings, and the insulation between the windings and the core.

Audio Transformers

Audio transformers are transformers that operate at frequencies that cover the range of audio frequencies. They bring together, or *couple*, circuits that carry audio frequencies. The range of audio frequencies is 30 to 15,000 Hz—that is, this range assumes that the average human ear can perceive sounds as low as 30 Hz and as high as 15,000 Hz. By way of contrast, note that power transformers operate on only one frequency: 60 Hz. Audio transformers, on the other hand, respond to any frequency within the audio range.

Audio transformers used as output transformers are usually rated in watts of power delivered at the secondary and in impedances that the primary and secondary windings are connected to. (Recall that impedance is the resistance to current flow in an AC circuit.) A typical set of values for an audio system's output transformer is 10 watts output into an impedance of 8 ohms and a primary winding matching a source that has 4,000 ohms impedance.

Transistors used as audio amplifiers usually have low-impedance outputs. Transformers are not normally used between the amplifier and the loudspeaker because the transistor's output can be designed to match the low impedance of the loudspeaker without use of the transformer. On the other hand, transformers are frequently used in solid-state circuits to couple one stage of amplification to another.

FILTER CHOKE COILS

A *filter choke coil* is not a transformer, but it resembles one and it operates on the principle of inductance, just as a transformer does. A filter choke has a single winding on a laminated iron core. Because it has only two leads or terminals, it is easily distinguished from a transformer. Filter chokes are rated according to their inductance, measured in henries (H), and their current-carrying capabilities.

Figure 82 shows three filter choke coils. They are commonly and erroneously called transformers; in reality, they are filter chokes. Figure 82 also shows a schematic of the audio output transformer, which is a special type of filter choke. (See the previous section on audio transformers.) Note that it only has two input leads; the secondary winding, however, has five taps. Also note the single straight line between the primary and secondary windings. This single line indicates that it is actually a single winding—that is, the primary and secondary arise from the same winding. Compare the schematic in figure 82 with the schematic in figure 81. The schematic in figure 81 shows two straight lines between the windings, which indicate that the primary and secondary are separate windings.

Once very popular, filter chokes were placed in the power supply of circuits that used *electron tubes*.

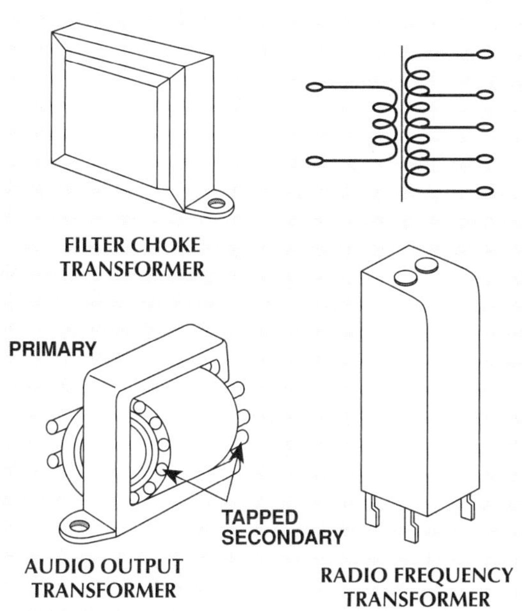

***Figure 82** A filter choke, two types of transformers, and a schematic diagram for the audio output transformer*

(An electron tube is an almost obsolete electronic device in which electrons move through a vacuum or gaseous medium within an airtight glass, or other material, container. It is also called a radio tube; a tube; or, in Great Britain, a valve. Today, transistors have virtually replaced tubes.)

Filter chokes are not commonly used today because of their cost and comparatively heavy construction. However, some *silicon-controlled rectifiers* (*SCRs*) may use large filter chokes to reduce the *ripple* on the DC *bus* before being used by an *inverter*. A silicon-controlled rectifier is a solid-state device that changes AC to DC. Ripple is the AC component in the output of a DC power supply, which arises in the power supply because of incomplete filtering or from a direct current generator. A bus is an assembly of electrical conductors that collect current from several sources and distribute it to feeder lines so that it will be available where needed. An inverter is an electronic device that converts DC to AC. So, a filter choke reduces or eliminates undesirable alternating current components in a direct current circuit before it flows to an inverter where it is converted to alternating current.

Modern power supplies for solid-state circuits eliminate the need for filter chokes. Today's power supplies provide excellent filtering of rectified alternating current with capacitors, resistors, and transistors connected as regulators.

INDUCTION MOTORS

Induction motors are the most popular electric motor in the world today. An induction motor is an alternating-current motor in which the primary winding on a part (usually the *stator*) is connected to a source of power, and the secondary winding (usually on the *rotor*) carries current induced by the magnetic field of the primary winding. The stator is the stationary part of the motor, and it contains parts of the magnetic circuit and associated windings. The rotor is the rotating part of the motor. Industry employs a considerable number of induction motors; moreover, many home appliances also use them. Reasons for the popularity of induction motors include—

1. They operate with alternating current, either single phase or *polyphase*. AC is widely available through the U.S. and the world.
2. They do not use brush-*commutator* or brush-slip ring systems. Because brushes

wear out, eliminating them from a motor reduces maintenance costs.

3. They operate at a constant speed. Constant speed operation keeps their construction simple.
4. They are efficient.
5. With polyphase induction motors, the bearings that support the rotor are the only parts that wear.
6. They cost less than brush-commutator motors.
7. They are safer in areas where flammable gases or liquids exist. Because they do not use brushes, no electrical arcing occurs, which can be an ignition source in flammable atmospheres.

In an induction motor, the stator coils and the alternating current set up a rotating magnetic field in the stator's core. The rotating magnetic field of the stator induces currents into the rotor. The rotor's induced currents set up magnetic fields that are attracted to the rotating magnetic field in the stator and cause the rotor to turn (fig. 83).

Rotor Construction

In the rotor of an induction motor, copper bars are embedded in the rotor's laminated iron core (fig. 84; the laminated iron core is removed to better reveal the bars). A shorting ring at each end of the rotor shorts the bars together—that is, the rings conduct electricity so that it can flow between the bars. Because the structure resembles a squirrel's exercise cage, rotors of this type are called *squirrel-cage rotors*. When the magnetic lines of force, or flux, cut across the copper bars

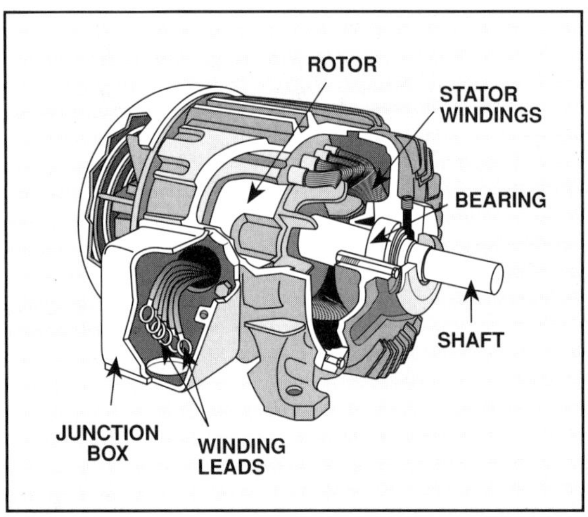

Figure 83 AC squirrel-cage induction motor

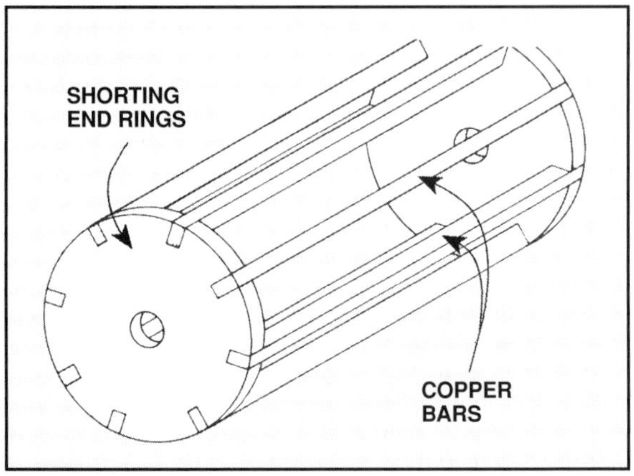

Figure 84 Squirrel-cage rotor

of the cage, currents flow in the cage. Because the iron core is laminated, eddy currents that would otherwise flow in an unlaminated iron core are minimized. Thus, practically all the current that flows in the rotor flows in the copper bars of the squirrel cage.

Rotor Action

As the squirrel-cage rotor rotates within the stator, the flux in the stator's core passes through the rotor and the embedded copper bars. The flux induces currents in the rotor bars. The induced current in the bars produces a magnetic field whose polarity is opposite that of the stator field. The result is an attractive magnetic force that exerts *torque* (turning force) on the rotor.

The instant that current is connected to the stator coils, the magnetic field of the two-pole stator begins to rotate at a constant speed of 60 revolutions a second (3,600 rpm). The induced current and resulting magnetic field on the rotor creates a starting torque and the rotor begins to turn, trying to catch up with the attractive magnetic poles on the rotating stator field.

The rotor in an induction motor never quite reaches the speed of the stator's rotating field, which is desirable. If the rotor did turn at the same speed as the rotating field, the rotor bars would move at the same speed as the rotating field's flux lines. The result would be no relative velocity between the stator and the rotor, and the flux lines could not cut the bars of the rotor. Consequently, no current would be induced in the rotor cage, no torque would be developed, and the rotor would stop turning. The rotor must, therefore, run at a speed slightly less than the rotating stator field.

The maximum speed of a two-pole induction motor operating on 60-Hz current is 3,600 revolutions per minute. An induction motor's maximum speed is called its *synchronous speed*. The actual rotating speed of an induction motor whose synchronous speed is 3,600 rpm is about 3,450 rpm. This rating is with the rated horsepower load applied to the motor. The difference between the synchronous speed and the actual speed of the motor under full load conditions is the *slip* of the motor.

Squirrel-Cage Motors

Polyphase motors are used only with AC power supplies of the particular frequency and number of phases for which they are designed. The polyphase AC power supply most commonly used in the United States is a three-phase 60-Hz system. However, a number of three-phase, 50 Hz, motors are also in use, especially outside the U.S.

Three basic types of three-phase AC motors are available: squirrel-cage induction, wound-rotor induction, and synchronous. However, wound-rotor motors are seldom used. Of the various types of electric motors, the squirrel-cage induction AC motor is the most popular. Because of their popularity, this manual discusses only squirrel-cage motors.

Construction

A squirrel-cage motor is simple in construction, rugged, reliable, and its speed is practically constant—regardless of load—within its normal working range. The stator, or field, of a squirrel-cage induction motor is composed of a yoke, a core, and the primary windings (fig. 85). The stator's yoke forms the

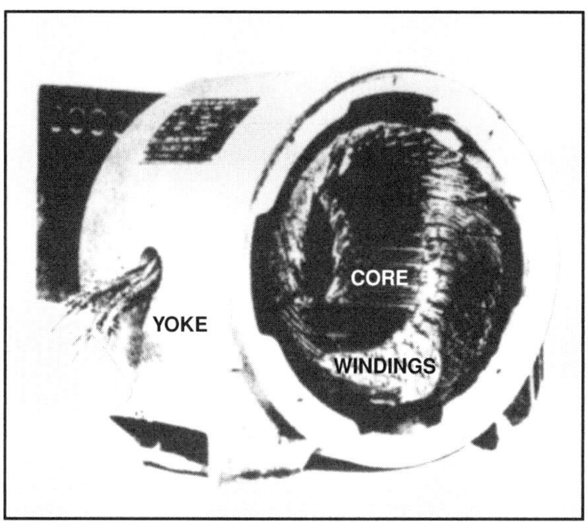

Figure 85 A squirrel-cage induction motor's stator

Figure 86 Stator core of an induction motor

cover for the motor and is the supporting frame for the stator core and coils, or windings. The stator core of a squirrel-cage induction motor is laminated. An insulator separates thin sheets of die-punched steel. The laminations are stacked under pressure to insure a permanently tight core (fig. 86). Figure 87 shows the coils (windings) being placed through die-punched slots in the stator core.

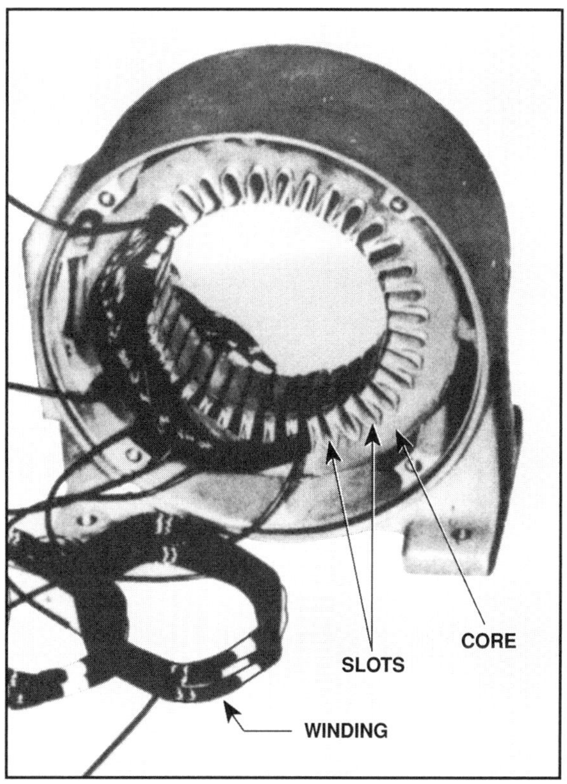

Figure 87 Partially-wound stator for a squirrel-cage motor

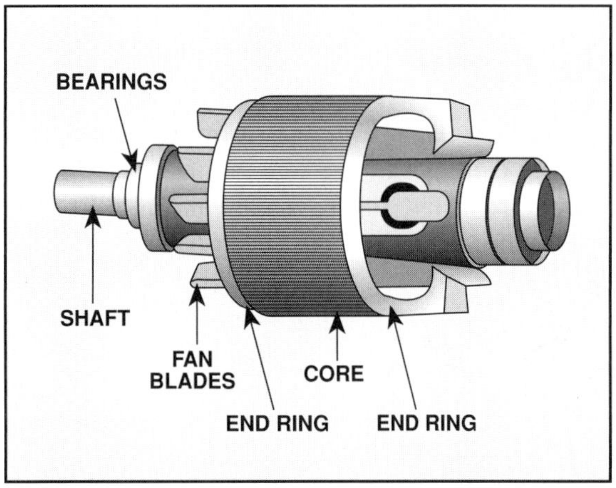

Figure 88 A typical squirrel-cage rotor, or armature

The rotor, or armature, of a squirrel-cage induction motor consists of a shaft, a core, and a secondary, or cage, winding (fig. 88). The core consists of thin, slotted, and laminated steel sheets that are mounted directly on the shaft or are supported by a *spider* mounted on the shaft. (A spider is simply a structure on the shaft that supports the core. It has a hub, spokes, and a rim.) The secondary winding conductors are usually made of copper, brass, or aluminum, and are wound through rotor slots. The winding is short-circuited by end rings or resistance rings. In figure 88, the end rings are the light-colored metal pieces from which the fan blades protrude.

Connections

The primary winding in the stator (see fig. 87) is connected to a power supply by leads that run to a *conduit* box. (Conduit is tubing made of flexible metal or other flexible material through which the insulated leads run.) The coils in the primary winding are arranged to produce several magnetic poles that are distributed around the core. The poles are formed by the way the individual slot coils of the stator winding are grouped and connected. They are not easily distinguished. The stator winding usually appears to be a continuous, uniform series of coils all around the stator's core (see fig. 85).

In a three-phase induction motor, the primary winding is wound and connected so that three separate phase windings exist, with each containing the number of groups of coils required to produce the magnetic poles. These phase windings are placed in the stator core so that the magnetic fields they produce are 120 electrical degrees apart.

No electrical connection exists to the rotor winding. Voltage and current are induced in its conductors by electromagnetic action, hence the name induction motor.

Rotating Field

Because the power supply to the motor is three-phase AC, during each half cycle, the voltage builds up to a maximum value in one polarity—first in one phase, then in the second, and finally in the third. Then it begins repeating. The currents that these voltages cause to flow in the primary-phase windings build up in the same order, and the flux produced by the currents fluctuate in the same manner. Because of the way the primary winding pole groups are arranged and connected, the combined field that is produced has the effect of a constant maximum value that rotates around the stator core.

This rotating field effect can be compared to the effect produced by a round electric sign where the light bulbs are lighted momentarily and then turned off, one after another, in succession, around the edge of the sign. The light rotates around the sign in the same way that the magnetic fields rotate around the stator. Figures 89 and 90 show two complete cycles of three-phase voltage and the condition in the armature at specific intervals during the cycles.

Torque Development

With the rotor at a standstill, the revolving lines of force cut the rotor conductors, which induce voltage in the rotor conductors. Since the rotor conductors are short-circuited through the end rings, the induced voltage causes current to flow in the conductors. This induced current in the rotor conductors situated in the rotating field is at least partially in phase with the rotating field. Therefore, this induced current reacts with the rotating field and exerts force on the conductor in the same direction as the field is rotating. The sum of the forces on the rotor conductors multiplied by the radius of the rotor produces torque in the rotor. If the rotor is free to turn, it rotates in the same direction as the rotating field.

Single-Phase Effect

A three-phase motor supplied with only single-phase power cannot develop starting torque. The stator windings produce only a pulsating field instead of a rotating field that enables it to develop starting torque.

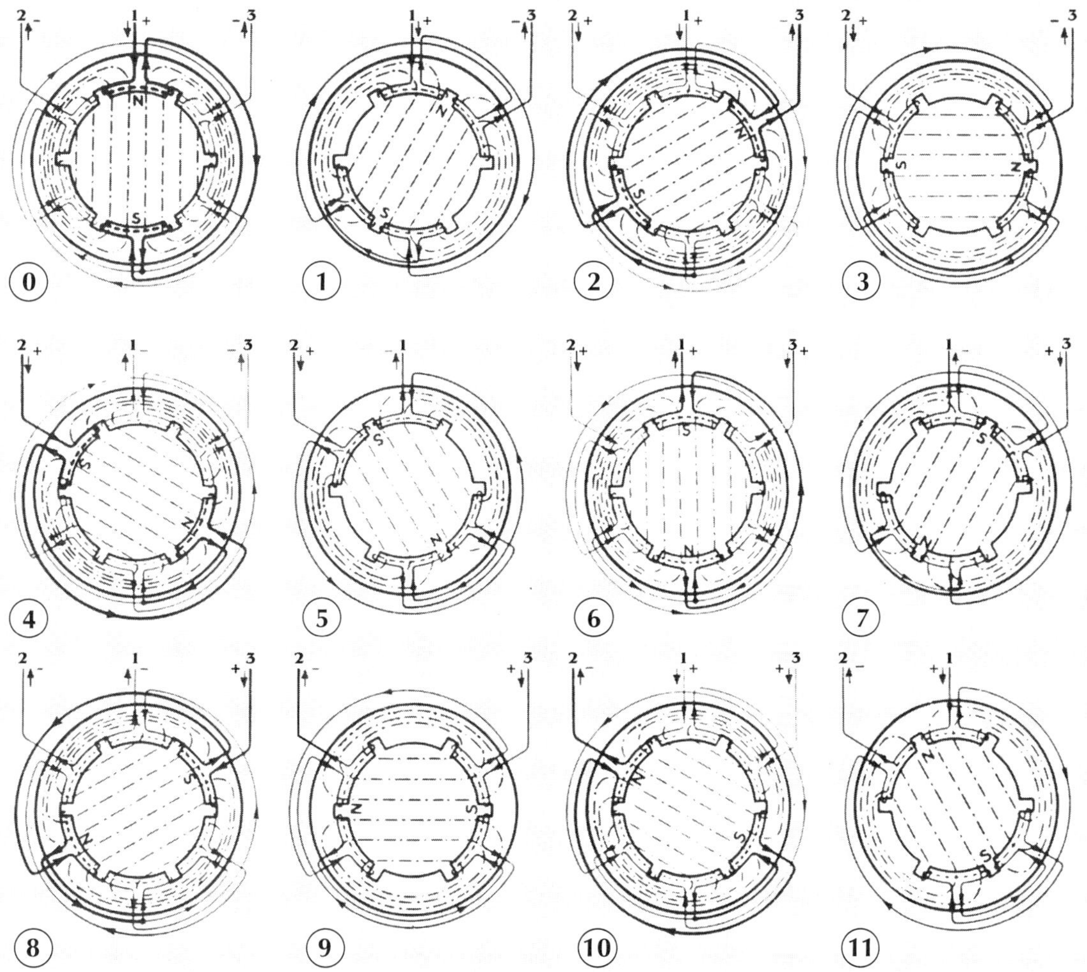

THESE DIAGRAMS SHOW HOW A REVOLVING MAGNETIC FIELD IS PRODUCED. AT INSTANT 0, THE CURRENT ENTERING THE MOTOR ON LINE 1 DIVIDES EQUALLY AND LEAVES THE MOTOR ON LINES 2 AND 3. AT INSTANT 1, THE CURRENT ENTERING ON LINE 1 LEAVES ON LINE 3; THERE IS ZERO CURRENT IN LINE 2. AT INSTANT 2, THE CURRENT IN LINE 1 IS SMALL: IT JOINS WITH LINE 2 CURRENT AND LEAVES ON LINE 3.

Figure 89 Revolving magnetic fields in the stator of a three-phase induction motor

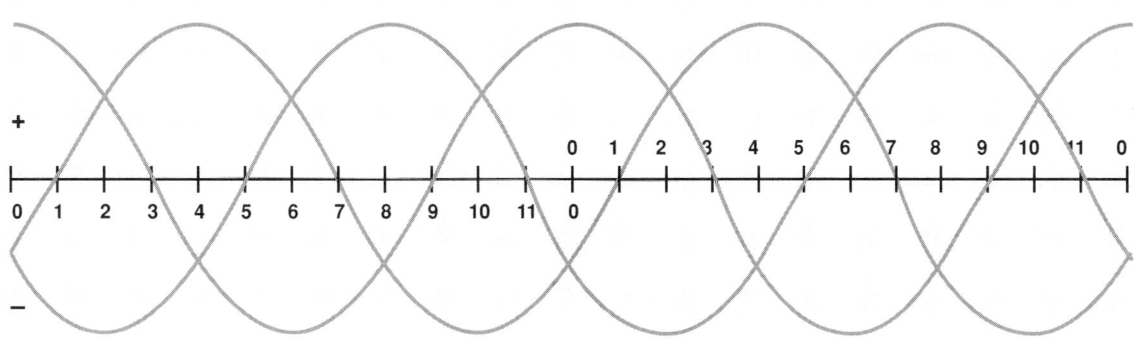

THIS DIAGRAM COVERS TWO CYCLES.

Figure 90 Three-phase currents in the armature of an induction motor

Nevertheless, if a single-phase condition of a three-phase power supply occurs while a three-phase induction motor is running at rated speed, the motor will continue to run as a single-phase motor. However, the torque capabilities of the motor will be drastically reduced, and it may stall. If it continues to run, some parts of its winding probably will become overheated and may burn out if it is carrying a load. Therefore, a three-phase motor should be provided with protective relays to prevent it from operating when a single-phase condition occurs.

Direction of Rotation

The direction of the force on the rotor conductor depends on the instantaneous direction of flow of rotor current produced by one phase and the instantaneous direction of field of the other two phases as determined by the direction of the phase rotation of the power supply. Therefore, reversing the phase connection of the power supply to the motor can reverse the direction of rotation of a polyphase induction motor. Simply interchanging any two of the three power-supply leads can reverse the motor's direction.

Synchronous Speed

The speed of the rotating field set up by the primary winding depends on the frequency of the power supply and the number of poles per phase that the primary winding is designed to produce. (Motors are commonly referred to by the number of poles per phase—for example, a six-pole motor produces six poles per phase.) The speed of the rotating field, which is called the synchronous speed of the motor, is determined by the equation—

$$S_s = \frac{120 \times f}{p_n} \qquad \text{(Eq. 43)}$$

where,

S_s = motor's synchronous speed, rpm
120 = a constant (for conversion to rpm)
f = frequency of the supply voltage, Hz
p_n = number of stator poles per phase.

As an example of using equation 43, determine the synchronous speed of a six-pole, 60-Hz motor. Thus,

$$S_s = \frac{120 \times 60}{6} = \frac{7{,}200}{6}$$

$$S_s = 1{,}200 \text{ rpm.}$$

Slip

As the rotor begins to turn and comes up to speed, the rotor conductors, which are turning in the same direction as the rotating field, cut the revolving field flux less and less frequently and at lower and lower speeds. This action causes less and less current to be induced in the rotor conductors. If the rotor could come up to synchronous speed or to the speed of the rotating field, no current would be induced in the rotor conductors, and no torque would be developed. For this reason, an induction motor does not quite attain its synchronous speed even at no load, and, as the load is increased, its speed further drops below its synchronous speed. This difference between synchronous speed and speed under load is called slip and is usually expressed in percent of synchronous speed.

For example, a six-pole, 1,200-rpm, synchronous-speed induction motor with a full-load speed of 1,140 rpm has 5-percent slip at full load because 5 percent of 1,200 rpm is 1,140 rpm (1,200 × 0.05 = 60 and 1,200 – 60 = 1,140).

Torque Characteristics

Induction motors exhibit two important torque characteristics. One is the maximum torque the motor develops as load is increased, starting from full speed at no load until the motor begins to stall. This characteristic is called *breakdown*, or *pullout, torque* and is defined as the greatest torque under which an electric motor can operate without sharply losing speed. The motor's design—particularly its field strength and leakage reactance—determines its pullout torque. The air gap, or clearance, between the rotor's core and the stator's core affect the field strength and leakage reactance. The greater the air gap, the less is the value of pullout torque. Therefore, manufacturers make an induction motor's air gap as small as possible. The resistance of the motor's winding does not appreciably affect the value of the pullout torque; however, it does determine the speed at which it occurs. The greater the resistance, the greater is the slip and the lower is the speed at which pullout torque occurs.

The second torque consideration of induction motors is the *starting*, or *breakaway, torque*. The value of starting torque, like pullout torque, is determined by field strength and leakage reactance. However, the resistance of the rotor's winding also affects starting torque. The rotor's winding resistance determines how close to starting condition that pullout torque occurs. Therefore, the greater the rotor winding resistance, the greater the starting torque up to the point at which pullout torque occurs.

The *National Electrical Manufacturers Association (NEMA)* classifies squirrel-cage induction motor designs and has a table of minimum values of pullout and starting torque for each classification. These are the maximum values for motor application purposes. Motor-torque values specified by NEMA are given in Table 16. Motors of recent manufacture carry their NEMA design classification

TABLE 16
NEMA Specified Motor Torque Values
(Percentage of Full-Load Torque)

NEMA Design	A & B			C			D
No. of poles	4	6	8	4	6	8	4, 6, or 8
Horsepower	Starting (Breakaway) Torques						
1	275	175	150				275
1	265	175	150				275
2	250	175	150				275
3	250	175	150		250	225	275
5	185	160	130	250	250	225	275
7	175	150	125	250	225	200	275
10	175	150	125	250	225	200	275
15	165	140	125	225	200	200	275
20–75	150	135	125	200	200	200	275
	Pull-Up (Accelerating) Torques						
1	193	123	105				
1	185	123	105				
2	175	123	105				
3	175	123	105		175	157	
5	130	112	100	175	175	157	
7	123	105	100	175	157	140	
10	123	105	100	175	157	140	
15	115	100	100	157	140	140	
20–75	105	100	100	140	140	140	
	Pullout (Breakdown) Torques						
1	300	275	250				
1	300	275	250				
2	275	250	225				
3	275	250	225		225	200	
5	225	225	225	200	200	200	
7	215	215	215	190	190	190	
10	200	200	200	190	190	190	
15	200	200	200	190	190	190	
20–75	200	200	200	190	190	190	

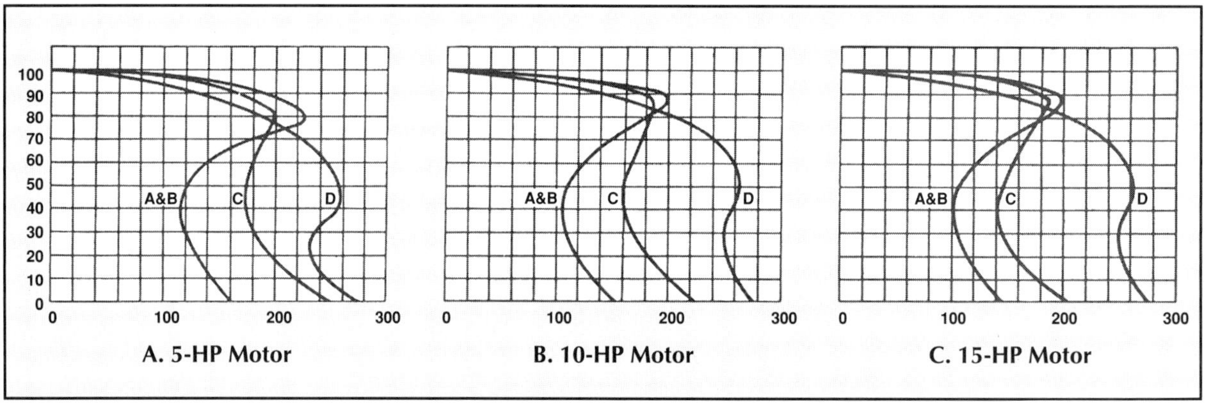

Figure 91 Speed-torque specifications for NEMA designs A, B, C, and D 6-pole motors

on the nameplate. Figure 91 shows torque characteristics for three typical oilwell pumping motors. Torque is plotted on the horizontal axis and speed in rpm on the vertical axis.

Changes in power-supply voltage greatly affect the pullout and starting torque capabilities of induction motors. The torque varies directly with the square of the voltage. For example, a 10-percent drop in voltage reduces the maximum starting or pullout torque to 81 percent of the torque a motor could develop at its rated voltage.

Efficiency

Squirrel-cage induction motors are generally quite efficient when compared to other types of electric motors and to most engines. Motors of average size and speed have efficiencies varying from zero at no load to about 75 percent at ½ load, 85 percent at full load, and 80 percent at 1¼ load. Motors designed with higher full-load slip have slightly lower full-load efficiency.

Starting and Inrush Current

When full-rated voltage is applied to the stator winding of a polyphase induction motor at standstill, a revolving field is set up instantly around the air gap, and the rotor slip is 100 percent. The revolving field cuts the rotor conductors rapidly, inducing high voltage in them. The rotor's winding resistance and reactance are the only factors that limit the current that flows in the rotor conductors as a result of this voltage. The heavy current in the rotor conductors causes a large current to be drawn by the stator winding because of transformer action.

When starting, squirrel-cage induction motors draw from five to ten times their full-load current across a line with full voltage. This heavy draw of current is termed *inrush current*. Inrush current occurs only until the motor accelerates to full speed. It usually decreases gradually as the motor's speed increases and then drops sharply when the motor reaches full speed. The magnitude of the inrush current depends only on electrical factors and not on the mechanical load on the motor. Put another way, the mechanical load on the motor does not affect the amount of inrush current. The duration of the inrush current depends on the time it takes the motor to accelerate. And, the time it takes for the motor to accelerate depends on the mechanical load on the motor.

POWER FACTOR

The power factor, as applied to an electric motor or other inductive circuit, is a measure of in-phase amperes compared with out-of-phase, or reactive, amperes. In-phase amperes do useful work. Reactive amperes magnetize the motor. The motor takes current from a power line to build up a magnetic field. However, as the current begins to decline and the magnetic field begins to collapse, an induced countervoltage, or back emf, in the motor drives current back into the power line. That is, when a motor is magnetized, the motor takes current from the line and then drives it back into the line. The net power taken from the line is zero; a wattmeter will not record this reactive power.

Reactive power must be considered in a motor's operation because reactive current loads and heats the power line and transformers just as useful in-phase current does. Reactive current causes voltage drops along the power line. It heats the line according to Watt's law and the following power equation—

$$P = I^2R \qquad \text{(Eq. 44)}$$

where

P = power, watts
I = current, amperes
R = reactance, ohms.

A wattmeter can record this power loss.

Reactive power also affects the transformer bank. For example, suppose a transformer's secondary is rated for 100 amperes and is supplying 100 amperes. In this case, the transformer is fully loaded. However, it could be that only 50 of the 100 total amperes are doing useful work. In such a case, the other 50 amperes are reactive. If the reactive power is not taken into account, an unknowledgeable person might install oversized transformers for a particular horsepower load and use oversized power-line conductors. It is important to know that oversizing and its increased costs can be avoided by improving the *power factor*.

The power factor is expressed as a decimal fraction ranging from zero to unity, or one. A power factor of 0.9 is quite good, while a power factor of 0.5 is too low to be acceptable for power circuits. The higher the power factor, the better it is for power circuits.

The power factor of squirrel-cage motors is generally moderately good at higher percentage loads and poor at no load or light loads. For motors of average size and speed, the power factor is from 10 to 20 percent at no load, about 60 percent at ½ load, about 75 percent at full load, and about 80 percent at 1¼ load. The power factor during starting is very poor, usually averaging about 20 percent. Motors designed with higher torques have a slightly lower full-load power factor.

Calculating the Power Factor

The power factor can be calculated for a large distribution system or for an individual motor by making several simple measurements. In the calculations that follow, the values are obtained from a running, unloaded, single-phase induction motor having a rating of ¼ horsepower. A wattmeter, a voltmeter, and an ammeter are used to take the measurements. Figure 92 is a schematic of the meters in a 120-volt AC circuit with the ¼-horsepower motor. The true power in the motor circuit is read by the wattmeter, while the product of the readings of the voltmeter and ammeter is the apparent power in the circuit. In other words, apparent power is obtained by multiplying the voltage by the amperage. The answer is apparent power in volt-amperes (VA). For example, 120 volts × 4.5 amperes = 540 VA.

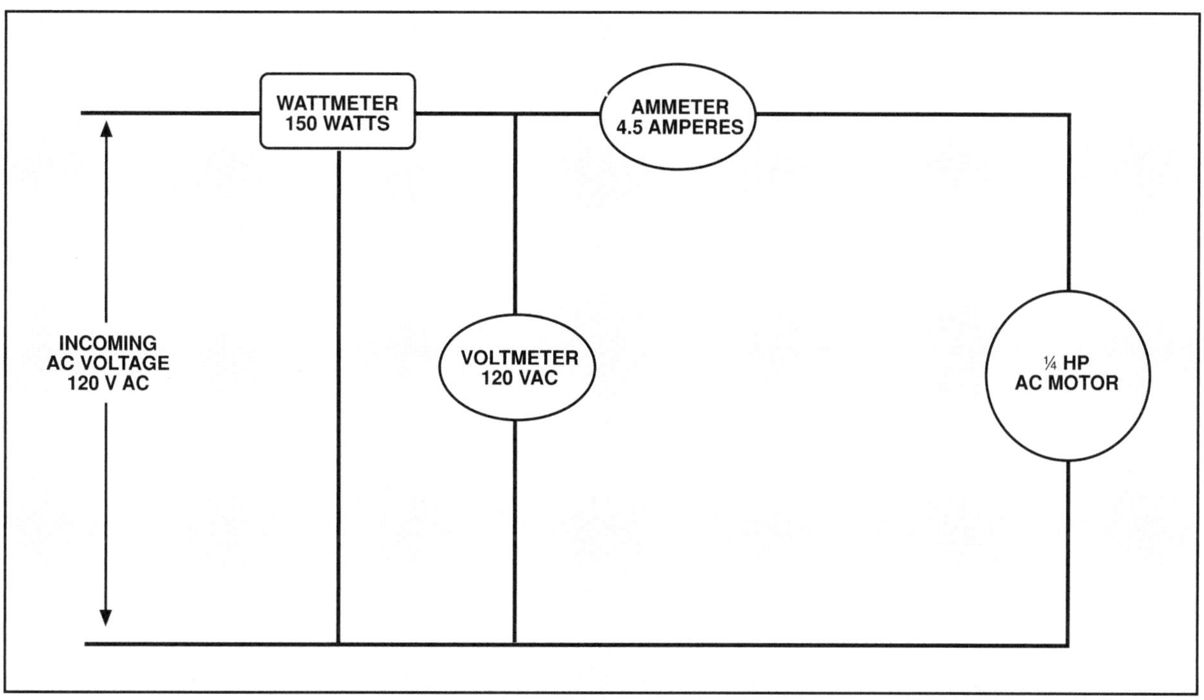

Figure 92 Connections for determining the power factor of an induction motor circuit

In simple terms, the power factor is the quotient obtained by dividing the real, or true, power in watts by the apparent power. The equation is—

$$pf = \frac{p_t}{p_a} \qquad \text{(Eq. 45)}$$

where

pf = power factor, volt amperes
p_t = true power, watts
p_a = apparent power.

A right triangle (fig. 93) shows the relationship between volts, amperes, VA, volt-ampere reactive (VAR), and watts. As an example, one leg of the triangle in figure 93 represents 150 watts, the other leg represents 519 VAR, and the hypotenuse represents 540 VA. Shortly, you will see how the value for VAR is determined, but first, some basic trigonometry must be discussed.

For example, the power factor can also be defined as the *cosine* of the angle by which the current either lags or leads the voltage in an AC circuit. A cosine is a *trigonometric function*. A cosine is one of six trigonometric functions that express relationships in a right triangle. Specifically, in a right triangle, the cosine is the ratio of the length of the side adjacent to an acute angle to the length of the hypotenuse. A trigonometric function is a function (such as the sine, cosine, *secant*, *cosecant*, *tangent*, and *cotangent*) of an arc or angle expressed as the ratios of the pairs of the sides of a right triangle. (A *function* is a mathematical rule between two sets that assigns to each member of the first set exactly one member of the second.)

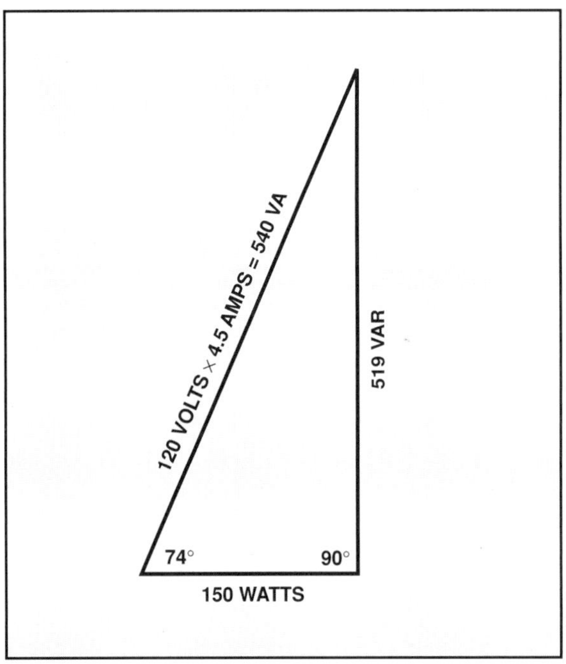

Figure 93 Power triangle for the circuit in figure 92

With trigonometric functions in mind, let's discuss the earlier example where a motor in a 120-volt, 4.5-ampere AC circuit, develops 150 watts of true power, and 540 VA of apparent power. To find the power factor, the first step is to use equation 45. Thus,

$$p_f = \frac{150 \text{ watts}}{540\ VA}$$

$$p_f = 0.2777.$$

This result must be converted to find the angle by which current lags behind voltage. To find the angle, refer to Table 17, which lists trigonometric functions for sine, cosine, tangent, and cotangent. (The functions secant and cosecant are not included because they are not needed in power calculations.) Then, because the power factor is the cosine of the angle by which current lags behind voltage, find the closest value in the cosine column of Table 17 to find the angle. In this case, 0.2756 in the table is closest to 0.2777. So, the angle with a cosine of 0.2777 is 74 degrees. That is the angle by which the current lags behind the voltage in this example.

The reactive power, or VAR, side of the triangle is calculated by using the equation—

$$VAR = VA \times \sin A \qquad \text{(Eq. 46)}$$

where

VAR = volt-ampere reactive
VA = volt-amperes
$\sin A$ = sine of angle A.

As an example, assume that angle A is 74 degrees (see fig. 93), true power is 150 watts, and apparent power (VA) is 540 watts. From table 15, the sine of a 74-degree angle is 0.9613. Plugging these values into equation 46 results in—

$$VAR = 540 \times \sin 74$$
$$= 540 \times 0.9613$$
$$VAR = 519.$$

Incidentally, the power factor of 0.2777 in the example is poor. However, because the example induction motor is running without a load, it exhibits a poor power factor. Indeed, any induction motor running without a load has a poor power factor. On the other hand, large induction motors operating under full load have power factors of about 0.8, which is quite good. Be aware, however, that small motors have inherently low power factors.

TABLE 17
Trigonometric Functions

Angle, °	Sine	Cosine	Tangent	Cotangent	Angle, °	Sine	Cosine	Tangent	Cotangent
0	0.0000	1.0000	0.0000	Infinity	45	0.7071	0.7071	1.000	1.0000
1	0.0175	0.9998	0.0175	57.29	46	0.7193	0.6947	1.036	0.9657
2	0.0349	0.9994	0.0349	28.636	47	0.7193	0.6829	1.072	0.9326
3	0.0523	0.9986	0.0524	19.081	48	0.7431	0.6691	1.111	0.9004
4	0.0698	0.9976	0.0699	14.301	49	0.7547	0.6561	1.150	0.8693
5	0.0872	0.9962	0.0875	11.430	50	0.7660	0.6428	1.192	0.8391
6	0.1045	0.9945	0.1051	9.514	51	0.7772	0.6293	1.235	0.8098
7	0.1219	0.9926	0.1228	8.144	52	0.7880	0.6157	1.280	0.7813
8	0.1392	0.9903	0.1405	7.115	53	0.7986	0.6018	1.327	0.7536
9	0.1564	0.9877	0.1584	6.314	54	0.8090	0.5878	1.376	0.77265
10	0.1737	0.9848	0.1763	5.671	55	0.8192	0.5736	1.428	0.7002
11	0.1908	0.9816	0.1944	5.145	56	0.8290	0.5592	1.483	0.6745
12	0.2079	0.9782	0.2126	4.705	57	0.8387	0.5446	1.540	0.6494
13	0.2250	0.9744	0.2309	4.332	58	0.8481	0.5299	1.600	0.6249
14	0.2419	0.9703	0.2493	4.011	59	0.8572	0.5150	1.664	0.6009
15	0.2588	0.9659	0.2680	3.732	60	0.8660	0.5000	1.732	0.5774
16	0.2756	0.9613	0.2868	3.487	61	0.8746	0.4848	1.804	0.5543
17	0.2924	0.9563	0.3057	3.271	62	0.8830	0.4695	1.881	0.5317
18	0.3090	0.9511	0.3249	3.078	63	0.8910	0.4540	1.963	0.5095
19	0.3256	0.9455	0.3443	2.904	64	0.8988	0.4384	2.050	0.4877
20	0.3420	0.9397	0.3640	2.748	65	0.9063	0.4226	2.145	0.4663
21	0.3584	0.9336	0.3839	2.605	66	0.9136	0.4067	2.246	0.4452
22	0.3746	0.9272	0.4040	2.475	67	0.9205	0.3907	2.356	0.4245
23	0.3907	0.9205	0.4245	2.356	68	0.9272	0.3746	2.475	0.4040
24	0.4067	0.9136	0.4452	2.246	69	0.96	0.3584	2.605	0.3839
25	0.4226	0.9063	0.4663	2.145	70	0.9397	0.3420	2.748	0.3640
26	0.4384	0.8988	0.4877	2.050	71	0.9455	0.3256	2.904	0.3443
27	0.4540	0.8910	0.5095	1.963	72	0.9511	0.3090	3.078	0.3249
28	0.4695	0.8830	0.5317	1.881	73	0.9563	0.2924	3.271	0.3057
29	0.4848	0.8746	0.5543	1.804	74	0.9613	0.2756	3.487	0.2868
30	0.5000	0.8660	0.5774	1.732	75	0.9659	0.2588	3.732	0.2680
31	0.5150	0.8572	0.6009	1.664	76	0.9703	0.2419	4.011	0.2493
32	0.5299	0.8481	0.6249	1.600	77	0.9744	0.2250	4.332	0.2309
33	0.5299	0.8481	0.6249	1.600	77	0.9744	0.2250	4.332	0.2309
34	0.5592	0.8290	0.6745	1.483	79	0.9816	0.1908	5.145	0.1944
35	0.5736	0.8192	0.7002	1.428	80	0.9848	0.1737	5.671	0.1763
36	0.5878	0.8090	0.7265	1.376	81	0.9877	0.1564	6.314	0.1584
37	0.6018	0.7986	0.7536	1.327	82	0.9903	0.1392	7.115	0.1405
38	0.6157	0.7880	0.7813	1.280	83	0.9926	0.1219	8.144	0.1228
39	0.6293	0.7772	0.8098	1.235	84	0.9945	0.1045	9.514	0.1051
40	0.6428	0.7660	0.8391	1.192	85	0.9962	0.0872	11.430	0.0875
41	0.6561	0.7547	0.8693	1.150	86	0.9976	0.0698	14.301	0.0699
42	0.6691	0.7431	0.9004	1.111	87	0.9986	0.0523	19.081	0.0524
43	0.6820	0.734	0.9325	1.072	88	0.9994	0.0349	28.636	0.0349
44	0.6947	0.7193	0.9657	1.036	89	0.9999	0.0175	57.290	0.0175
					90	1.0000	0.0000	Infinity	0.0000

Effect of Load on Power Factor

As just stated, as mechanical load is applied to an induction motor, the power factor improves. Improvement continues even to the point where overload occurs. As an example, suppose a brake is applied to the motor setup in figure 92, and the load is increased until the wattmeter reads 500 watts. The ammeter reading rises to 6 amperes, and, of course, voltage remains the same. The power factor has now changed according to power factor equation 45. Thus—

$$p_f = \frac{500}{120 \times 6}$$

$$= \frac{500}{720}$$

$$p_f = 0.694.$$

The power factor of 0.694 corresponds to a cosine of 46 degrees. So, with this information, a right triangle with an acute angle of 46 degrees can represent the conditions in the example (fig. 93), where VA = 720 (because 120 × 6 = 720), VAR = 519, and true power = 500 watts.

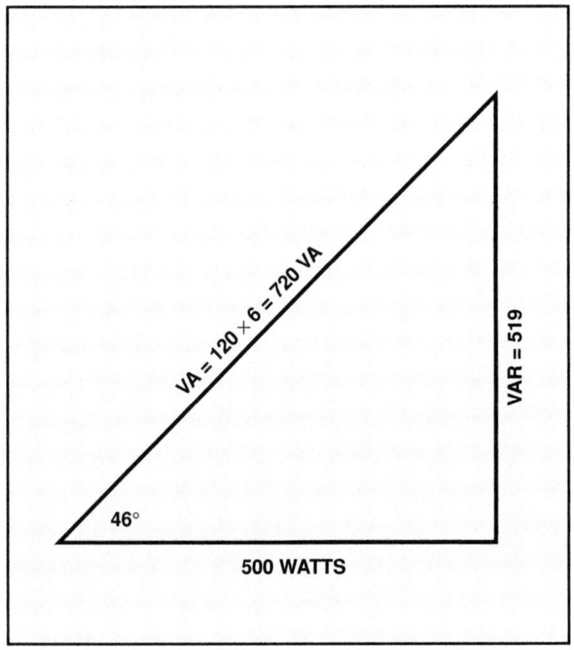

Figure 94 Power factor improvement with an increased load on the motor shown in figure 92

Correcting the Power Factor with Capacitors

When a capacitor is connected across an AC line, a current flows. This current is reactive current, just as the magnetizing current flowing in a motor is reactive. The difference between the two reactive currents is that the reactive current in a motor lags behind the applied voltage while the reactive current in the capacitor leads the voltage. When inductors and capacitors are in the same circuit, their reactance tends to neutralize one another. This neutralizing tendency can be used to correct for low power factors in inductive circuits.

A properly sized capacitor connected across an induction motor improves the power factor. Moreover, this improvement is reflected throughout the entire circuit. Thus, installing a capacitor can prevent large reactive currents that could cause power losses in transformer banks and in the conductors of the power line.

Capacitors used for power-factor correction are not rated in microfarads (the usual way) but are rated in kilovolt-amperes, reactive (kVAR) at a specified voltage. For example, a capacitor rated at 0.519 kVAR exactly corrects the power factor to unity (1.0) in figure 94. Because of the cost of capacitors, however, power factors are not normally corrected to better than 0.91.

SOLENOIDS

Basically, a solenoid is a coil of wire wrapped around a tube through which a rod of iron can travel. A spring mechanism usually holds the rod in position at one end of the coil. Current at the rated voltage is passed through the coil of wire to establish a magnetic field that moves the rod to the opposite end of the tube.

Solenoid valves are *needle* or *orifice valves* that allow the flow of liquid or gases when the needle or orifice is in one position but do not allow flow when in the other position. The needle or orifice is attached to the rod so that when current passes through the coil, the valve is in the flow position. When the current stops, a spring moves the valve to the no-flow position. Opposite action is also possible.

The solenoid valve shown in figure 95 is a type of relay that activates pneumatic or hydraulic systems. In this case, the spring keeps the valve closed under normal operating conditions. When the coil is energized, the magnetic core attached to the inner valve is pulled upward, thus opening the valve. When the current to the coil is interrupted, the valve returns to its normal, closed position. The opening or closing of the valve can perform many different functions, such as bringing the level of the liquid in a vessel under control.

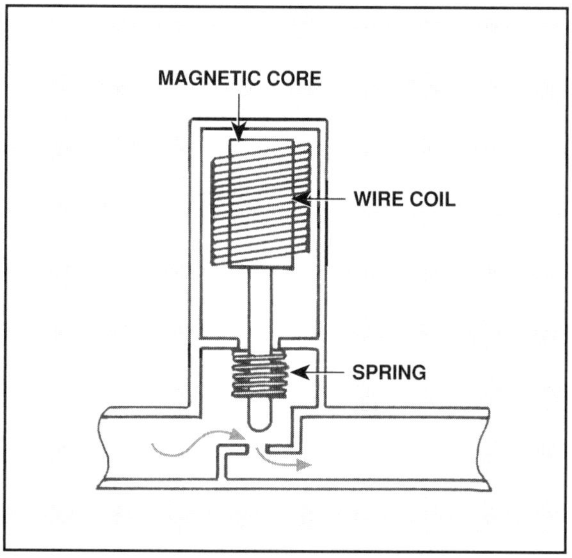

Figure 95 Solenoid valve

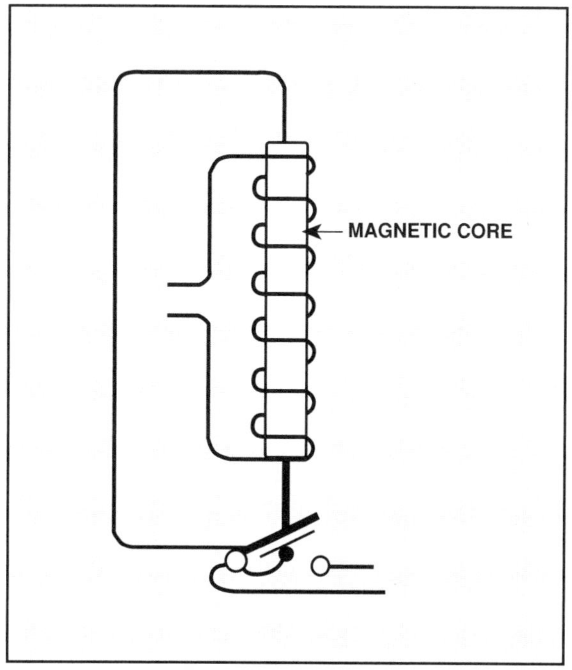

Figure 96 Single-pole, single-throw (SPST) relay

ELECTROMECHANICAL RELAYS

An *electromechanical relay* is probably the most widely used application of the solenoid principle. The iron rod of the solenoid is mechanically attached to an arm that bridges a set of contacts. When current is passed through the solenoid, this arm either breaks the contact bridge, as a switch does, or breaks one contact and bridges another set, using a double-throw switch. More than one arm may be attached to the rod, as in the case of multiple-pole relays. Many variations of relays are available in which the number of poles, coil ratings, contact ratings, switching speed, and so forth differ. Figure 96 is a simple sketch of a single-pole, single-throw (SPST) relay. Figure 97 is a schematic of a double-pole, double-throw (DPDT) relay.

REVIEW QUESTIONS

1. A practical generator can produce voltage by having a coil rotate in a stationary magnetic field. What other arrangement can be used to allow the generator to produce voltage?
2. What components of a generator are used to make connections to the rotating coils?
3. In a two-pole generator, a complete revolution of the coil produces one cycle. How many revolutions per minute produce a 60-Hz current?
4. What is the name of the waveform produced by commercial 60-Hz generators?
5. Why is it desirable to transmit electricity at high voltage and low current?
6. What are two conditions that tend to limit the maximum voltage that can be transmitted?
7. What is a single word that best describes transformer action?
8. A step-down transformer has a turn ratio of ten to one. If 120 volts are applied to the primary, what is the secondary voltage?
9. If the current in the secondary of the transformer in question 8 is 10 amperes, what is the primary current?

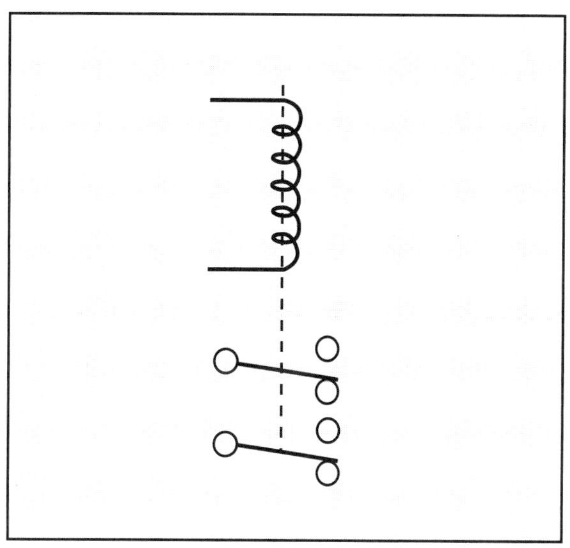

Figure 97 Double-pole, double-throw (DPDT) relay

10. Large transformers are rated according to primary maximum capacity. What is the symbol usually used to designate the rating?
11. The primary and secondary windings of large distribution transformers are identified by the use of the letters H and X. How are the windings of power transformers in electronic equipment identified?
12. A squirrel-cage rotor is used in what kind of electric motor?
13. Applying full line voltage to most induction motors starts them. At the moment this line voltage is applied, how much current does a motor draw?
14. Write the equation for the power factor, using true power and apparent power.
15. How does a poor power factor cause a loss in a transmission (power) line?
16. What electrical component can be used to improve the power factor of an induction-motor circuit?
17. Name two principal applications of the solenoid.

Semiconductors and Applications

In the years since their invention in the late 1940s by physicists at Bell Laboratories, transistors and similar semiconductor components have virtually replaced vacuum tubes and other conventional devices in electronic equipment. Terms such as *vacuum tube*, *thyratron*, *triode*, and the like are virtually unheard today.

Semiconductors, which include *diodes*, transistors, and *light-emitting diodes* (*LEDs*), among many other such components, offer several advantages over vacuum tubes. Semiconductors are small, lightweight, rugged, and efficient. What is more, they require no warm-up time and can operate under severe environmental conditions.

BASIC SEMICONDUCTOR THEORY

Semiconductors are *solid-state* devices that depend on the flow of electric charges in a solid for their operation. A semiconductor's resistance to current flow is less than the resistance of insulators, or nonconductors. However, a semiconductor's resistance is greater than a conductor's. Stated another way, the conductivity of a semiconductor is higher than a nonconductor but is lower than a conductor.

The conductivity of a material depends on the number of free electrons contained in its atomic structure. For example, a conductor such as copper has 1.64×10^{24} free electrons per cubic centimetre, while an insulator such as rubber has only three or four free electrons per cubic inch. Interestingly, the most common materials used to make semiconductors—*silicon* and *germanium*—are nonconductors in their pure states. They have no free electrons and become semiconductors only after manufacturers add impurities.

Semiconductor Materials

Manufacturers make most semiconductors from silicon. They rarely use germanium, except in special applications, because germanium cannot operate at high temperatures. Silicon, on the other hand, can operate at high temperatures.

Four electrons in the outer ring, or shell, orbit the nucleus of a silicon atom. Electrons in the outer ring of an atom are *valence electrons*. Valence electrons in one atom are available to combine with valence electrons of other atoms. When valence electrons of atoms combine, they form a new substance. For example, when the valence electrons of silicon combine with the valence electrons of oxygen, silicon dioxide (SiO_2) is formed, which is one of the most common substances on earth in that it is the main component of sand.

Besides valance electrons in one atom of one element combining with the valance electrons of a different atom, the valence electrons can also combine with the valence electrons of the same atom. When valence electrons of one element combine with the valence electrons of the same element, they form *electron-pair bonds*. An electron-pair bond is the combining of atoms (the bonding) in which each atom of a bond pair contributes one electron to form a pair of electrons.

Silicon has a crystal structure. That is, silicon is a homogenous solid whose atoms are arranged in a regularly repeating pattern. When a substance occurs in such a pattern, it is a *crystal* or has a crystalline structure. The four valence electrons of silicon are tightly bound in its crystalline structure by electron-pair bonds. This type of structure is called a *lattice*. (A lattice, in this sense, is the regular geometrical arrangement of points or objects over an area in space.) In a lattice, the four electrons of each atom form electron-pair bonds with one valence electron from each of the four adjacent atoms (fig. 98).

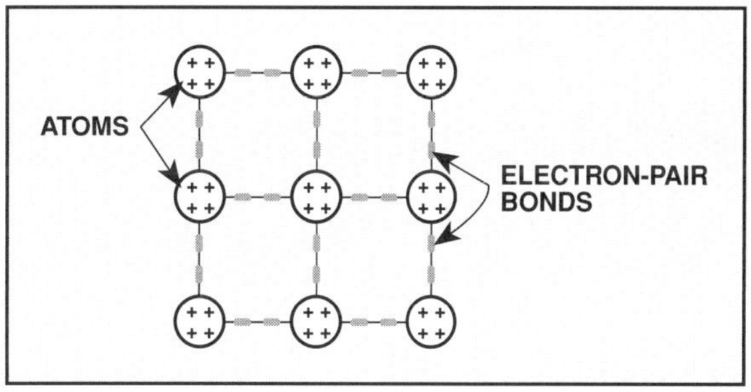

Figure 98 Lattice structure of semiconductor crystals

Because all of the valence electrons combine to form electron-pair bonds, no free electrons are in the crystal, which means that its conductivity is low.

Manufacturers increase the conductivity of a semiconductor material, such as silicon, by adding impurities. The impurities have a different number of valence electrons so that when the electron-pair bonds form, a few electrons are left over. Only a small amount of impurities is required to give silicon the needed conductivity.

Two types of semiconductor structures are required for semiconductor devices. The two structures are made with different types of impurities. One structure is called *n-type*; the other is called *p-type*. The n-type, which is short for negative type, possesses a negative charge because free electrons exist in it. The p-type, or positive type, possesses a positive charge because it has too few electrons. This condition creates positive charges called *free holes*.

Manufacturers form an n-type structure (fig. 99) by adding an impurity, such as *arsenic*, which has five valence electrons. (Arsenic is a chemical element that occurs as a brittle, steel-gray hexagonal mineral.) Each impurity atom displaces an atom in the lattice structure. The valence electrons in the impurity form electron-pair bonds with valence electrons of the four adjacent atoms. Because the semiconductor material atom has four valence electrons and the impurity atom has five, only four electron-pair bonds are formed. Thus, one excess, or free, electron is available. The existence of this free electron causes the structure to have a negative charge.

The p-type structure (fig. 100) is obtained by adding an impurity such as *gallium* that has only three valence electrons. (Gallium is a silvery-white metallic element whose chemical symbol is Ga.) In this

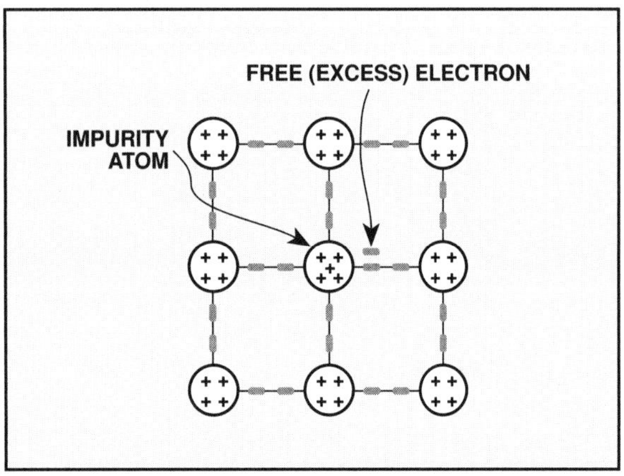

Figure 99 Lattice structure of n-type material

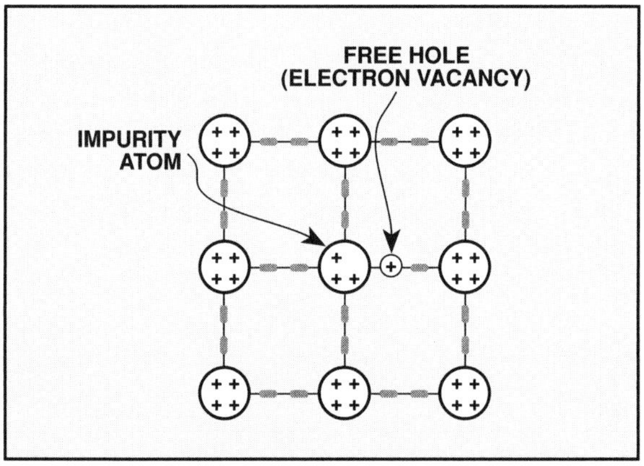

Figure 100 Lattice structure of p-type material

case, only three complete electron-pair bonds can be formed because the impurity atom has only three electrons available for pairing. The lack of a fourth electron creates a vacancy, or free hole, in the lattice structure. This free hole causes the structure to have a positive charge.

Current Carriers

Free electrons and free holes serve as current carriers in semiconductors. That free electrons, which are negative charges, conduct current is a basic concept of electricity. However, that free holes, which are positive charges, conduct current is a relatively new concept. Current flow that depends on free holes occurs only in semiconductor devices and is opposite in direction to that of free electron flow.

As stated previously, free holes are formed in p-type semiconductor material because the impurity atom that displaces a semiconductor atom has only three valence electrons available for forming electron-pair bonds. When these electrons form pairs with the electrons of the surrounding semiconductor atoms, not enough are available to go around. The semiconductor atom has four valence electrons, so the fourth electron is left without an electron to pair with. This action forms a *hole-electron pair*, and this hole, or vacancy, means the atomic structure of the crystal is unbalanced.

The hole in the lattice exerts a pull on surrounding electron pairs and causes an electron to break its bond and move to fill the hole. When this movement occurs, a hole is created in the electron pair that the electron broke away from. In effect, the hole moves from its original position to the position that was occupied by the electron that broke away and filled the original hole. This action is repeated over and over as shown in figure 101.

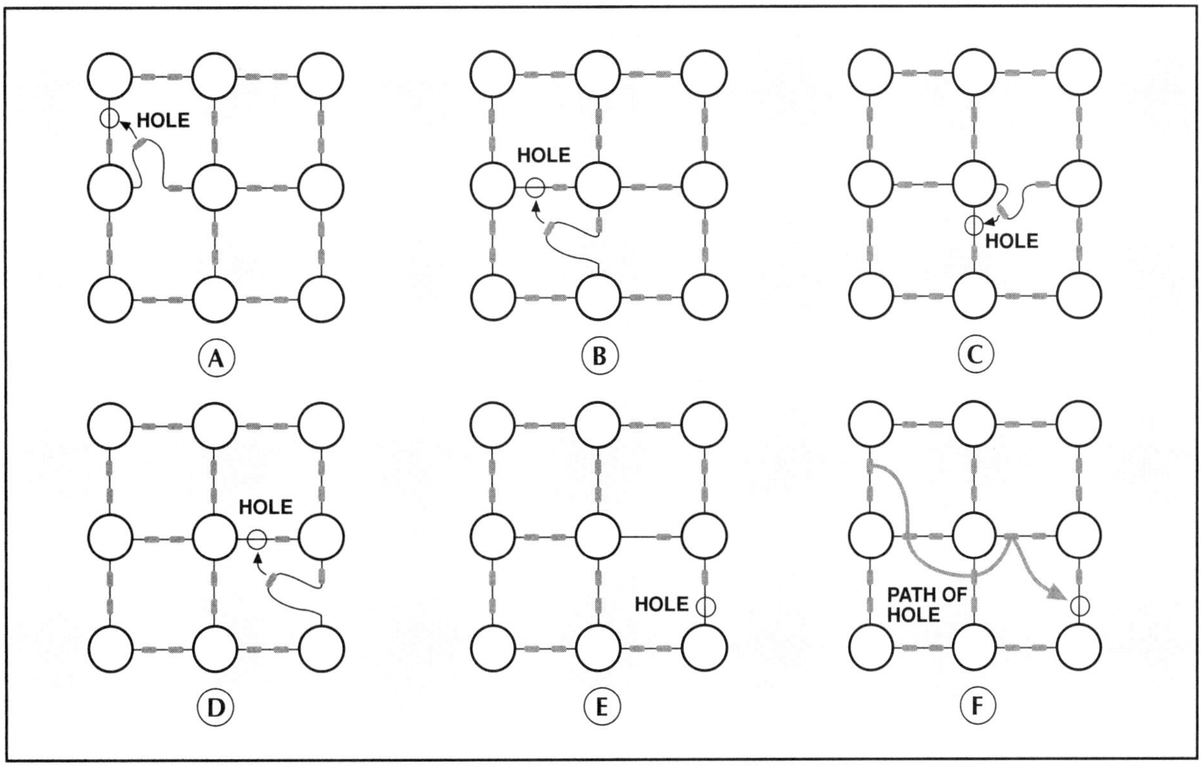

Figure 101 Movement of free holes through semiconductor crystal

Each time an electron breaks away and fills the hole, a new hole is created in its place. In figure 101A, the free hole is on the upper left side of the crystal lattice. An electron has broken its bond and is moving to fill the hole. In figure 101B, a new hole occurs where the electron was. This action occurs again and again until the hole is in the lower right side of the lattice (fig. 101E). Bear in mind, however, that the hole itself does not move. Instead, a new hole is created each time an electron breaks its bond. Nevertheless, the effect is the same as if the free hole had moved along the path shown in figure 101F.

Free electrons move through an n-type semiconductor crystal in the same manner as free holes except that a free electron displaces another electron from a nearby electron pair. No holes are created because an excess of electrons is available. Each time an electron is displaced by a free electron, it, in turn, becomes a free electron and moves to displace an electron from another pair.

Free electrons and holes move randomly in a crystal to which no voltage is applied. They haphazardly displace other electrons and exchange positions. But, when voltage is applied, it causes free electrons and holes to move in a relatively straight line. Applied voltage causes free electrons to flow across the crystal from the negative terminal to the positive terminal. Applied voltage causes free holes to flow from the positive terminal to the negative terminal.

Figure 102 is a schematic of an n-type crystal, in which free electrons carry current, and to which

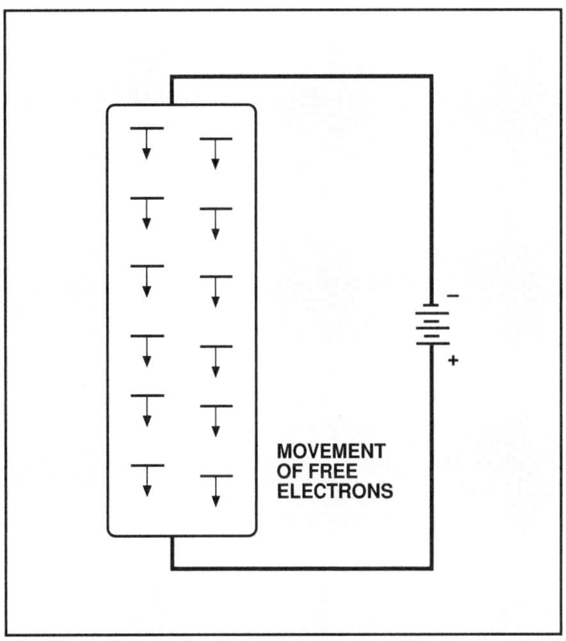

Figure 102 Electron flow in n-type crystal

DC voltage is applied. Voltage makes the free electrons move from the negative terminal toward the positive terminal. Each electron displaces an electron from an electron pair and the displaced electron becomes a free electron, which, in turn, displaces an electron from another pair. With voltage applied, the free electrons displace others in the direction of the positive terminal. In this way, the free electrons move across the crystal from the negative to the positive terminal.

Figure 103 is a schematic of a p-type crystal, in which free holes carry current, and to which voltage is applied. The flow of free holes within the crystal is opposite that of electron flow. The holes move toward the negative terminal; however, electrons flow toward the positive terminal. When a hole reaches the negative terminal, an electron from the DC supply terminal cancels the hole. Simultaneously, an electron from one of the bonds enters the positive terminal and the new hole moves toward the negative terminal.

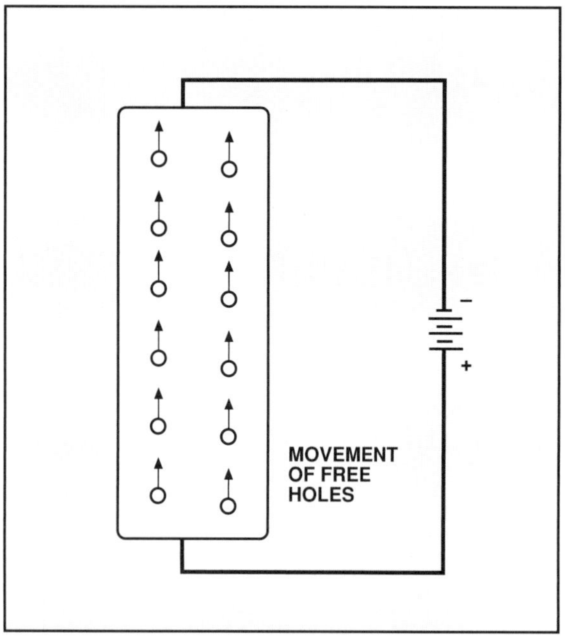

Figure 103 Hole flow in p-type crystal

Current Flow

Hole flow from the positive to the negative terminal has the same effect as electron flow in the opposite direction. An n-type or a p-type material conducts current in both directions—that is, regardless of which way the DC supply is connected, current will flow.

It is interesting to note that the concept of hole flow began only when semiconductors were developed. Prior to then, current flow was thought to be only the flow of electrons from the negative to the positive of the power supply. However, as semiconductors became prominent, physicists had to change the way to visualize current flow. Today, industry has accepted the concept that current flows from the positive terminal to the negative terminal of the supply. Flow from positive to negative is the direction of hole flow, and not electron flow. In short, conventional current flow is now thought of as hole, or positive-to-negative, flow, rather than electron, or negative-to-positive, flow. (When symbols are used to depict semiconductors, arrows depict the direction of this conventional current flow.)

Because of the modern concept of conventional current flow, technicians and engineers no longer consider the negative terminal of a power supply as the hot terminal. Instead, they consider the positive terminal to be the hot terminal. Further, they think of the negative terminal as being connected to earth, or ground, which is *zero potential.* (Zero potential is the actual potential of the surface of the earth taken as a point of reference.) Moreover, throughout this manual, current means conventional current flow from positive to negative of the DC power supply.

SEMICONDUCTOR DIODES

A *diode* is a device that has two electrodes—a *cathode* and an *anode*. A diode passes current in one direction—from anode to cathode—and blocks current flow in the opposite direction.

Manufacturers make semiconductor diodes by placing n-type and p-type materials together to form a junction. When the materials are joined, an interaction occurs between the free holes of the p-type material and the free electrons of the n-type material near the junction (fig. 104). This interaction enables semiconductor devices to function.

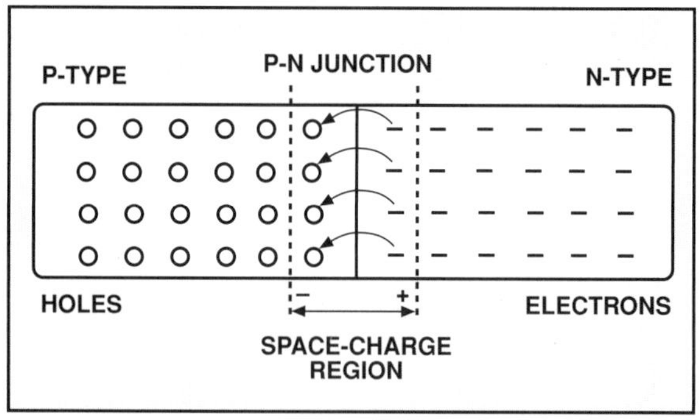

Figure 104 Interaction of free electrons and holes at a p-n junction

The free holes of the p-type material attract a few electrons from the n-type material across the junction. After a few electrons have crossed the junction and filled holes in the p-type material, a *space-charge region* is set up at the junction. Next to the junction, the n-type material becomes slightly positive, and the p-type material becomes slightly negative. When the space charge has been established, the junction is stabilized. No more electrons can flow across the junction because they are now repelled by the slight negative charge in the p-type material.

This space charge acts as an energy barrier, or *potential hill*. If current is to flow across this potential hill, external voltage must be applied to the semiconductors. (This voltage must be about 0.4 to 0.7 volts for silicon semiconductors.) The potential hill preserves the differences between the two types of material. It prevents a total interaction between the holes in the p-type and the free electrons in the n-type. Total interaction would neutralize the charge in the two types and make the diode useless.

As stated before, current can flow in both directions through a piece of semiconductor material provided the semiconductor is not joined to form a p-n junction with the opposite type of material. Current, however, cannot flow in both directions across a p-n junction. When external voltage is applied, current can flow from the p-type material to the n-type; however, it cannot flow from the n-type to the p-type. This characteristic enables a p-n junction to be used as a diode—that is, it blocks current flow in one direction, but allows current flow in another.

Figure 105 shows a p-n junction that has the positive terminal of a battery connected to the n-type material and the negative terminal connected to the p-type material. This reversal of polarities is *reverse bias* and prevents current from flowing through the junction. Reverse bias in a p-n junction attracts free electrons from the n-type material toward the negative terminal. Free holes in the p-type are attracted away from the junction and toward the positive terminal. This action causes the free electrons and holes to bunch up toward the ends of the crystal. This bunching up of free electrons and holes leaves a depleted area at the junction that offers high resistance to current flow. Any current that may flow is small and is sometimes referred to as leakage current.

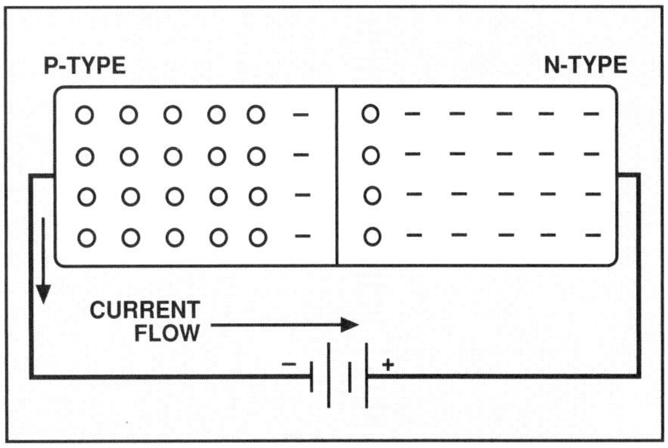

Figure 106 Forward bias of a p-n junction

Figure 106 shows a *forward biased* p-n junction. Here the polarities are matched; the negative supply terminal is connected to the n-type material and the positive terminal is connected to the p-type material. Free electrons in the n-type are attracted across the junction toward the positive terminal while holes in the p-type are attracted across the junction toward the negative terminal. A forward-biased p-n junction offers low resistance and current flows freely across it.

Figure 107 shows two symbols used in circuit diagrams for a semiconductor diode. The connection made to the p-type material is called the anode

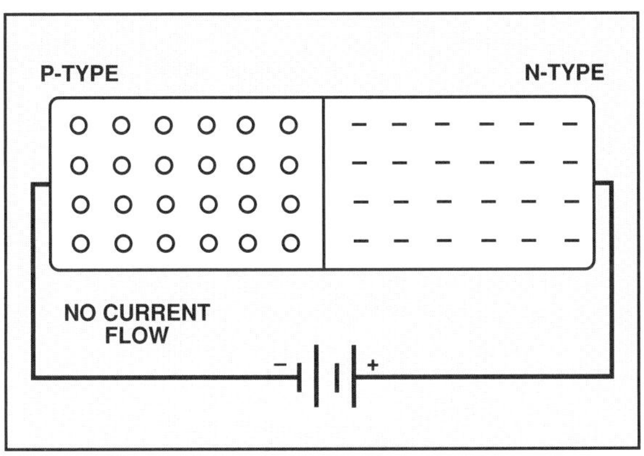

Figure 105 Effect of reverse bias on a p-n junction

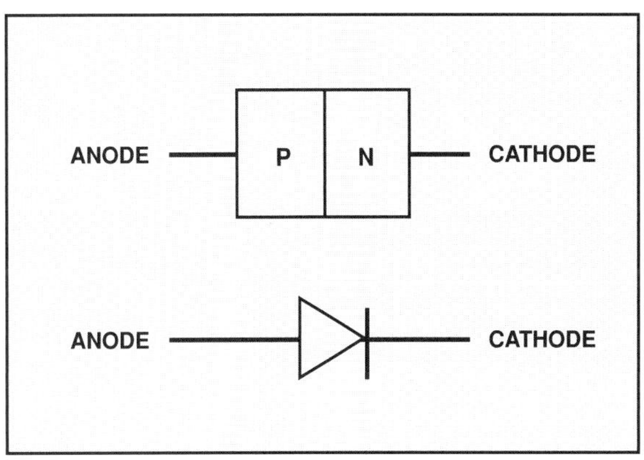

Figure 107 Symbols for diodes

and the connection to the n-type is called the cathode. The direction of current flow is from the anode to the cathode, which is indicated by the bottom drawing in figure 107. Notice that the symbol for the anode is a large arrow whose head points in the direction of current flow.

When the anode (p-type connection) is negative with respect to the cathode (n-connection), current flow cannot occur. When the anode is positive with respect to the cathode, current is permitted to flow without restriction except for the junction voltage of 0.4 to 0.7 volts.

Figure 108 depicts a graph with a curve that shows the response of a diode to both forward and reverse voltage. The horizontal axis of the graph is labeled –V and +V, which indicates negative and positive voltage. The graph's vertical axis is labeled –I and +I, which indicates negative and positive current. In the forward-biased region, where voltage and current are positive, the diode conducts freely with a forward voltage drop of about 0.7 volts under full conduction. When biased in the reverse direction, where voltage and current are negative, no current flows (except leakage current) until *reverse*, or *Zener, breakdown* occurs. Zener breakdown occurs when the electric field across the space-charge region becomes high enough to produce a form of field emission that suddenly increases the number of carriers in this region. This breakdown is typically several hundred to several thousand volts, depending on the semiconductor's construction, and is not destructive. Indeed, Zener breakdown allows diodes to be employed as voltage regulators, as you will learn later in this section. (Zener breakdown is named for the American physicist Clarence M. Zener, who first proposed the phenomenon.)

Bipolar is a descriptive term applied to devices constructed of p-n materials. It means that the device has two poles or polarities. Diodes, transistors, SCRs, and similar devices feature bipolar construction.

A diode operates somewhat like a check valve (fig. 109). A disc-shaped flap in the check valve fits inside a pipe carrying fluid. The valve's flapper swings up when liquid flows in one direction, but when the

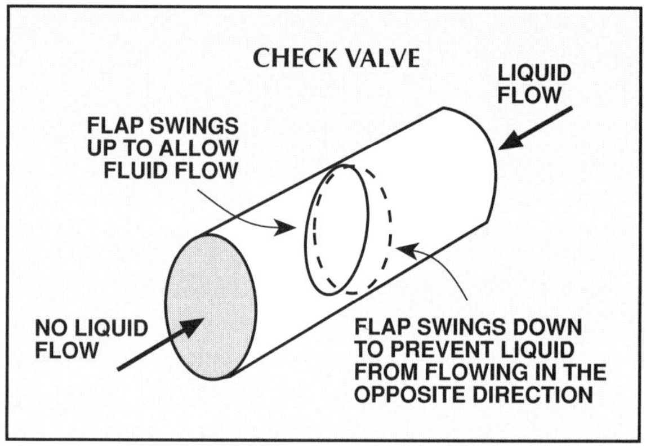

Figure 109 Check valve analogy of a diode

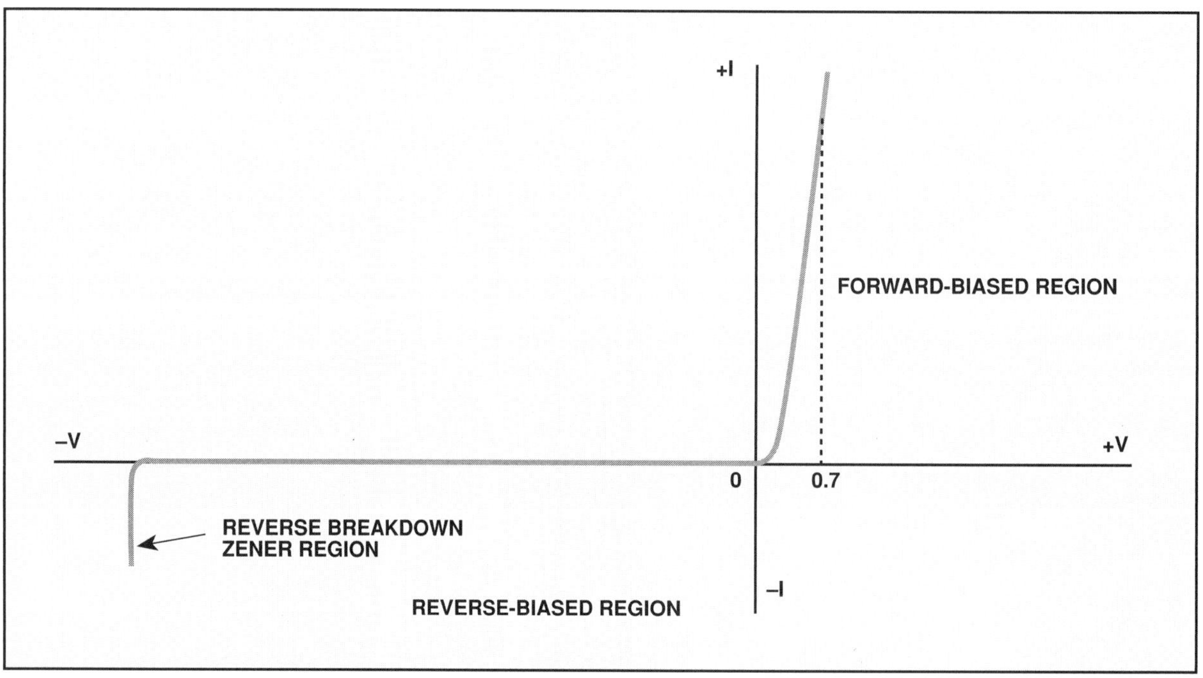

Figure 108 Forward-reverse characteristics of a diode

fluid tries to flow in the opposite direction, the flapper swings down to prevent fluid flow. Similarly, in a diode, when forward-biased voltage is applied in one direction, current flows; however, when reverse-biased voltage is applied in the opposite direction, current flow is blocked.

Diodes are used in a variety of applications and DC or AC can be used depending on the desired result. For example, diodes are used in logic circuits as *steering diodes*, which guide logic-level voltages from one point to another. Diodes can be used as *blocking diodes* to restrict current from flowing in an undesirable direction. In AC circuits, diodes can be used to convert AC to pulsating DC by only allowing the positive voltage waveform to pass while restricting the negative portion of the AC (or vice versa). Circuits using diodes to convert AC to DC are normally referred to as rectifier circuits and the diodes used in such circuits are commonly called rectifiers (fig. 110).

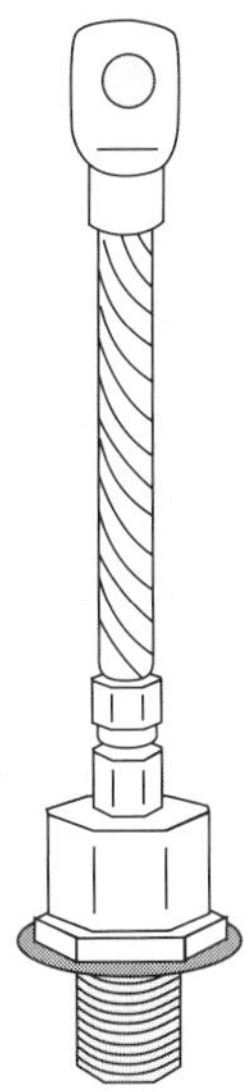

Figure 110 Power rectifier

Diode Ratings

Users should carefully review a diode's rating before installing it in a circuit. A diode's primary ratings include its voltage, current, surge, and temperature characteristics.

A diode's *peak reverse*, or *inverse*, *voltage* is the primary consideration when applying a diode in an AC or DC circuit. This peak reverse, or inverse, voltage is the maximum voltage that might appear across the diode in the reverse direction at any instant. Theoretically, it can be as high as twice the peak voltage that appears at the input to a filter section or it can be exposed to voltage transients in other parts of the circuit. In general, once the peak reverse voltage has been determined for normal operation, most technicians select a diode that has a peak voltage rating that is twice the normal value.

Diodes are also rated in terms of their *average current rating*. This rating is the average of the maximum and minimum current expected in a circuit and can be read on a DC ammeter. However, in some DC circuits, the current may intermittently fluctuate (irregularly go up and down). These fluctuations are called *discontinuous DC current pulses*. These pulses create heat. Enough heat may be created to overheat the diode beyond its temperature rating. Thus, it is important to check with the diode manufacturer in cases where discontinuous pulses may occur in a circuit to ensure that the selected diode is suitable.

The *current surge rating* of a diode is a measure of the diode's ability to continue to operate properly when an occasional and momentary current pulse is applied to the circuit. An application where current may surge is in a power supply that uses diodes as rectifiers and uses a capacitor to filter the input current. During the initial charging of the capacitor, a surge current passes through the diodes into the capacitor. This surge can be 10 to 15 times higher than the average current. Testing and analysis of such a circuit is needed to assure that the surge current is not beyond the rating of the diode.

Permissible operating temperature is a rating of the ability of a diode to dissipate heat while it operates. Heat results as current passes through the diode's connection to, or junction with, the circuit. About 0.7 volts occur at the diode junction when current passes through it. Multiplying current times *diode junction voltage* determines the wattage produced during operation. The wattage takes the form of heat and the heat must be removed through conduction, convection, or forced cooling to maintain the diode temperature within its rating.

A diode can be as small as an integrated circuit chip or as large as a 5-in. hockey puck. The greater the current rating is, the larger the diode. In general, a diode's voltage rating does not affect its size, except in high-voltage applications.

Diode Protection

Diodes must be protected from current surges. Protection can be achieved in several ways. One is a brute-force approach, whereby a diode is selected whose current surge rating is well over the maximum surge expected in the circuit. Another way to protect diodes from current surges is to insert current-limiting resistors in series with the diode. The resistors can either be left in permanently or a time-delay contact can be used to short them out.

Besides current surges, voltage surges can also adversely affect diode operation. Protection against voltage surges requires voltage suppression devices such as *metallic oxide varistors* (*MOVs*), *thyrectors*, or similar voltage-limiting devices. MOVs and thyrectors are solid-state devices that absorb voltage spikes above a certain level. By absorbing the spikes, the devices restrict the voltage from being applied to the diodes. Another way in which to protect diodes from voltage surges is to install capacitors in parallel with the diodes. Such an installation allows the capacitors to absorb fast-rising voltage surges.

Special Diodes

Earlier, this manual discussed Zener breakdown, which, remember, occurs in diodes when the electric field across the space-charge region becomes high enough to produce a form of field emission that suddenly increases the number of carriers in this region (see fig. 108). Through proper construction to take advantage of the voltage breakdown region, a diode can be used as a *voltage regulator*. (A voltage regulator maintains a given voltage within required limits in spite of variations in voltage or load.) The reason such a diode can be used is because the voltage that occurs during breakdown is relatively constant and its output is therefore relatively constant. Diodes using voltage breakdown as a voltage regulator are *Zener diodes* (fig. 111).

Manufacturers can modify diodes to produce other useful components. For example, by adding certain impurities to the p-n materials to change their atomic structure, they can produce light emitting diodes (LEDs). When energized with current, LEDs produce visible light. Also, they can add special impurities to the basic p-n structure and make a diode that produces DC voltage when subjected to sunlight. These are *photovoltaic diodes* or *PVDs*. As you can guess, PVDs are often used to produce electricity in solar cells.

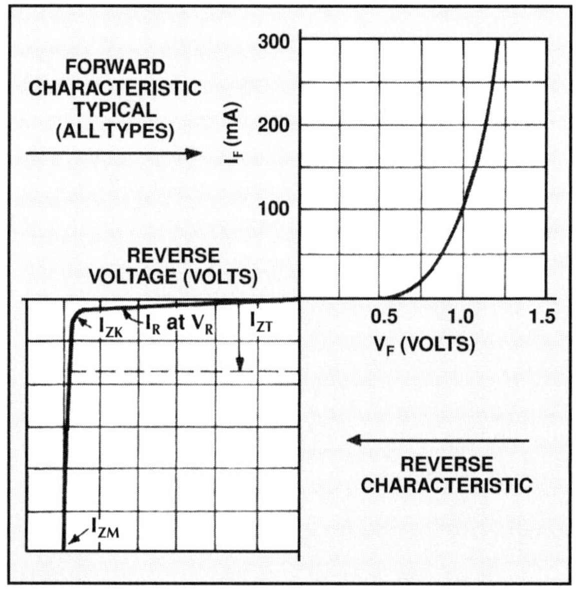

Figure 111 Zener diode characteristic

Also, manufacturers can produce other special diodes such as *tunnel diodes* and *Schottky diodes*. A tunnel diode is a diode with a special impurity that causes the diode to have negative resistance at low voltages in the forward bias direction and a short circuit in the reverse bias direction. Tunnel diodes are used as oscillators and as microwave amplifiers. (They are also known as Esaki tunnel diodes after the Japanese physicist, Leo Esaki, who discovered the tunnel effect.)

A Schottky diode is one formed by contact between a semiconductor layer and a metal coating. It is used as a special rectifier. Electrons or holes are emitted from the area where the semiconductor and metal coating contact each other and move to the metal coating. It provides very fast switching speeds. They are named after Walter Schottky, a German physicist whose research in solid-state physics and electronics yielded many effects and devices that now bear his name.

BIPOLAR TRANSISTORS

In solid-state electronics, a *junction* is a region of transition between two semiconducting regions in a semiconducting device where the p and n materials come into contact. A *junction transistor*, then, is a transistor in which emitter and collector barriers are formed between the semiconductor regions of opposite conductivity (the region where the p and n materials come into contact). Manufacturers can construct junction transistors using bipolar technology in much the same way as they make diodes.

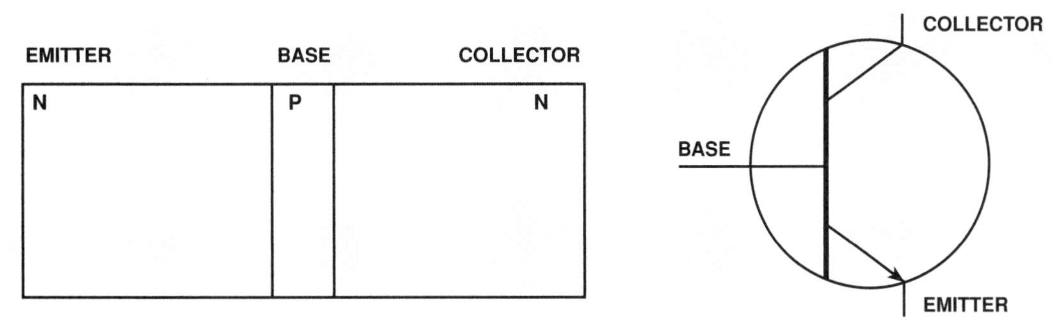

A. Crystal Arrangement

B. Circuit Symbol

Figure 112 Npn transistor

Keep in mind that a transistor is a resistor in transition when placed in an operating circuit. That is, the resistance of the transistor changes from high levels to low levels as current levels flowing through it change. One function of a transistor is to amplify control current to its output terminal. To accomplish this function, two combinations of connecting the p-n materials together are available.

NPN Transistors

One type is an *npn transistor* (fig. 112). To form it, manufacturers place n-type material on each side of a p-type material (fig. 112A). While this sandwiched construction is shown for clarity, manufacturers actually commingle the p-n materials in a space of about 0.001 in. (0.025 mm).

An npn transistor has three terminals: the *collector*, *emitter*, and *base* (fig. 112B). The base is the control terminal and receives current from an external source into its region. The collector is the transistor's output terminal and it is connected to positive voltage from a supply through a resistor. Base and collector currents are combined in the emitter region and it serves as the outlet for the total current in the transistor. An arrow on the symbol for the emitter shows the direction of current flow.

With the emitter connected to a negative source of voltage, the emitter gives off, or emits, electrons that the collector, which is connected to the positive voltage source, gathers, or collects. The current delivered from the base's control terminal determines the amount of electrons being transferred from the emitter to the collector. Therefore, base current controls the current in a transistor.

PNP Transistors

Another type of junction transistor is a *pnp transistor* (fig. 113). It is formed by placing p-type material on

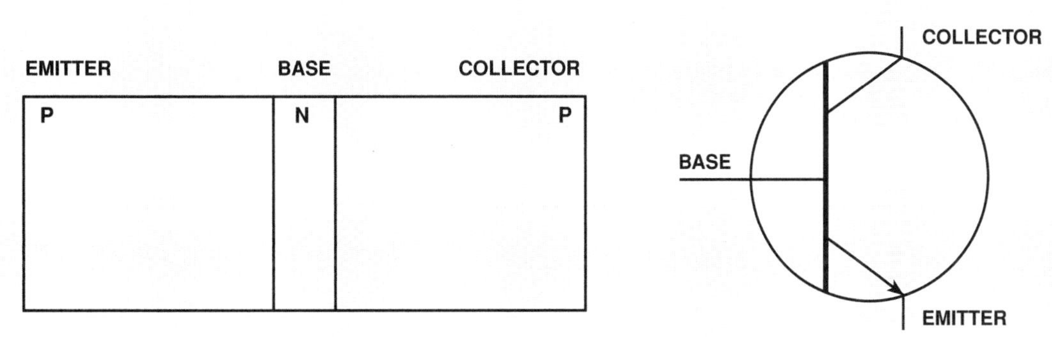

A. Crystal Arrangement

B. Circuit Symbol

Figure 113 Pnp transistor

each side of n-type material (fig. 113A). Again, this sandwiched construction is shown for clarity. Actually, most transistors are constructed with commingled p-n materials and are about 0.001-in. (0.025-mm) thick.

Like an npn transistor, a pnp transistor also has three terminals: the collector, emitter, and base (fig. 113B). The base is the control terminal and delivers current to an external source out of the base region. The collector is the output terminal and is connected to negative voltage through a resistor. The emitter is connected to a positive source of current. Emitter current combines base and collector currents. An arrow on the emitter symbol shows the direction of current flow.

With the emitter connected to a positive source of voltage, the collector, which is connected to a negative voltage source, gathers, or collects, electrons that the emitter gives off, or emits. The current delivered from the base control terminal determines the amount of electrons transferred from the emitter to the collector. Therefore, base circuit current controls current in a pnp transistor.

Transistor Biasing

Recall that bias is DC voltage applied to a transistor's control electrode to establish the desired operating values. So, voltage biasing of a transistor is important for proper operation. Also recall that forward bias is the condition that occurs when the positive terminal of a battery is connected to the lead on the p-type semiconductor and the negative terminal of the battery is connected to the lead on the n-type semiconductor. Reverse bias is a condition in which no current flows in the semiconductors because the negative terminal of a battery is connected to the lead of the p-type material and the positive terminal of the battery is connected to the lead of the n-type material.

In an npn transistor, the emitter-base junction is forward biased (fig. 114)—that is, the base is positive (fig. 114A). The collector-base junction is reverse bias—that is, the base is negative (fig. 114B). To achieve this biasing, the positive terminal of the supply voltage is connected to the collector and the negative terminal of the supply voltage is connected to the emitter (fig. 114C).

Figure 115A–C shows pnp transistor biasing. The positive terminal of the supply voltage is connected to the emitter while the negative terminal of the supply is connected to the collector.

Biasing ensures that current from the base and collector are always combined in the emitter, regardless of current direction. This phenomenon is true in both npn and pnp transistors. While each type has different performance characteristics and are used in different circuit applications, both perform amplification in much the same way.

Operation of NPN Transistors

Recall that a potential hill is an energy barrier created by the semiconductor's space-charge region (see fig. 104). Potential hills occur when no external voltage is applied to a transistor. The potential hills are established along each junction. The electrons in the

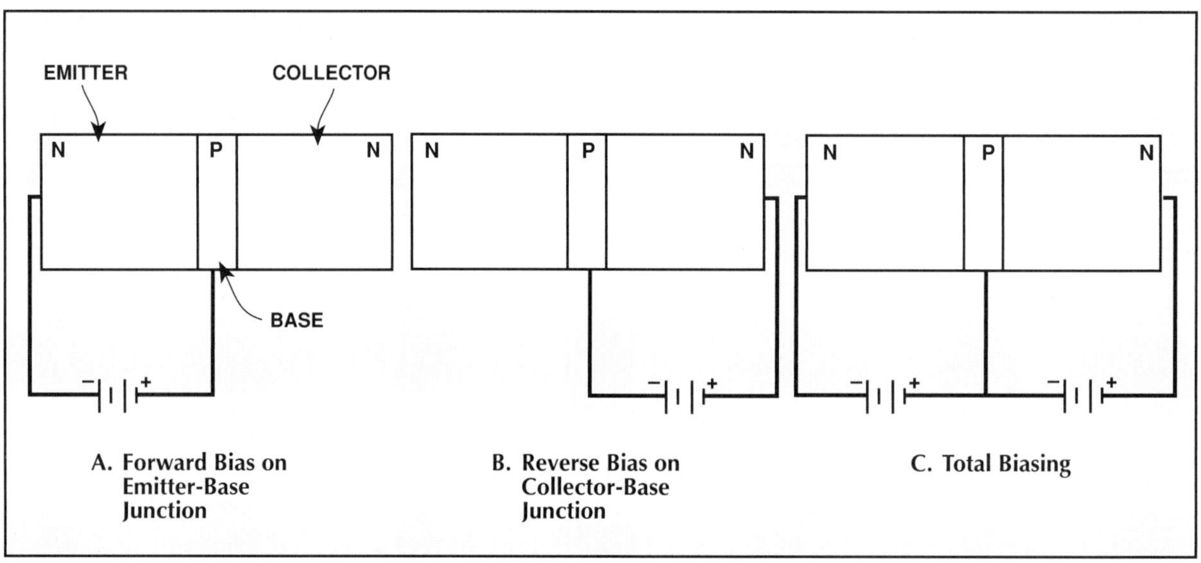

Figure 114 Npn transistor biasing

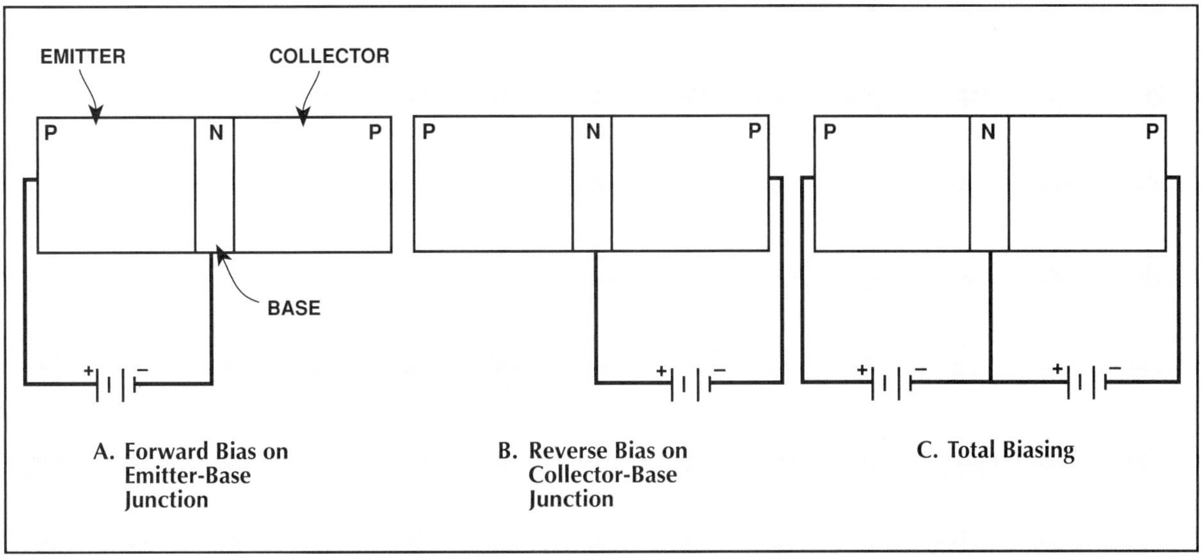

Figure 115 Pnp transistor biasing

n-type region interact with the holes in the p-type region in the same way that they interact in a junction diode. That is, a few electrons in both n-type regions of an npn transistor are attracted by holes in the p-type region. The electrons diffuse across the two junctions and fill holes in the p-type region. The p-type region becomes slightly negative along with each junction because of the extra electrons that were acquired from the n-type regions. The n-type regions become slightly positive along with the junctions because holes now exist where the electrons had been in the n-type regions.

As shown in figure 114 of npn biasing, the emitter-base junction of an npn transistor is biased in the forward, or low-resistance, direction. Electrons flow freely from the emitter across the emitter-base junction and into the base when external voltage is applied. The collector-base junction is reverse biased in the high-resistance direction. The high resistance presented by the reverse-biased junction prevents any usable current from flowing across the junction. (Current would also flow across the collector-base junction if the base region were not made extremely thin.)

To understand the operation of the npn transistor, first visualize the transistor as having voltage applied across the emitter-base junction (shown in fig. 114A). With the voltage applied across the emitter-base junction, electrons leave the negative terminal of the supply voltage and enter the emitter section. They diffuse across the junction and fill holes in the p-type base section. Each time an electron leaves the negative supply voltage terminal, a replacement electron must enter the positive battery terminal. The electrons that enter the positive terminal in this case must come from the base section. As they leave the p-type base section, holes are created in the base. These holes move toward the emitter-base junction and combine with electrons that come across the junction from the emitter.

When voltage is applied only across the emitter-base junction, electrons from the base section replace all electrons that leave the negative supply terminal and enter the emitter. No current flows across the collector-base junction. No current flows across the collector-base junction even when voltage is also applied across the collector-base junction (as shown in fig. 114C). The collector-base junction is reverse biased and therefore presents high resistance to electron flow. Since the forward-biased emitter-base junction presents a very low resistance to electron flow, while the collector-base junction presents a very high resistance, it would seem that almost all the electrons would return to the supply directly from the base section. This situation would occur if the base section were thick. If the base section were thick, no usable current could flow between the base and collector and the transistor could not function as an amplifier.

The base section, however, is extremely thin. As pointed out before, the base of a typical transistor is only 0.001-in. (0.025-mm) thick. Because it is so thin, the p-type base section has comparatively few free holes. Of the great number of electrons that enter the base section from the emitter, only a few are able to combine with holes in the base. Consequently, almost all of the electrons travel across the collector-base junction and enter the collector.

Once an electron enters the collector, it is strongly attracted to the positive connection on the collector. Because of its positive charge (with respect to the emitter), the base attracts electrons from the emitter. Most of the electrons escape into the collector because the base has so few holes available and because of the much greater positive charge on the collector (with respect to the base).

The collector has a greater positive charge than the base because the bias voltages have different levels. For example, the voltage from emitter to base is typically 1.5 volts. From the base to collector to base, voltage is 4.5 volts. Therefore, the voltage between the emitter and collector is 6 volts. Because of the higher voltage, electrons leaving the emitter are much more strongly attracted to the collector than to the base.

Only about 5 percent of the electrons leaving the emitter combine with holes in the base. On the other hand, about 95 percent of the electrons pass through the base and into the collector. Though the number of electrons leaving the emitter is determined entirely by the emitter-base voltage, the collector is the element that receives almost all of the electrons. The current that flows through the collector is important because it flows through the output circuit.

A transistor amplifies a signal because the emitter-collector current (the current in the output circuit) can be controlled and varied by changing the emitter-base voltage. A small change in emitter-base voltage produces a large change in emitter-collector current. This small emitter-base voltage change produces a large emitter-collector current change because the base current, or number of electrons combining with holes in the base, is quite small.

Operation of PNP Transistors

Pnp transistors operate very much like npn transistors. The basic difference is that the holes carry current in pnp transistors while free electrons carry current in npn transistors. Figure 115 shows that the junctions of a pnp transistor are biased in the same direction as npn junctions. That is, the emitter-base junction is biased in the forward, or low-resistance, direction. The collector-base junction is biased in the reverse, or high-resistance, direction.

Because of the different internal structure of pnp transistors, the polarities of the biasing voltages must be opposite those used to bias npn transistors. To forward bias the emitter-base junction of a pnp transistor, the positive supply voltage must be connected to the emitter and the negative terminal to the base. To reverse bias the collector-base junction, the negative supply voltage must be connected to the collector and the positive terminal to the base. The base must be made negative with respect to the emitter and positive with respect to the collector.

When external voltage is applied across the emitter-base junction of a pnp transistor, the positive terminal of the supply drives the holes (positive carriers) in the p-type emitter section toward the base. At the same time, the negative supply voltage drives electrons from the n-type base section toward the emitter. Each time an electron from the base combines with a hole in the emitter, a new hole is created and the new hole travels toward the base. Also at the same time, negative supply voltage drives another electron into the base, and an electron leaves the emitter and enters the positive terminal.

With the collector-base junction also biased (shown in figure 115C), most of the holes from the emitter section pass through the base section and enter the collector. Only about 10 percent of the holes combine with electrons in the base because the extremely thin base has comparatively few electrons and because there is a much larger potential difference between the emitter and collector. Thus, about 90 percent of the holes enter the collector section.

Once in the collector, a strong negative force in the collector attracts the holes toward the collector's terminal. When a hole reaches the collector's terminal, an electron from the supply terminal combines with the hole. This action neutralizes the hole at the edge of the collector section; so, the holes do not travel outside of the transistor.

At the same time that an electron leaves the negative terminal of the collector supply on its way to neutralize a hole in the collector, a replacement electron leaves the emitter. The replacement electron travels through the emitter connection, through the emitter battery, and enters the positive terminal of the collector supply.

Because the emitter-base current is small in comparison to the emitter-collector current, a pnp transistor functions as an amplifier in the same way that an npn transistor does. A small change in emitter-base voltage produces a large change in collector current, and collector current is supplied to the output circuit.

Transistor Gain

Because the current required in the base region is small compared to that in the collector circuit, significant control can be achieved with a transistor. For example, when an npn transistor is operating in a circuit, only a small amount of current is required in the base to change the collector current a significant amount. A typical transistor whose *current gain* is 100 means that 1 milliampere of current is required in the base to change the collector current by 100 milliamperes. The equation for determining current gain is—

$$h_{FE} = \frac{I_C}{I_B} \qquad \text{(Eq. 47)}$$

where

h_{FE} = transistor current gain, unitless ratio
I_C = collector current, milliamperes or amperes
I_B = base current, milliamperes or amperes.

Typically, manufacturers specify transistor current gain on the specification sheet accompanying the transistor. In most cases, transistor current gain values range from 50 to 500.

When transistors are used in circuits with resistors, *voltage gain* occurs—that is, the transistors increase the voltage in the circuit. Multiplying the change in base current with its base resistance and multiplying the change in collector current with its collector resistance can determine voltage gain. The equation is—

$$VG = h_{FE}\frac{R_C}{R_B} \qquad \text{(Eq. 48)}$$

where

VG = circuit voltage gain, volts
h_{FE} = transistor current gain, unitless ratio
R_C = collector resistance, ohms
R_B = base resistance, ohms.

For example, if a transistor circuit has a base resistor of 10 kohms and a collector resistor of 2 kohms with a transistor whose current gain is 100, the voltage gain is—

$$VG = 100 \times \frac{2{,}000}{10{,}000}$$

$$= 100 \times 0.2$$

$$VG = 20 \text{ volts.}$$

Insulated-Gate Bipolar Transistors

A special transistor is the *insulated-gate bipolar transistor (IGBT)* (fig. 116). An IGBT combines low *forward voltage drop*, *gate-controlled turnoff*, and high switching speed. Forward voltage is voltage that the transistor carries in a forward, rather than a reverse, direction. A gate in a transistor switches a signal going to it on or off. Because of the manner in which they operate, IGBTs are used in high-power converters and in inverters. (A converter changes DC to AC or AC to DC. An inverter changes DC to AC.) Large motor drives that require inverters and power converters use IGBTs to perform the conversion.

Figure 116 IGBT power transistor

Thyristors, Silicon-Controlled Rectifiers, Triacs

Another product using bipolar technology is a solid-state switch that is referred to as a *thyristor*. These devices are used to switch current to high levels from a low-level control signal in AC and DC circuits. Thyristors that conduct current only in one direction are called silicon-controlled rectifiers (SCRs). Thyristors that conduct current in both directions are called *triacs*. Thyristors arose from diode-rectifiers that allow current flow only in one direction while blocking voltage in the opposite direction. Thyristors add another feature in that they control current in a forward direction.

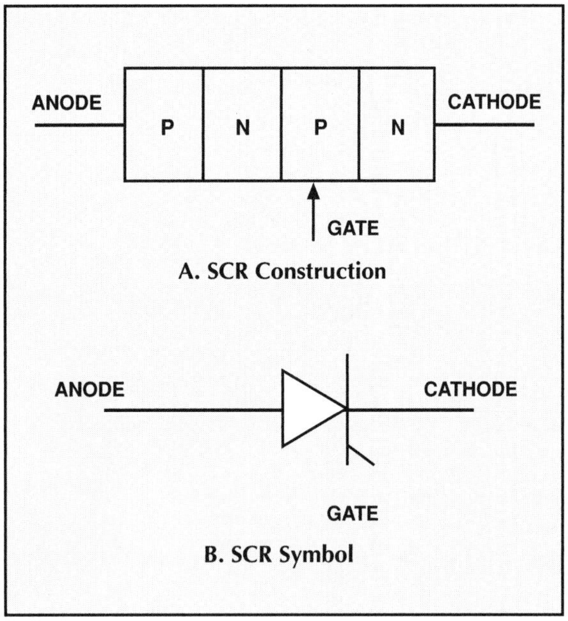

Figure 117 SCR construction and terminals

Silicon-Controlled Rectifier

The most popular thyristor is the silicon-controlled rectifier (SCR). Figure 117A shows that an SCR consists of four bipolar elements—two type p materials in junction with two type n materials, which are alternately connected. It has three terminals: an anode, a cathode, and a gate. Figure 117B shows the symbol for an SCR that is used in schematics. Figure 118 shows several power SCRs, which are used in high-powered circuits to rectify AC to DC.

The SCR works on the principle that when the anode is forward biased with a positive voltage with respect to the cathode, it blocks current flow. However, if current is applied to the gate terminal while the anode is forward biased and current flows into the gate p-material through the p-n junction connected to the cathode, and returns to the gate current supply source, the SCR switches to a conducting condition. This action occurs because the gate current through the gate's p-n junction causes an increase in current from the anode to the cathode. The result is analogous to an avalanche of hole flow that is initiated with initial gate current. Once the current level from the anode to cathode reaches a level referred to as the *holding current*, the gate current can be removed and the SCR remains in conduction because its current passes through the p-n gate junction. (Holding current is the minimum amount of current required to hold the SCR in a conductive state.)

To turn the SCR off, and cease conduction, the anode-cathode current must be reduced below holding current level. To reduce current, the anode-cathode voltage polarity is reversed, which sweeps out the holes in the SCR. Polarity remains reversed until the current drops below holding current level. When sufficient current drop occurs, the SCR ceases to conduct and resumes its blocking state.

Figure 119A plots a curve of voltage (*V*) and current (*I*) characteristics of an SCR. Figure 119B is the schematic symbol of an SCR. In figure 119A,

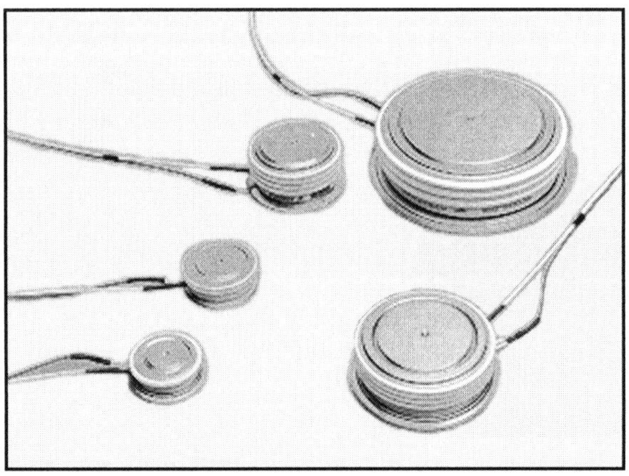

Figure 118 Power SCRs

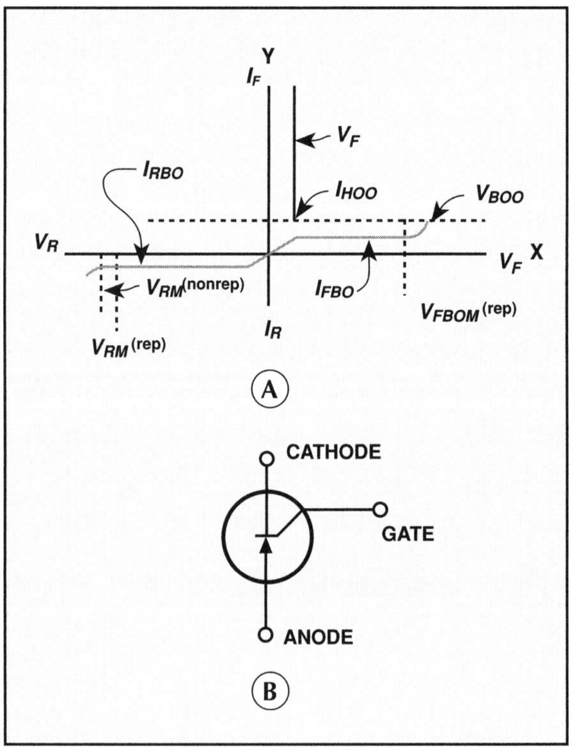

Figure 119 SCR characteristics p-n junction field-effect transistor

reverse bias voltage (V_R) and forward bias voltage (V_F) form the *x* axis of the graph. Forward current (I_F) and reverse current (I_R) form the *y* axis. The curve shows the small values of leakage current that exist for forward and reverse bias before heavy conduction occurs. I_{RBO} represents reverse-bias leakage, and it remains a fairly uniform small value until the peak reverse voltage (V_{RM}) is exceeded. (V_{RM} may be repetitive, or rep, and nonrepetitive, or nonrep.) Reverse current increases drastically for higher values of reverse bias.

For forward bias, leakage current is small before heavy conduction. This leakage current is abbreviated as I_{FBO}, which is short for forward blocking current. I_{HOO} stands for holding current (the minimum amount of current required to maintain the SCR in a closed, or conducting, state after it is energized). As forward bias increases, the forward breakover point is reached at V_{BOO}. Beyond this point, the voltage across the anode-cathode elements drops to a small value (V_F), and the current of conduction increases rapidly for any additional forward bias. In this state of heavy conduction, the SCR is triggered, or in the on state.

A popular application for SCRs is in AC circuits where they convert AC to variable voltage DC. When connected in a single device or in a multiple SCR bridge circuit, the SCR's forward-biased voltages are synchronized with AC line frequency and the gate current signals are applied at appropriate timing intervals. Selecting the timing intervals allows SCRs to conduct over small intervals or at intervals that cover the complete half-cycle for full conduction for each SCR. Consequently, variable DC output is available from zero to maximum voltage in an infinite series of steps. When the AC voltage completes its half cycle, the polarity across the SCR is reversed, which turns it off.

Special SCRs are available that can be turned off with reverse gate current; thus, they do not depend on AC voltage for commutation (current reversal). These devices are *gate-turnoff silicon-controlled rectifiers*, or *GTO-SCRs*, for short.

Triacs

SCRs can be connected in an inverse-parallel, or back-to-back, manner. When so connected, they convert fixed AC voltage to variable AC voltage. This arrangement requires two SCRs and two sources of gate-control current. A triac is a special thyristor that uses bipolar technology. Bipolar technology allows the triac to achieve conduction in both directions with a single gate source. Triacs are used in variable heating controls, lamp dimmers, and small, variable-speed motor drives.

FIELD-EFFECT TRANSISTORS

Another transistor is a *field-effect transistor* (*FET*). It operates in a significantly different way from bipolar npn and pnp transistors. An FET is so called because current flow through it is controlled by an electric field rather than by an electric current. Electric current controls npn and pnp transistors.

Figure 120 is a simplified drawing of a *junction-type field-effect transistor*, or *JFET* for short. It is called a junction type (JFET) because the two types of materials, n and p, form junctions where they contact one another. Another type of FET, an insulated gate-type, is called a *MOSFET* for *metallic-oxide-semiconductor field-effect transistor*.

Figure 120 also shows the names that solid-state scientists have assigned to FET elements. *Gate* is another name for the transistor base. *Source* is equivalent to the emitter. *Drain* resembles the function of the collector. Current in an FET normally flows from source to drain and is therefore a *unipolar transistor*, and not a bipolar transistor. Note that current in an FET can flow in either direction; in a bipolar transistor, it can flow only in one direction.

When a source of voltage is connected between source and drain, current flows through the n-type material from source to drain. The amount of current depends on several factors, including the voltage and

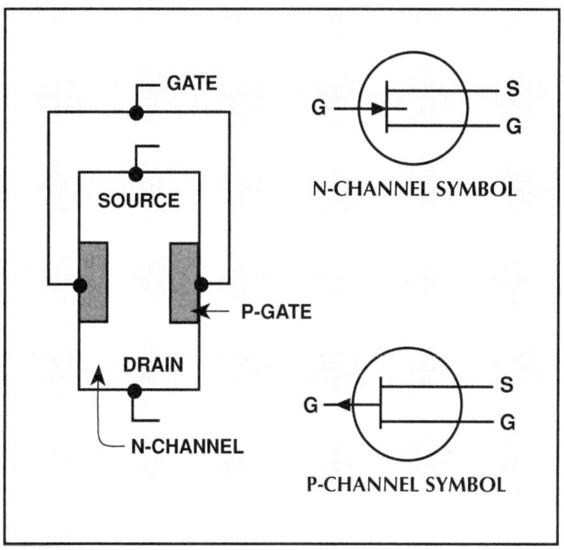

Figure 120 P-n junction field-effect transistor

the density of *doping* in the n-type material. (Doping is the adding of impurities to n-type or p-type materials to give them the desired semiconducting characteristics.)

Figure 121 shows two schematics of the same circuit, which illustrate source-drain current characteristics as a function of gate bias. The circuit has a power source, a potentiometer, two FETs, and a meter. The potentiometer provides a way to reverse bias the p-type gate from zero to a negative voltage source with respect to the source. As shown in the top schematic in figure 121, practically no current flows in the gate circuit when it is reverse biased. However, since each gate and channel junction forms a diode, depleted areas form near the junctions. Depletion near a junction results in a high-resistance condition. Note that the meter indicator (the arrow) is deflected to the right to indicate this condition.

If the potentiometer is advanced to make the gate more negative (fig. 121, bottom schematic), the

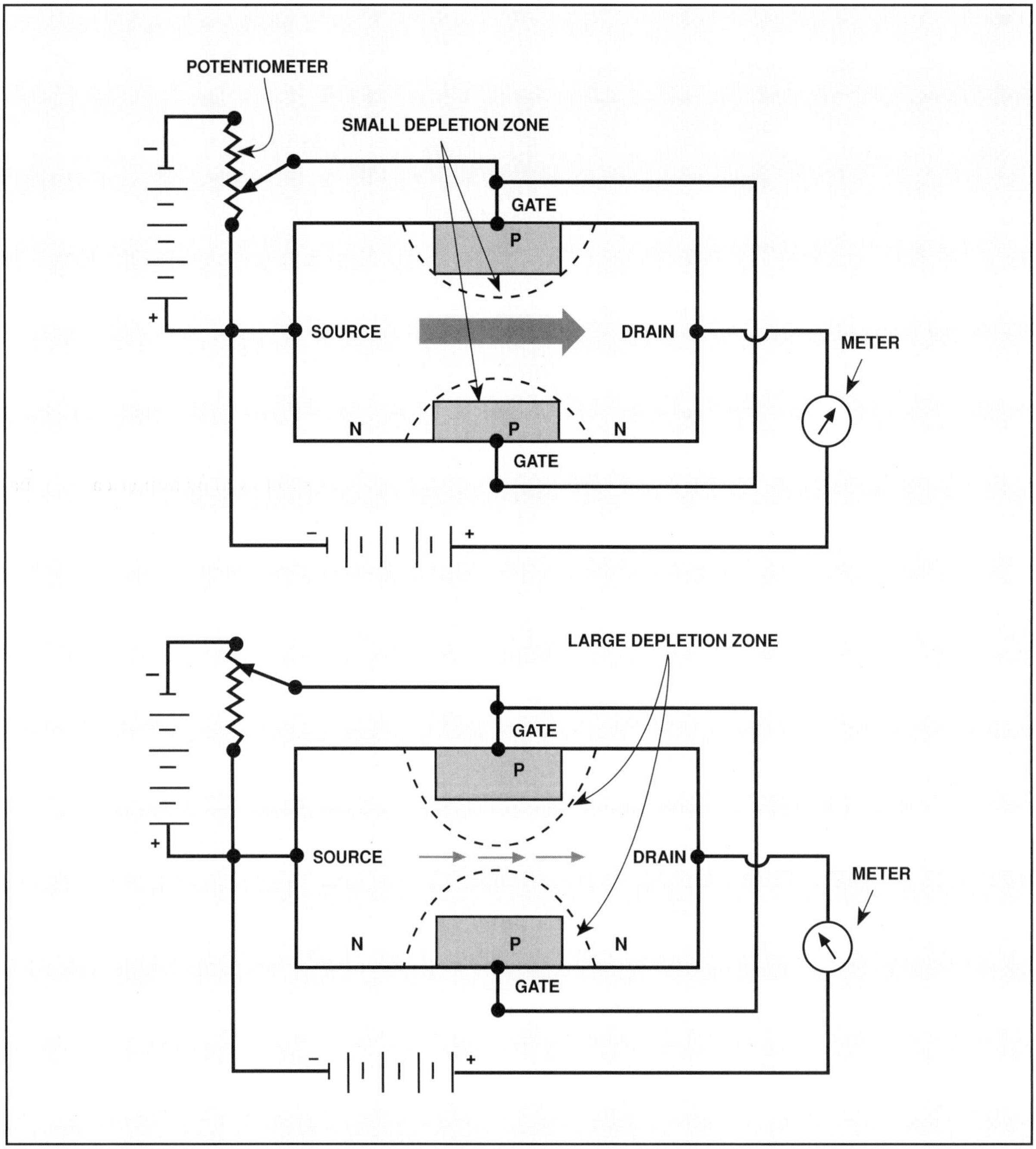

Figure 121 Source-drain current characteristics as a function of gate bias

depletion zones become even larger and resistance to current flow between the source and drain becomes greater. With proper design, the arrangement can be made to pinch off virtually all current flow.

The amount of voltage applied to the gate to control a wide range of current through the channel is relatively small. Thus, the transistor performs well as an amplifier because it increases current with small voltage increases. The input impedance of the FET is very high, which gives it a real advantage over bipolar transistors. Because of the high input impedance, less power is required to operate an FET. The output impedance of the FET is also quite high.

The FET shown in figure 121 is an n-channel type, which means that the material that forms the path between the source and the drain is n-type material. A p-channel type of FET is also available. The two types are quite similar. In the p-channel FET, the gate is reverse biased by placing positive voltage on it with respect to the cathode.

MOSFETS

A junction-type FET uses a reverse-biased semiconductor junction between the gate and channel. The junction is like that of a diode, in that current flows if forward bias is applied. A metallic oxide semiconductor field-effect transistor (MOSFET) contains an insulating channel between its gate and channel, usually in the form of a dielectric material such as silicon dioxide (SiO_2). The MOSFET gets its name from its use of a metal gate, an oxide layer for gate-channel insulation, and a semiconductor material as the channel.

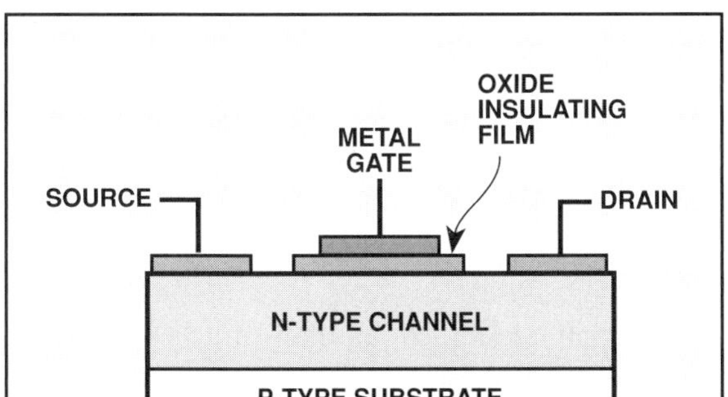

Figure 122 MOSFET construction

In a junction FET, reverse bias is always used between gate and source and the input impedance is very high. Consequently, if a forward bias is attempted, the input impedance will drop drastically, current will flow in the circuit and the power gain will diminish.

Figure 122 shows an example of a MOSFET. It uses a metal gate insulated from the n channel. Like the p-n junction, the insulated gate can cause a depletion area in the channel when properly biased. The insulated gate feature can also increase conductivity of the channel when biased in the opposite direction and, in doing so, does not draw current or affect the power gain.

SOLID-STATE DESIGNATIONS

The *American National Standards Institute* (*ANSI*) has established standard designations for diodes, transistors, SCRs and other semiconductors that have fixed specifications. The standards are promulgated by the *Joint Electron Device Engineering Council (JEDEC) Solid State Technology Association*, which represents the solid-state electronic industry. (It is usually simply called JEDEC.) An alphanumeric code identifies semiconductors. For example, JEDEC has assigned diodes and other two-terminal devices a 1N prefix, which is followed by numerical designations of 2 to 4 digits. Thus, a typical low-current, low-voltage diode of 1 ampere is designated as 1N4001. A higher voltage version is designated as 1N4004.

Three-terminal devices, such as transistors, SCRs, and triacs, have designations that begin with 2N followed by numbers. For example, a popular power transistor is the 2N3055. Keep in mind that manufacturers produce unique solid-state devices that have their own design characteristics. Because they are not standard, the manufacturer uses its own designation method.

INTEGRATED CIRCUITS

An *integrated circuit* is a device constructed on a substrate in much the same manner as a transistor. An integrated circuit, however, combines many devices into a complete circuit. Typically, an integrated circuit contains diodes, resistors, transistors, and other components. The components are mounted on a *chip*. A chip is a shaped and processed semiconductor die that is mounted on a substrate. Thousands of components can be placed on a chip and thus are economical to produce.

Typical integrated devices include *logic gates*, such as *AND, OR, NOR, NAND, exclusive or, buffer,* and *inverter circuits*. An AND gate has two or more input-signal ports that deliver an output only if and when every input port is simultaneously energized. An OR is an instruction that performs the logical operation "or" on a bit-by-bit basis. In a NOR circuit, output voltage appears only when a signal is absent from all of its input terminals. A NAND circuit is a logic circuit whose output signal is a logical 1 if any of its inputs is a logical 0 if all of its inputs are logical 1. An exclusive OR is an instruction that performs on a bit-by-bit basis. A buffer is a component that prevents undesirable electrical interaction between two circuits or components. An inverter circuit (also called a NOT circuit) is a logic circuit with one input and one output that inverts the input signal at the output—that is, the output signal is a logical 1 if the input signal is a logical 0 and vice versa.

Other integrated devices include operational amplifiers (op amps), *Schmitt triggers*, SCRs, diode arrays, LED drivers, *arithmetic converters*, and *EEPROMs*. A Schmitt trigger is a pulse generator capable of assuming either of two stable states (a bistable pulse generator) in which an output pulse of constant amplitude exists only as long as the input voltage exceeds a certain value. An arithmetic converter changes electronic signals into units that additional circuitry can add, subtract, multiply, divide, or compare. EEPROM stands for electrically erasable programmable read-only memory, which is an integrated-circuit memory chip that has an internal switch to permit a user to erase the contents of the chip and write new contents into it by means of electrical signals.

The most popular and productive integrated circuit is the *microprocessor*, which is used in personal and laptop computers, programmable logic controllers (PLCs), flow controllers, and the like. A microprocessor is a single silicon chip on which all the arithmetic and logic functions of a computer are placed. A microprocessor is a highly complex device that integrated circuit technology has made economical.

Operational Amplifiers

An operational amplifier is a special linear circuit that consists of multiple transistors. The transistors are arranged not only to provide high amplification of signals, but also to provide—

- an inverting input,
- a noninverting input,
- an output,
- high input resistance, and
- low output resistance.

An op amp is used as an amplifier with adjustable gain, as a *timing circuit*, as a *differential signal* producer, as a *signal integrator*, as a device to sum up signals, and so on. A timing circuit provides accurately timed pulses to synchronize a digital computer or to provide a time base in a transmission system. (It is also called a clock.) A differential signal is an indicator (a signal) that consists of the voltage difference between two junctions (nodes), neither of which is at ground potential. A signal integrator senses electronic signals in a circuit and electronically arranges the signals so that they function together in an efficient and logical way. Figure 123 is a diagram of a typical op amp circuit that uses transistors and resistors.

Manufacturers usually package operational amplifiers as an integrated circuit. They connect the circuit's input and output terminals to pins, which users employ to connect the component into the circuit of the device in which it is installed. Figure 124 shows the symbol for an op amp. It has two inputs: an *inverting*, or *minus*, *input* and a *noninverting*, or *plus*, *input*. The terms inverting and noninverting refer to the change in output signal when an input signal is changed. For example, if the inverting input increases in voltage, the output signal will be inverted, or will decrease in value. The opposite is true for the noninverting input. If the noninverting input increases in voltage, the output signal is not inverted; it also increases in value.

A typical application for an op amp is a simple amplifier. Without any external feedback resistors, but using the device as an *open-loop amplifier*, the voltage gain of the device can be as high as 100,000. (An open-loop amplifier is an electronic device that increases, or amplifies, a signal without any feedback from the rest of the circuit.) For example, an op amp used as an open-loop amplifer can amplify an input voltage of 1 millivolt 100,000 times to a level of 100 volts if the power supply's range permits it. In circuits that require less voltage gain, a feedback resistor can be connected from the op amp

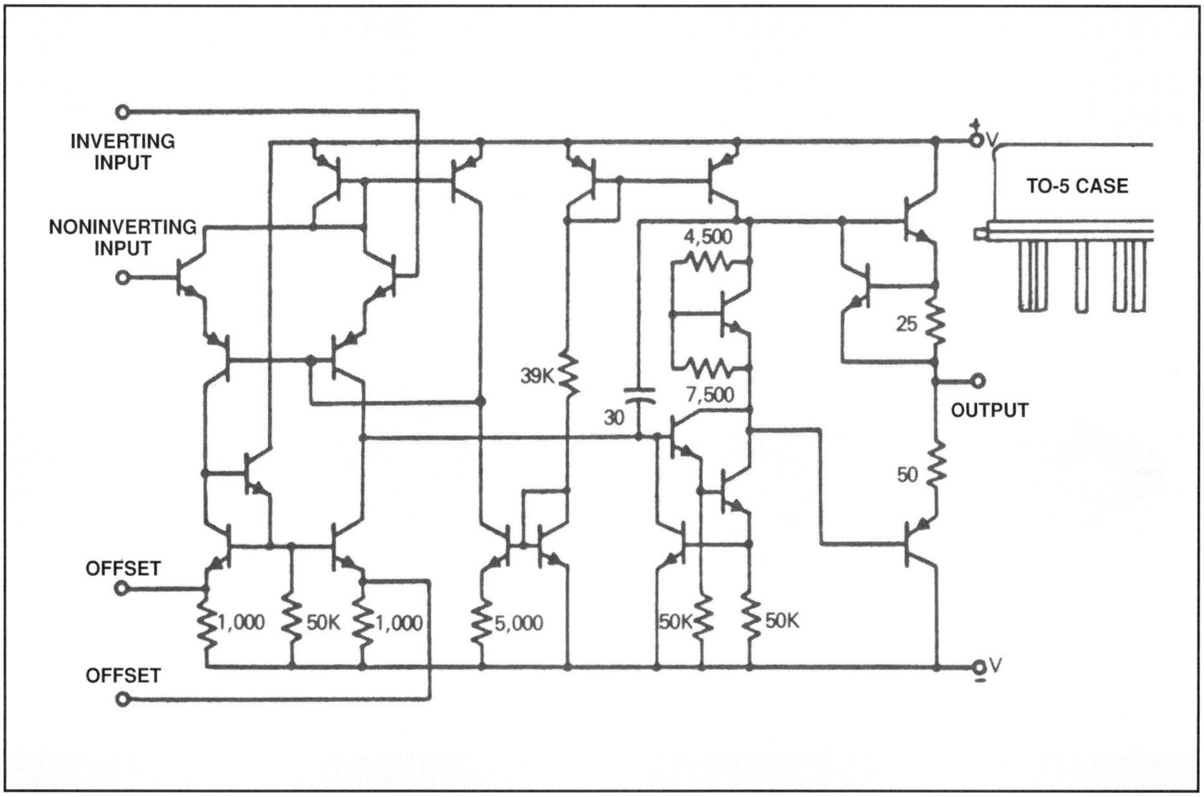

Figure 123 Operational amplifier circuit diagram

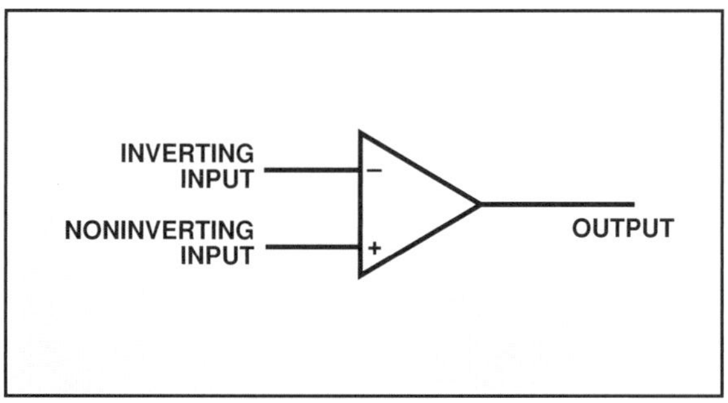

Figure 124 Symbol for an op amp

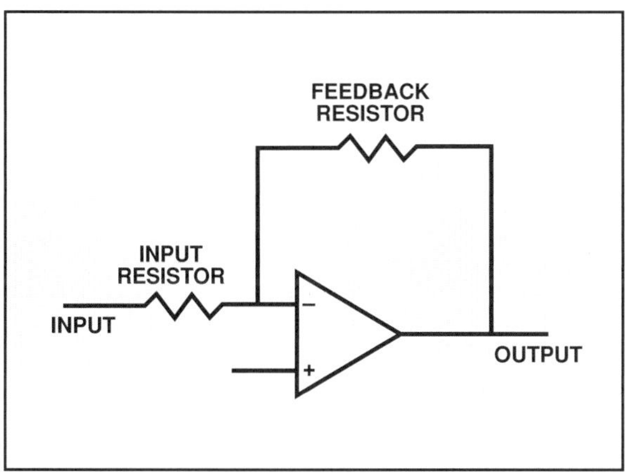

Figure 125 Op amp with feedback resistor

output to its inverting input (fig. 125). By adjusting the ratio of the feedback resistor to the input resistor, the voltage gain can be established. The equation is—

$$VG = \frac{R_{FB}}{R_{In}} \quad \text{(Eq. 49)}$$

wherewhere

VG = voltage gain, volts
R_{FB} = feedback resistance, ohms
R_{In} = input resistance, ohms.

For example, if the feedback resistor is 100 kohms, and the input resistance value is 1 kohm, the voltage gain is—

$$VG = \frac{100}{1}$$

$$VG = 100 \text{ millivolts.}$$

In this case, a 1 millivolt input signal change produces 100 millivolts at the output. This input signal can be applied to either the inverting or noninverting input with the same amplifying result. Only the output's signal change in direction is affected.

Often, it is desirable for a circuit's output voltage to be proportional to the rate of change of the

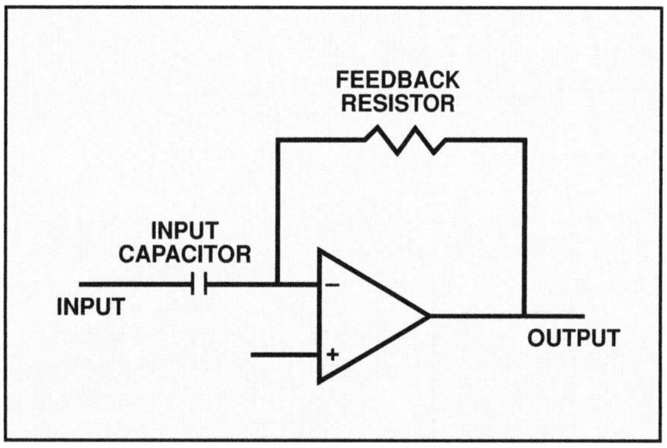

Figure 126 Op amp differentiating circuit

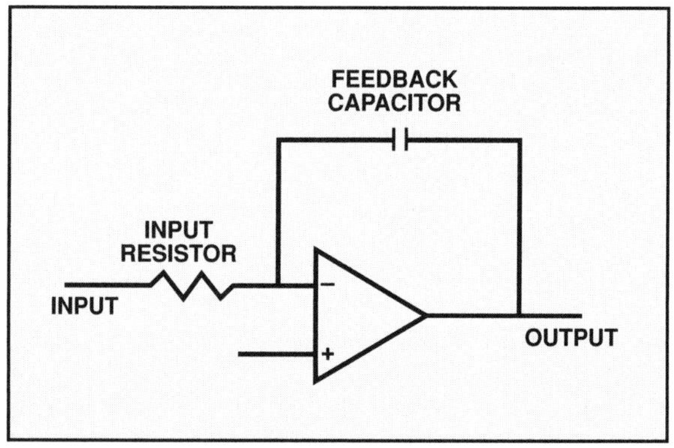

Figure 127 Op amp as an integrating amplifier

input voltage. Such a circuit is a *differentiating circuit*, and op amps are used in them (fig. 126). In a differentiating circuit, an input capacitor is added instead of an input resistor; however, a feedback resistor is included. When a sudden change in input voltage is applied to the input, the charging current of the capacitor causes the op amp output to change abruptly and remain at that level for a time that depends on the values of the input capacitor and feedback resistor. At the end of the time interval, the output resumes its original state.

An op amp can also perform an integrating function by placing a resistor in the input and a capacitor in its feedback loop (fig. 127). The output of the amplifier is proportional to the charge that accumulates from the input current flow into the feedback capacitor. The amount of time delay involved, which is called integration, or summing, is proportional to the product of the input resistor and feedback capacitor.

Circuits Using Semiconductors

Several integrated circuits use the basic functions of a rectifier, Zener diode, and transistor to convert AC voltage to DC voltage and also provide filtered regulation of this voltage.

Half-Wave Rectifier

Figure 128 is a schematic of a half-wave transformer-rectifier circuit. The transformer steps down the AC input voltage to a voltage level compatible with the desired output voltage. A single diode then rectifies the low-voltage AC and allows current flow only during the positive half-cycle of the AC voltage. The resulting output is a pulsating voltage waveform whose average value represents the DC voltage. The value of the DC voltage is about 32 percent of the peak voltage or 45 percent of the root mean square (rms) value. This circuit is relatively inefficient since it uses only half of the available voltage and few circuits can use the pulsating, discontinuous voltage.

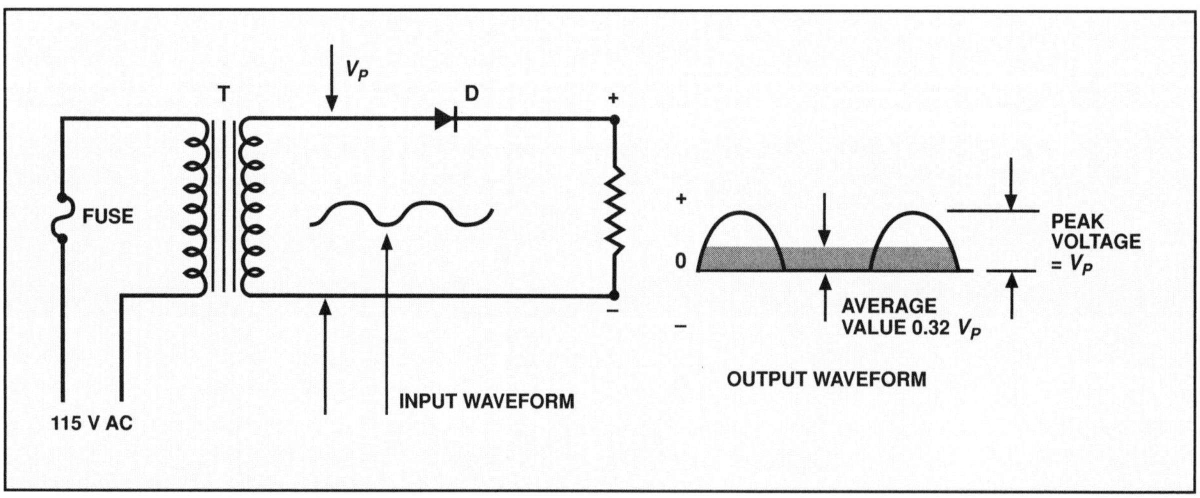

Figure 128 Half-wave transformer-rectifier

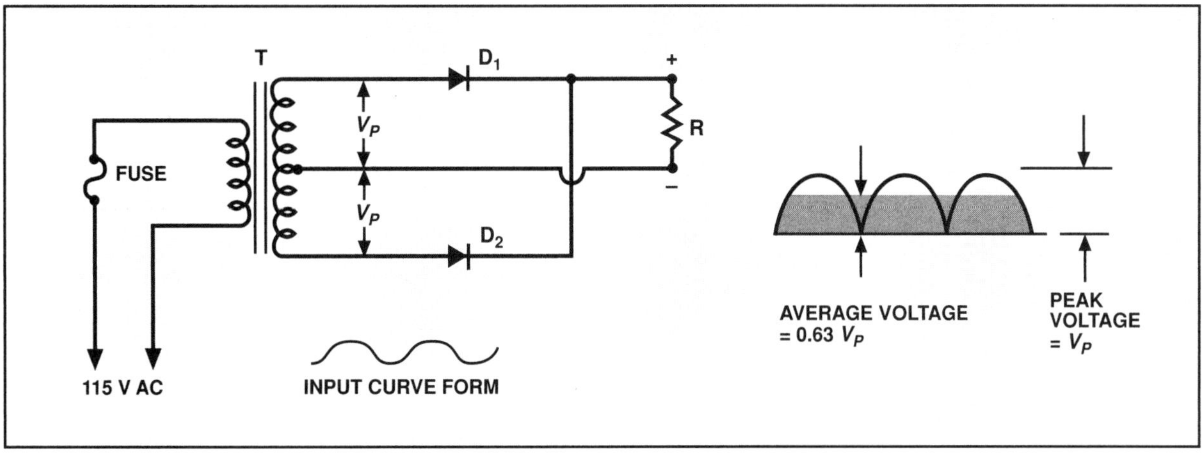

Figure 129 Transformer center-tap full wave rectifier

Full-Wave Circuits

Figure 129 is a schematic of a full-wave rectifier that uses the center tap of a transformer. This circuit improves operation over the half-wave circuit by using two diodes to rectify the output voltage. With a common connection to the center tap of the transformer, the diodes are connected to the opposite ends of the transformer winding. During the positive half-cycle, current can flow through diode D_1 and produce half-wave voltage output while diode D_2 blocks current flow. During the negative half-cycle, diode D_2 conducts the voltage to the output while diode D_1 blocks current flow. The resulting output is full-wave voltage with both half-cycles being rectified to a DC level.

The average voltage when measured with a DC voltmeter is about 63 percent of the peak or about 90 percent of the rms value of the AC voltage. These figures are a significant improvement over the half-wave circuit. The circuit requires a center-tapped transformer and an additional diode. Nevertheless, the voltage is still pulsating DC voltage whose frequency is twice that of the AC supply frequency.

Figure 130 is a schematic of a circuit that improves the waveform produced by the full-wave, center-tapped, AC-to-DC rectifier circuit. The storage effect of capacitor C_1 filters the pulsating DC output. The gaps between the pulsations remain at a higher level because voltage is stored. The high-level gaps produce continuous DC voltage that still retains some ripple voltage. However, the *choke*, or DC inductor, in the circuit further reduces ripple voltage by keeping the current relatively constant to the output. The output is further filtered with capacitor C_2 to produce a smooth DC voltage with little ripple voltage. This type of filter is an *LC filter* or a pi (π) filter because the two capacitors and choke roughly take the form of the Greek letter π. It is very effective in attenuating any ripple appearing at its input.

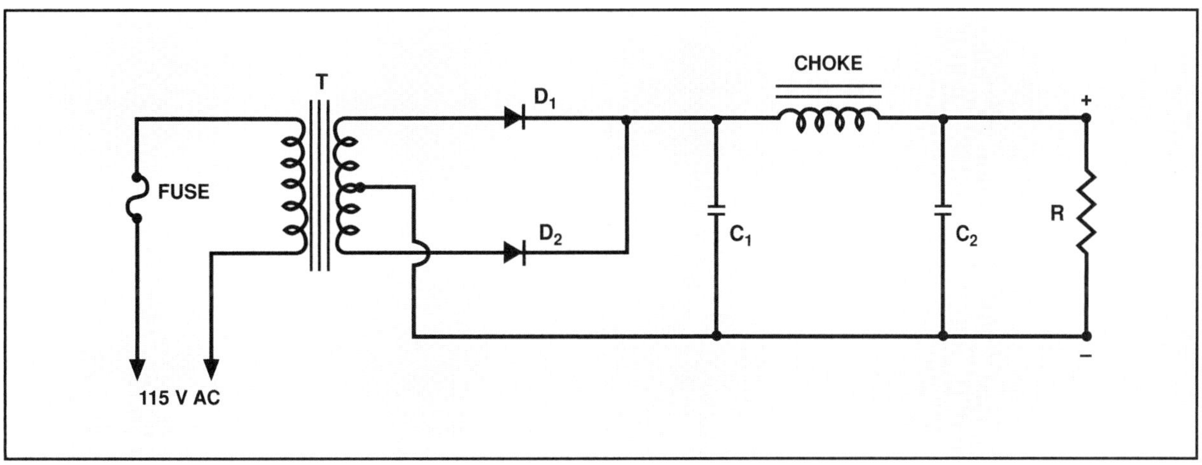

Figure 130 Full-wave rectifier with LC filter

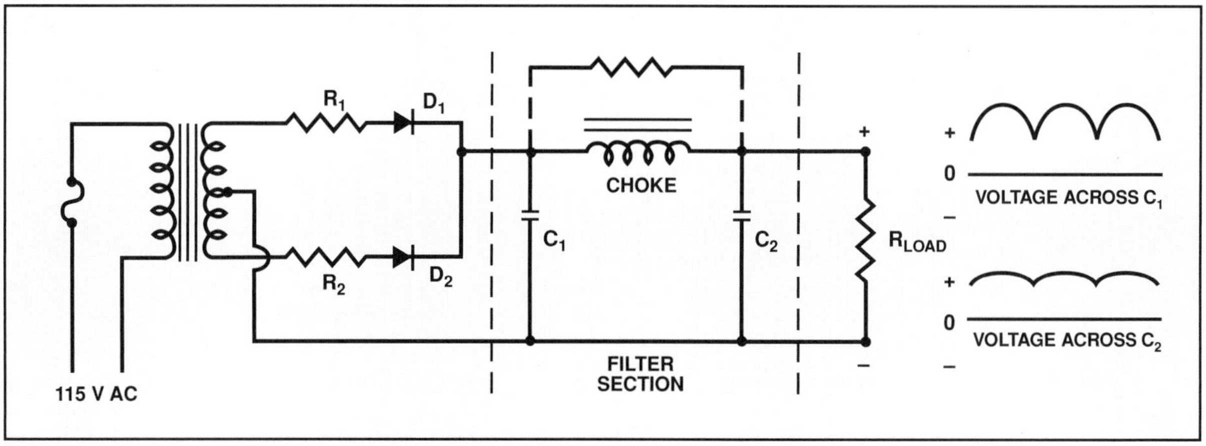

Figure 131 Full-wave rectifier with RLC filter

Figure 131 is a schematic of a similar circuit with additional improvements. It has an RLC filter. The resistors that are wired in series with each rectifier provide current limiting so that the diodes do not reach peak-current limitations. In a capacitor-input filter such as this, the current during capacitor charging can be significant and the resistors limit the charging current to an acceptable level. In addition, a dampening resistor is added in parallel to the DC choke to limit the peak voltages that might develop when a sudden current change results in the load.

Figure 132 is a schematic that shows a variation of a full-wave rectifier. This circuit uses a rectifier bridge. A bridge circuit is better than center-tap circuits because the transformer winding is completely used during both cycles of rectification. Although this circuit requires four diodes to achieve full-wave output, the result is the same as with the center-tapped circuit. Filtering can be added to the output as required by the load.

Power Supply Regulation

As noted before, the DC voltage on the output can be filtered or unfiltered, depending on the design of the system. In addition, the voltage in the previously discussed circuits is unregulated, which means that any change in AC supply voltage or load current will cause the output DC voltage to be affected. Most applications for DC power require that the voltage not only be smooth and ripple-free, but relatively constant. Electronic regulation is the primary method used to provide the constant voltage required by most electronic loads.

Zener Voltage Regulator

In a Zener voltage regulator (fig. 133), unregulated voltage is applied to the input on the left. By using a

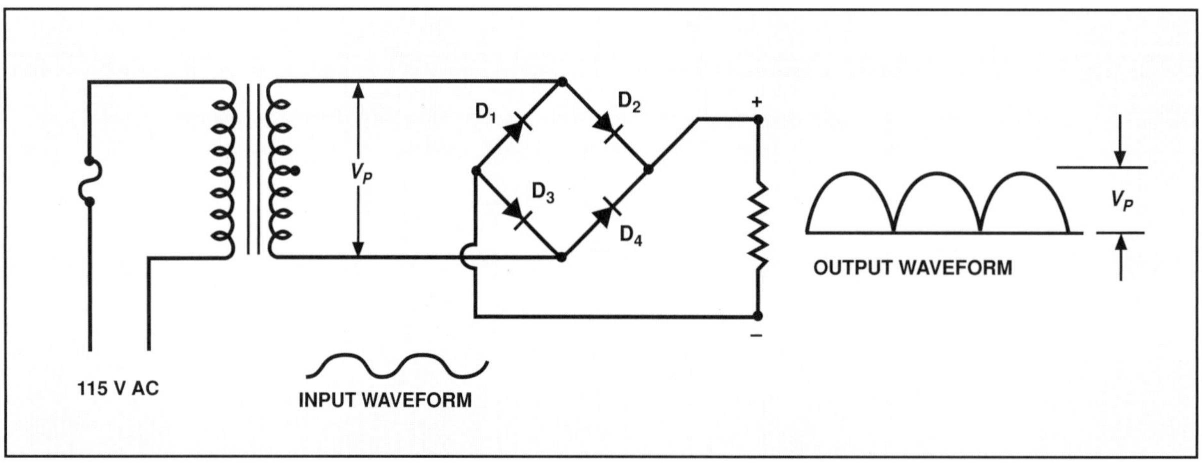

Figure 132 Rectifier bridge full-wave rectifier

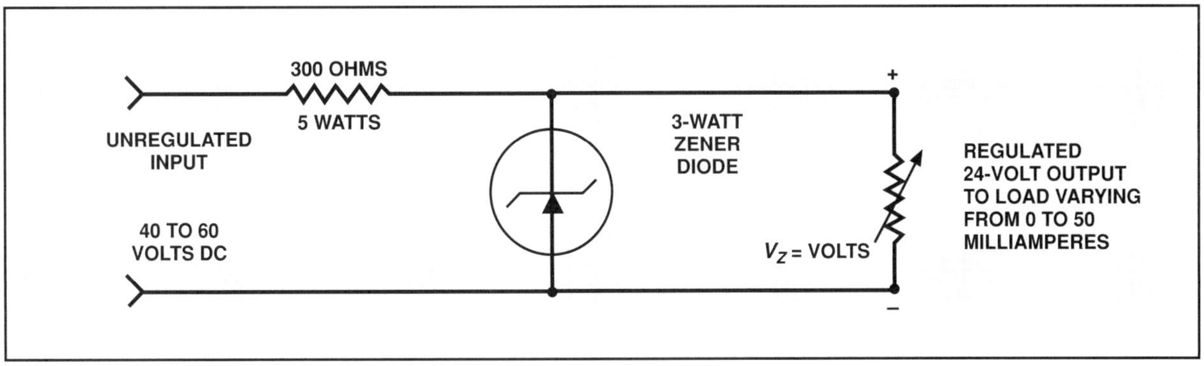

Figure 133 Zener regulated circuit

series resistor to drop the undesirable voltage, the Zener diode provides a fixed voltage and low resistance path for current to flow. By connecting the load resistance to the output, the voltage remains relatively constant as long as the load current does not exceed the current flowing through the Zener diode.

A regulator with a Zener diode is limited by the amount of power it can control because it uses a shunt type of regulation. That is, the maximum load variation is the current permitted to flow in the Zener diode from no-load to full-load. Once the current of the load exceeds this level, the circuit becomes unregulated since the Zener diode is no longer conducting and maintaining a constant voltage.

A Zener diode circuit can be improved by using a series form of regulation with power transistors. In the schematics in figure 134, unregulated, but filtered, DC voltage is provided to the input of the voltage regulator. In the first example (fig. 134A), the filtered DC voltage has a negative polarity with respect to the common connection. The transistor's base is connected to a Zener diode that provides the desired voltage reference. When the input voltage or load current changes, the transistor base-emitter circuit detects this change against the Zener voltage and provides the required changes from collector to emitter to keep the output voltage constant.

In this type of circuit, the transistor regulator provides additional filtering if the filtered input voltage contains any undesirable ripple content. Because of the fast response of the transistor and Zener diode, the transistor regulator attenuates any ripple, or change, on the input.

While the circuit in figure 134A produces a negative output with respect to its common, the circuit shown in figure 134B provides a positive output voltage. The functions of regulation are basically the same, but circuit 134A uses a pnp transistor regulator

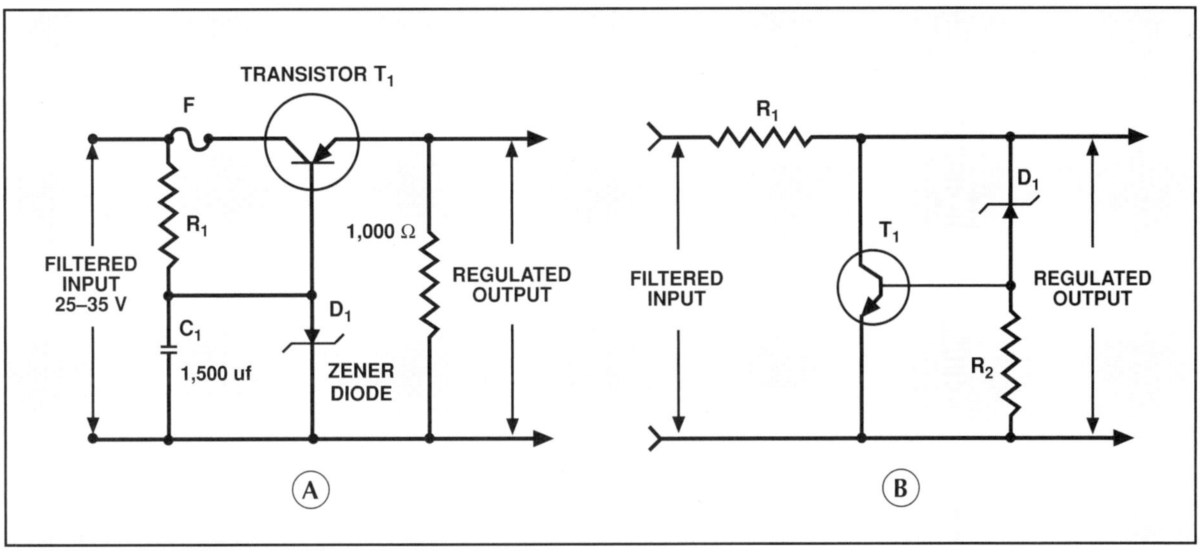

Figure 134 DC regulators

while circuit 134B uses an npn transistor. Circuit 134B also uses shunt regulation with the transistor operating in parallel with the load in much the same manner as the Zener diode did in the circuit shown in figure 133. The transistor, however, has more current-handling capability and can provide the capacity for most loads.

Series regulation is the most efficient form of regulator because the power handled by the transistor is the difference in input and output voltage times the current of the load, while the shunt regulating transistor must handle the total-load voltage as well as the total-load current change.

Sophisticated regulators (fig. 135) are available that provide excellent voltage regulation with load changes as well as rejecting any incoming ripple voltage from the unregulated rectifier. In addition, these types of regulators provide a means for adjusting the output voltage over a limited range.

Three-Phase Rectifiers

In high-power applications where three-phase exists, rectifiers can convert the three-phase from AC to DC for power supplies. Figure 136 is a schematic of a three-phase half-wave rectifier. The three-phase voltage provided by a wye-type transformer has a single diode connected to each winding. (It is called a wye-type transformer because it resembles the letter Y.) Each diode conducts when its anode becomes positive with respect to its cathode. Each diode blocks reverse current flow when its cathode becomes positive with respect to the anode. Unlike a single-phase circuit, where voltage appears for conduction every 180 degrees, three-phase voltage appears for conduction every 120 degrees. Therefore, overlap in the voltages occurs and the resulting DC output has continuous voltage even though it retains some ripple.

Further improvements can be made with three-phase rectification by installing three additional rectifiers to produce three-phase, full-wave, supply. Figure 137 shows a printed wiring board voltage regulator of this type. The output ripple is quite low, generally in the neighborhood of 5 percent of the total DC voltage. Filtering of this type of voltage is quite small and simple.

REVIEW QUESTIONS

1. What is the basic difference between conductors, semiconductors, and nonconductors?
2. How are n-type and p-type semiconductor materials made?

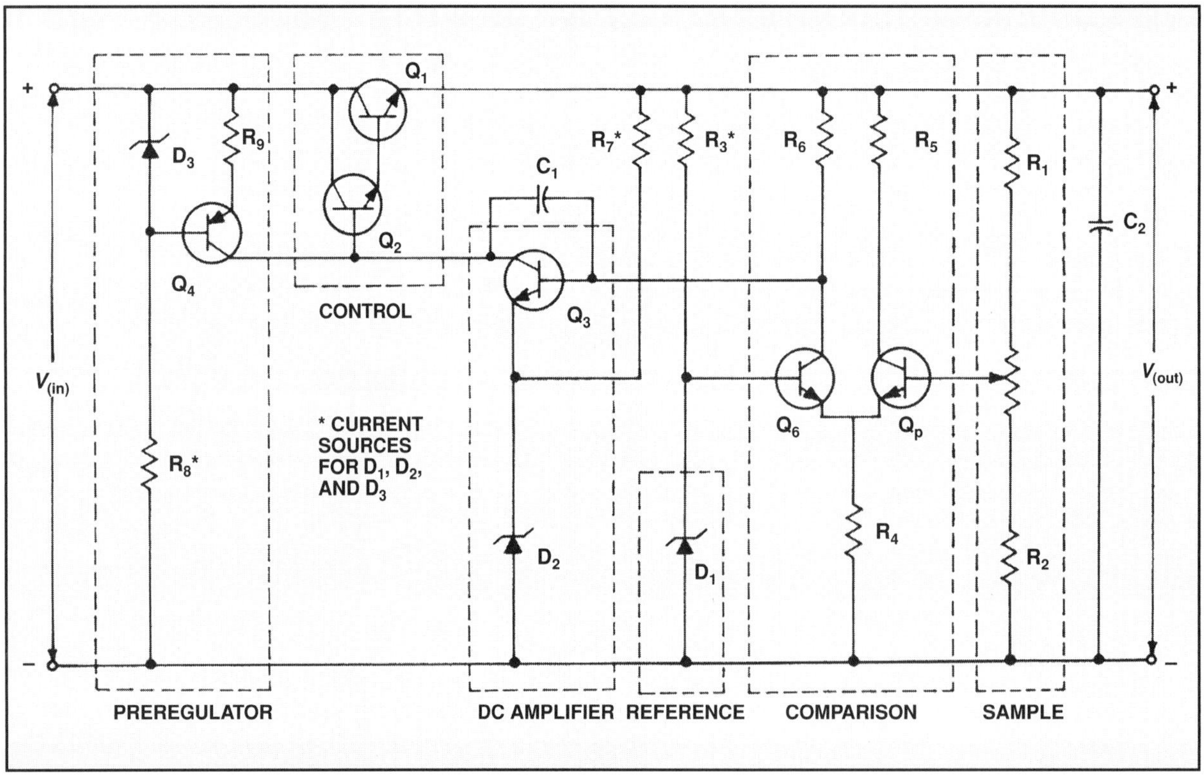

Figure 135 Sophisticated adjustable voltage regulator

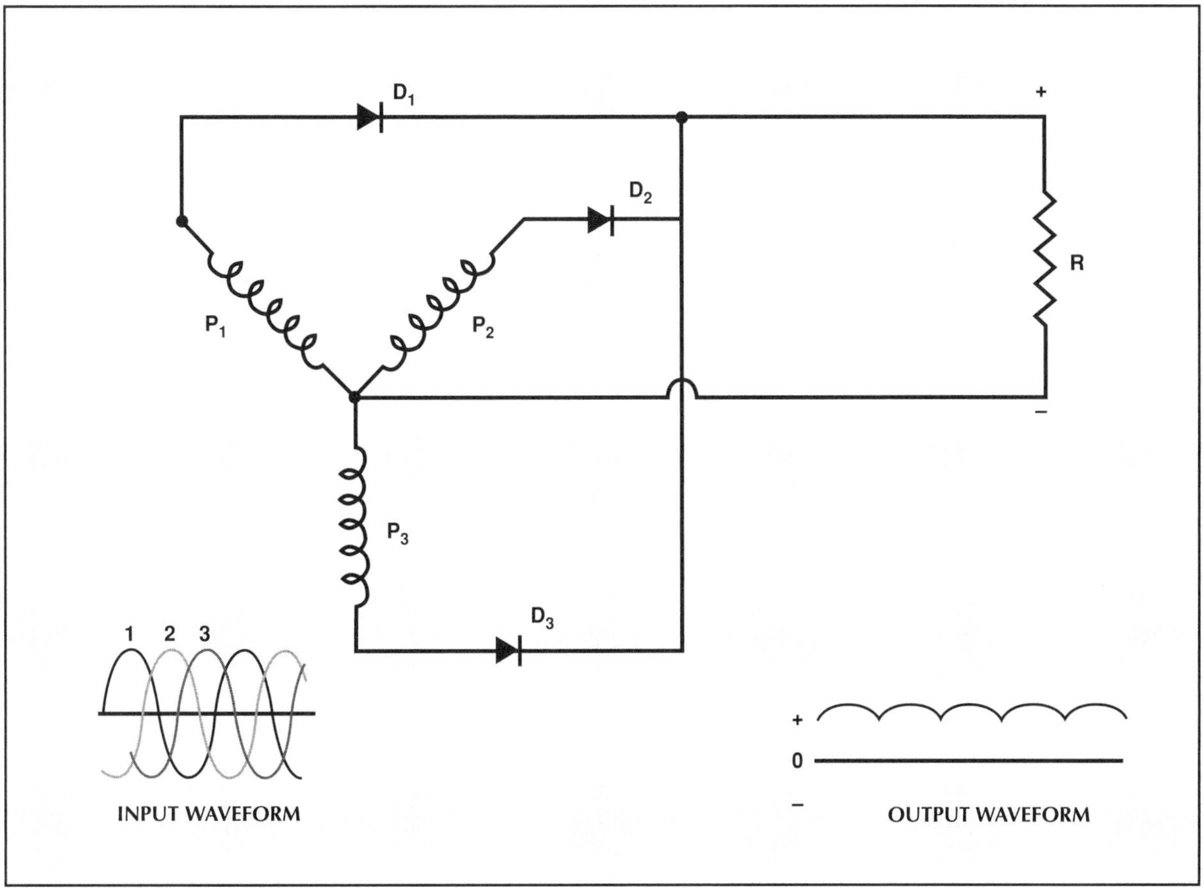

Figure 136 Three-phase half-wave rectifier

Figure 137 Printed wiring board voltage regulator

3. What serves as the current carriers in n-type and p-type material?
4. How is a junction diode formed?
5. What prevents a total interaction between the free holes and free electrons in a p-n junction?
6. What is forward bias and reverse bias? How does the direction of the bias affect current flow?
7. With an alternating current across a semiconductor diode, during what part of the cycle does the diode conduct current?
8. What are the two basic types of junction transistors?
9. What are the basic elements of transistors?
10. Are npn and pnp transistors biased in the same direction? Are the polarities of the external connections the same for both types?
11. What is the direction of current flow through each of the two types of transistors?
12. Why is the base section of bipolar transistors made extremely thin?
13. Are the two biasing voltages connected in series or parallel?
14. What is the current gain relationship in a transistor?
15. An SCR has three terminals. Name them and their function.
16. Name three applications of an op amp.
17. To create a DC regulated power supply from AC, what are the basic functions required?

11 Digital Logic

We are firmly entrenched in a world where *digital logic* equipment and systems have become the dominant means of computation and communication. Several years ago, analog equipment dominated because digital devices had not been developed to the point of widespread use. Development of solid-state electronics, logic functions, and integrated circuits became commercial in the late 1960s and became fully developed in the 1970s. Logic functions, such as AND, OR, NAND, NOR, and *flip-flop circuits*, became available at low cost and led to the replacement of electromechanical relays. (A flip-flop circuit is a type of electronic circuit in which either of two active devices may remain conducting, with the other nonconducting, until the application of an external pulse.) However, replacement was passive—that is, the change from electromechanical relays to logic circuits was made only when it became necessary to replace a bad relay.

In the 1980s, a new integrated circuit was introduced: the microprocessor. Microprocessors allowed interaction between the user and the desired output. Further development of microprocessors led to production of calculators, digital phones, digital computers, and programmable logic controllers (PLCs). Today, supervisory control systems, digital telemetry devices, data acquisition equipment, and similar equipment are widely available. Because they are so widespread, these devices are usually well within economic reach of most consumers and companies.

BINARY NUMBERS

Because digital logic and the equipment that use it are based on the binary number system, this section further explores the concept. Earlier, basic concepts of digital and binary numbers, which use the symbols 1 and 0 to represent logic states or conditions, were discussed. This section builds on these concepts.

Binary Arithmetic

Binary numbers are different from decimal numbers in that binary numbers use base 2 and decimal numbers use base 10. All decimal numbers can be represented as the number 10 raised to some power, and all binary numbers can be represented as the number 2 raised to some power.

The decimal system uses digits ranging from 0 to 9, while the binary system uses only the digits 0 and 1. The use of only two digits fits in well with electrical systems, because 0 can represent the absence of voltage while 1 can represent the presence of voltage. Switches or solid-state devices can establish the presence or absence of voltage.

In all numbering systems, a number takes on a weighted value according to its position to a reference point. In binary systems, this reference point is the *binary point*. Decimal systems take on weight relative to the position of the *decimal point*. A binary point is similar to the decimal point. A decimal point is a dot placed between numbers to indicate the point where numbers take on positive or negative powers of 10 (positive to the left of the decimal point and negative to the right of the decimal point). Because binary numbers are used in the binary system, the point is called the binary point, rather than the decimal point. In the binary system, numbers to the left of the binary point are raised to positive powers of 2, while numbers to the right of the binary point are raised to negative powers of 2. For example—

Binary Point

↓

$$2^3 \mid 2^2 \mid 2^1 \mid 2^0 \mid 2^{-1} \mid 2^{-2} \mid 2^{-3} \mid 2^{-4}$$

In the base-10 decimal system, a number becomes 10 times larger each time it is moved one position to the left of the decimal point. For example, the number 3,747 can be represented in weighted columns as—

10^3	10^2	10^1	10^0
1,000	100	10	1
3	7	4	7

Thus, the number 3,747 also means 3,000 + 700 + 40 + 7 = 3,747.

In the base-2 binary system, a number also increases in value to the left of the binary point. Thus, the binary number 1111 can also be written as 1,000, 100, 10, and 1. And, it can also be written as $1(2^3) + 1(2^2) + 1(2^1) + 1(2^0)$. Writing the number as 1 times the power to which 2 is raised makes it easy to convert the binary number to its decimal equivalent. Thus, the binary number 1111 is the same as $1 \times 2^3 + 1 \times 2^2 + 1 \times 2^1 + 1 \times 2^0$, which solves to $8 + 4 + 2 + 1 = 15$. Similarly, the binary number 1010 is the same as $2^3 + 2^1$, or $8 + 2 = 10$, and the binary number 101 is the same as $2^2 + 2^0$, or $4 + 1 = 5$. (Since $1 \times 2^2 = 4$ and is the same as 2^2, or 4, by itself, it is not necessary to multiply each number by 1.)

Arranging binary numbers in rows and columns shows how the numbers increase in weight to the left of the binary point. In the following arrangement, the binary numbers 1111, 1010, and 0101 are put in columns and their decimal equivalents of 15, 10, and 5 are written to the right.

				Decimal Number
2^3	2^2	2^1	2^0	
8	4	2	1	
1	1	1	1	15
1	0	1	0	10
0	1	0	1	5

Binary numbers with negative powers of base two occur to the right of the binary point. The decimal numbers decrease by half as they move one column to the right. The following columns and rows show binary and decimal numbers to the right of the binary point and the decimal point.

				Decimal Number
2^{-1}	2^{-2}	2^{-3}	2^{-4}	
0.5	0.25	0.125	0.0625	
1	1	1	1	0.9375
1	0	1	0	0.6250
0	1	0	1	0.3125

Just as the decimal equivalents of binary numbers to the left of the binary point can be determined by adding the numbers that result when raised to their respective powers, decimal equivalents of binary numbers to the right of the binary point can be determined by adding them. For example, the binary number 0.1111 is the equivalent of $0.5 + 0.25 + 0.125 + 0.0625$, or 0.9375.

Binary numbers can be added, subtracted, multiplied, and divided just as decimal numbers can be added, subtracted, multiplied, and divided.

Binary Addition

Adding binary numbers is easier if the numbers are arranged in columns and rows. In the following columns and rows of numbers, the binary numbers are placed and added in six left-hand columns. The equivalent decimal numbers are in the far right-hand column. The binary numbers 110.11 and 011.01 are added, as are the equivalent decimal numbers 6.75 and 3.25. The binary number 110.11 and the decimal number 6.75 are *augends*. The binary number 011.01 and the decimal number 3.25 are *addends*. An augend is simply a quantity to which another quantity is added. An addend is one of a collection of numbers to be added.

Function			Binary				Decimal
No. carried		1	1	1	1		1 1 1
Augend		1	1	0.	1	1	6.75
Addend		0	1	1.	0	1	3.25
Sum	1	0	1	0.	0	0	10.00

To add the binary numbers, refer to the binary columns and the two rows labeled augend and addend. The operation begins with the right-most binary number column and works to the left, just as decimal numbers are added. Beginning at the right-hand binary column, the augend 1 is added to the addend 1, and the result is 2. However, because 1 is the highest number in the binary system, a 0 is written for the sum of 1 plus 1, just as 0 is put down for the sum of 9 plus 1 in the decimal system. The 1 is carried over to the next column and placed in the first row, which is labeled 'no. carried'.

To add the second column, the 1 in the augend is added to the 1 in the no. carried row; so, once again a 0 is written for the sum of 1 plus 1, and a 1 is carried to the third column in the no. carried row. When the third column is added, the result is again 1 plus 1, so a 0 is placed in the sum, and the 1 is carried to the fourth column.

The fourth column now consists 1 + 1 + 1. As before, 1 plus 1 gives 0 and 1 to carry; so, the remaining 1 and 0 are added to get 1 for the sum in the fourth column. Going to the fifth column, 1 plus 1 gives the sum of 0 and 1 is carried into the sixth column. The final sum is 1010.00. In decimal form, the answer is 10.00. In binary addition, keep in mind that 1 plus 1 equals 0 with 1 to carry to the next column.

Binary Subtraction

Subtraction of binary numbers is similar to subtraction of decimal numbers. In subtraction, as in addition, the numbers are arranged in columns and rows. The binary numbers subtracted in the five left-hand columns are equivalent to the decimal numbers in the right-hand column. The binary number 010.11 is subtracted from 110.10 and is compared with subtracting the equivalent decimal numbers of 6.50 and 2.75. The numbers 110.10 and 6.50 are minuends. A minuend is the number from which another number, the subtrahend, is subtracted. In this case, the binary subtrahend is 010.11 and the decimal subtrahend is 2.75.

Function			**Binary**			**Decimal**
No. borrowed	– 1	1	1	1		– 1 1 1
Minuend	1	1	0.	1	0	6.50
Subtrahend	– 0	1	0.	1	1	– 2.75
Difference		1	1.	1	1	3.75

Subtraction starts with the first right-hand column of the binary numbers, just as with decimal numbers. In the right-most binary number column, the subtrahend of 1 cannot be subtracted from the minuend of 0, so 1 is borrowed from the next column to the left. Since 1 is moved one column to the right the number becomes 2, so subtracting 1 from 2 leaves a 1 in the right-most column. This 1 is entered in the row labeled difference.

In the second column, –1 is in the subtrahend, 1 is the minuend, and a –1 is in the no. borrowed column. The numbers add up to –1. A 1 is borrowed from the third column, which is treated as a 2 when adding up the second column; so, 1 is placed in the difference row for the second column.

The operation proceeds in the like manner for the remaining columns. Remember: when subtracting binary numbers, the no. borrowed column has the same sign as the subtrahend, which is a minus (–). The final answer is 11.11 in binary notation and 3.75 in decimal notation.

Binary Multiplication

Multiplication of binary numbers is the simplest of the four mathematical operations because binary multiplication is the same as decimal multiplication. Even the placement of the binary point in the product is determined in the same way as placement of the decimal point. In the example, 10.11 is multiplied by 1.01 and its decimal equivalent of 2.75 is multiplied by 1.25. The numbers 10.11 and 2.75 are multiplicands and the numbers 1.01 and 2.75 are multipliers.

Function			**Binary**				**Decimal**
Multiplicand			1	0.	1	1	2.75
Multiplier			×	1.	0	1	× 1.25
			1	0	1	1	1375
Partial products		0	0	0	0		550
	1	0	1	1			275
Product	1	1.	0	1	1	1	3.4375

The final answer (product) in binary numbers is 11.0111; in decimal numbers, the product is 3.4375. The three rows of numbers labeled partial products are the figures that result from each multiplication step. Partial products are 1011, 0000, and 1011 in the binary multiplication steps, and 1375, 550, and 275 in the decimal multiplication steps.

Binary Division

Dividing binary numbers is also simple. Because a trial divisor must be either a 1 or a 0, there is little likelihood of failing on the first trial. In the following example, borrowing is done mentally during all the intermediate subtraction steps. The placement of the binary point and the placement of the decimal point are determined in the same manner. First, the binary number 11.0111 is divided by 1.01; then, the decimal number 3.4375 is divided by 1.25. In the binary operation, the divisor is 1.01, the dividend is 11.0111, and the quotient is 10.11 as determined by—

```
        10.11
1.01 ) 11.0111
       101
       -------
        111
        101
       -------
          101
          101
       -------
```

In the decimal operation, the divisor is 1.25, the dividend is 3.4375 and the quotient is 2.75 as determined by—

```
         2.75
1.25 ) 3.4375
       250
        937
        875
         625
         625
```

DIGITAL LOGIC

Statements or conditions that are either true or false, yes or no, or energized or deenergized can be described with digital logic. Since digital implies either one state or the other, binary devices such as signals, components, and the like can establish binary conditions and obtain results.

Classical Logic

Over 2,000 years ago, Greek philosophers studied the concepts of logical reasoning, or *logic*. One concept that came from this study was that all observations could be simplified until a statement about an observation was either true or false. We know today, however, that many observations are both true and false and that many statements are either true or false. Put another way, the Greeks were wrong to think that all observations could be simplified to be either true or false. However, unbeknownst to them, they laid the groundwork for today's digital revolution.

A fundamental concept of logic is that by using two or more logical statements as premises, a correct conclusion can be made. For example, if a car's gas gauge reads empty, and the gas gauge is accurate, then the tank is out of gas. In terms of classical logic, the preceding statements can be stated as—

- Major premise: the car's gas gauge reads empty.
- Minor premise: the car's gas gauge is accurate.
- Conclusion: the car's gas tank is out of gas.

Note that the conclusion is correct or true only if both conditions are met—that is, both statements must be true. Notice, too, that at least two premises must be used and both must be true.

Logical AND Operation

Digital logic depends on three basic logic operations, or relationships. The first logic operation is termed the *logical AND operation*. This operation is illustrated by the example concerning the car's gas tank. That is, both the statement regarding the gauge's reading empty AND the statement regarding the accuracy of the gauge must be true before the conclusion is true. (In a logical AND operation, AND is capitalized,)

Another example of logic AND reasoning concerns the statement, "If Tom is taller than Bill and Bill is taller than Dick, then Tom is taller than Dick." This statement is correct, because if Tom is taller than Bill AND Bill is taller than Dick, then Tom must be taller than Dick.

Logical OR Operation

Another logical operation is the *logical OR operation*. This operation also makes it possible to draw a conclusion from two or more statements. In this case, however, if either or both of the statements are true, then the conclusion is true. Thus, the logical OR operation differs from the AND operation where both statements had to be true for the conclusion to be true. As an example of a logical OR operation, consider the statement, "if the sun is shining OR if artificial lights are on, then it must be light." In this case, the logical OR operation holds that if the sun is shining, it is light OR if artificial lights are on, it is light. In this case, either the sun has to be shining OR the artificial lights have to be on for it to be light. On the other hand, both the sun can be shining and the artificial lights can be on for it also to be light.

Logical NOT Operation

Still another logical operation is the *logical NOT operation*. This operation means that a true statement is made false by the use of NOT. For example, consider the statement, "If it is light, and someone says that it is NOT light, then the second statement is false." In other words, if it is true that it is light, the statement that it is NOT light is false.

Boolean Logic

In 1850, the English philosopher and mathematician George Boole developed a technique of using numerical values to relate to the truth or falsehood of a logical statement. He assigned a mathematical symbol to logical operations. Using this technique, logical statements can be expressed mathematically. For example, Boole's system can show so-called *if-then operations* as algebraic equations. An if-then operation (also called an implication) is a statement that asserts that if a certain condition is true, then a certain other condition is also true. Stated algebraically, an if-then operation reads, "If p then q," which means that if p is true then q is also true. Boolean math makes it possible to manipulate equations just as any algebraic expression, with only a few exceptions, can be manipulated. The method of expressing logic as equations is called *Boolean logic.*

Boolean logic designates or labels each logical statement or premise. For example, the letter A can replace the premise, "Tom is taller than Bill." Similarly, B replaces, "Bill is taller than Dick," and C replaces, "Tom is taller than Dick." Now, the letters can replace the statements as, "if A AND B, then C." This shorthand notation greatly simplifies logic expressions.

The following symbols replace the three basic logic operations—

AND = • or ×, multiplication symbols;
OR = +, the addition symbol; and
NOT = a bar placed across the top of the symbol—for example, $\overline{A}$.

In many cases, as with conventional algebra, the AND symbol (• or ×) is left out, and two characters are simply written side by side. Using symbols and equations, the example, "If A AND B, then C," can be written in Boolean equation form as $A \bullet B = C$ or $AB = C$. This expression reads as, "A and B equal C."

In the same way, logical OR operations can also be written as a Boolean algebraic equation. For example, the previous logical OR operation, which stated that, "If the light is on OR the sun is shining, then it must be light," can broken down as—

- Statement A = the sun is shining,
- Statement B = the artificial lights are on, and
- Statement C = it is light.

The logical expression is written as $A + B = C$ and it reads as, "A or B equal C."

An example of a logical NOT operation is—

Statement A = it is light, and
Statement B = it is not light;
then, $\overline{A} = B$.

This expression reads as, "not A equals B."

Boole also assigned the numerical symbol 1 to a true statement and 0 to a false statement. These numbered symbols are states or conditions of logical statements and, in themselves, have no numerical value. Thus, to express a true statement in Boolean logic one simply writes $A = 1$. This statement can be spoken as, "A is true." Similarly, to express a false statement, one simply writes $B = 0$. This statement reads, "B is false."

Boolean logic contains many definitions and expressions. This manual does not list them all, but a few include—

$0 \bullet 0 = 0$, which reads, "a false statement and false statement are false";

$1 + 1 = 1$, which reads, "a true statement or a true statement is a true statement";

$0 \bullet 1 = 1 \bullet 0 = 0$, which reads, "a false statement and true statement are the same as a true statement and a false statement and both are false";

$0 + 1 = 1 + 0 = 1$, which reads, "a false statement or a true statement is the same as a true statement or a false statement and both are true";

$1 \bullet 1 = 1$, which reads, "a true statement and a true statement are true";

$0 + 0 = 0$, which reads, a false statement and false statement are false";

$\overline{0} = 1$, which reads, "a false statement is not a true statement";

$\overline{1} = 0$, which reads, "a true statement is not a false statement";

$A \bullet 1 = A$, which reads, "A and a true statement is A";

$A + 0 = 0$, which reads, "A or a false statement is a false statement";

$A \bullet 0 = 0$, which reads, "A and a false statement is a false statement"; and

$A + 1 = 1$, which reads, "A or a true statement is a true statement."

At the time Boolean logic was developed, few philosophers were interested in it. However, when the telephone was developed, engineers found that Boolean logic lent itself well to describing the operation of the complex relay switching gear that was necessary. A set of open relay contacts could be described by a logical 0 and closed contacts by a logical 1. Thus, racks of switches could be described by a few equations instead of several pages of electrical drawings. From that time, Boolean logic has been used mathematically to describe everything from simple relay systems to complex computer systems.

Truth Tables

In many cases, it is desirable to know the effect of all possible conditions on a logical statement. For example, consider the statement, "If Tom is taller than Bill and Bill is taller than Dick, then Tom is taller than Dick." However, what if the part of the statement that says, "Bill is taller than Dick," is false? To be absolutely sure of the conclusion, it is necessary to list all possible conditions. Other possibilities include—

1. If Tom is taller than Bill and Bill is not taller than Dick, then Tom is not taller than Dick.
2. If Tom is not taller than Bill and Bill is taller than Dick, then Tom is not taller than Dick.
3. If Tom is not taller than Bill and Bill is not taller than Dick, then Tom is not taller than Dick.

Now, label the first statement A, the second B, and the third C. With these labels, the following equations are possible.

1. $A = 0, B = 0$
   $C = AB = 0$

This Boolean equation reads, "If A is false and B is false, then C is A and B, which is false."

2. $A = 0, B = 1$
   $C = AB = 0$

This equation reads, "If A is false and B is true, then C is A and B, which is false." (If A is false and B is true, C is false, because in a logical AND operation, both premises must be true for the conclusion to be true.)

3. $A = 1, B = 0$
   $C = AB = 0$

This equation reads, "If A is true and B is false, then C is A and B, which is false." (Both premises must be true for the conclusion to be true.)

4. $A = 1, B = 1$
   $C = AB = 1$

This equation reads, "If A is true and B is true, then C is A and B, which is true. (In this case, both premises in the logical AND operation are true, so the conclusion is true.)

A tabular format called a *truth table* is a simple method of expressing a logic equation. Table 18 is a truth table for the previous example.

Table 18 shows four possible combinations for consideration. Since two possible conditions exist for each input statement, the number of possible combinations or considerations for a possible decision or conclusion is $2N$, where N is the number of input conditions. If $C = AB$, the number of possibilities is 2^2, or 4, since A and B are two inputs. The truth table shows these four conditions, which are: (1) if A is false and B is false, then C is false; (2) if A is false and B is true, then C is false; (3) if A is true and B is false, then C is false; and (4) is A is true and B is true, then C is true.

TABLE 18
Truth Table for C = AB

A	*B*	*C*
0	0	0
0	1	0
1	0	0
1	1	1

If the logic equation is $D = ABC$, which reads, "D is either true or false depending on the *truth values* of A and B and C." (That is, whether D is true or false depends on whether A, B, and C are true or false.) In this case, the possibilities are 2^3, or 8 since A, B, and C are three inputs. Table 19 shows the possible combinations. Note that in only one case can D be true.

TABLE 19
Truth Table for D = ABC

A	*B*	*C*	*D*
0	0	0	0
0	0	1	0
0	1	0	0
0	1	1	0
1	0	0	0
1	1	0	0
1	1	1	0
1	1	1	1

TABLE 20
Truth Table for D = A + B + C

A	*B*	*C*	*D*
0	0	0	0
0	0	1	1
0	1	0	1
0	1	1	1
1	0	0	0
1	0	1	1
1	1	0	1
1	1	1	1

Another truth table (Table 20) can be drawn from the equation $D = A + B + C$, which reads, "*D* is either true or false depending on the truth values of *A* or *B* or *C*." In this case, because it is a logical OR operation, *D* can be true in six out of eight possibilities.

Boolean logic can also be applied to describe electrical and electronic components used in electrical and electronic circuits. Symbols have been developed that portray the AND, OR, NOT forms of statements in addition to other statements, such as buffer, NOR, NAND, and XOR (exclusive or).

Electromechanical relays are devices that consist of a coil and one or more contacts that are magnetically and mechanically actuated when the coil is energized. The contacts of the relays can be normally open or normally closed; normally refers to the coil's being deenergized. The relays are used in control applications for turning on or off discrete devices such as lamps, *contactors*, *annunciators*, and other devices. (A contactor is a switch that opens or closes a circuit. An annunciator is an electromechanical device that signals when current flows in a circuit.) Schematic diagrams containing relays are usually referred to as *ladder logic* diagrams because they resemble the rungs and rails on a ladder. The components operate in a logical fashion to achieve certain and definite results.

Ladder logic diagrams developed from a need in the field to solve a particular control problem. Ladder logic diagrams are also used as troubleshooting aids, and help explain the operation and programming of PLCs. When creating and reading a ladder logic diagram, several rules must be followed. Rules include—

1. show all symbols for the devices in the diagram in their off-the-shelf, or deenergized, condition;
2. show input devices on the left part of the rung;
3. show output devices on the right part of the rung;
4. connect input devices in such a manner as to create logic to accomplish a desired function and produce an output; and
5. read the diagram from top to bottom and left to right.

To the left in figure 138 are the symbols of six relays: buffer, AND, OR, inverter (NOT), NAND, and NOR. They are labeled with Boolean logic letters *A*, *B*, and *C*. To the right in figure 138 is a ladder logic diagram that shows each of the relays in schematic form. They are also labeled with Boolean letters.

Truth tables for each of the individual logic rungs in figure 138 follow. Note that each relay is referred to as a gate, which means that when its input receives the proper logic conditions, the output changes its state. Table 21 is the truth table for the buffer gate, where $B = A$.

TABLE 21
Buffer Gate
Truth Table for B = A

A	*B*
0	0
1	1

Table 22 is the truth table for the AND gate, where $C = AB$.

TABLE 22
AND Gate
Truth Table for C = AB

A	*B*	*C*
0	0	0
0	1	0
1	0	0
1	1	1

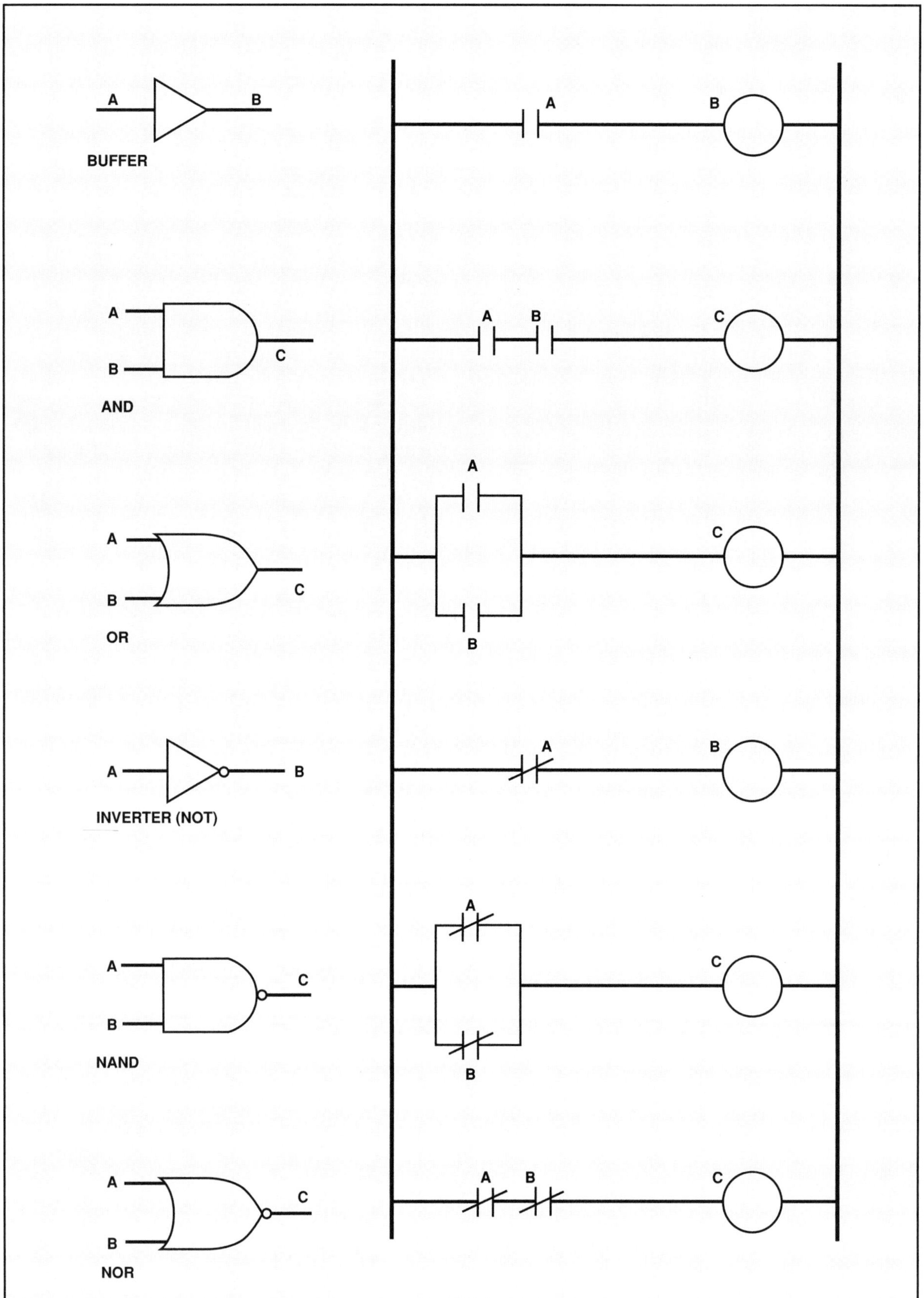

Figure 138 Logic-relay logic symbols

Table 23 is the truth table for the OR gate, where $C = A + B$.

TABLE 23
OR Gate
Truth Table for C = A + B

A	*B*	*C*
0	0	0
0	1	1
1	0	1
1	1	1

Table 24 is the truth table for the inverter (NOT) gate, where $B = A$.

TABLE 24
Inverter (NOT) Gate
Truth Table for B = A

A	*B*
0	1
1	0

Table 25 is the truth table for the NAND gate, where $C = A + B$.

Table 26 is the truth table for the NOR gate, where $C = A + B$.

While the truth tables are similar, it is important to examine them carefully to understand the logic functions of each gate. For example, a Boolean logic equation for the buffer gate is the same as for the inverter (NOT) gate. However, the truth tables are different.

TABLE 25
NAND Gate
Truth Table for C = A + B

A	*B*	*C*
0	0	1
0	1	1
1	0	1
1	1	0

TABLE 26
NOR Gate
Truth Table for C = A + B

A	*B*	*C*
0	0	1
0	1	0
1	0	1
1	1	0

SOLID-STATE LOGIC GATES

Besides relays, devices such as diodes, transistors, and other switching devices can be used for logic gates. Figures 139, 140, 141, 142, 143, 144, and 145 are circuit diagrams that use diodes and transistors as

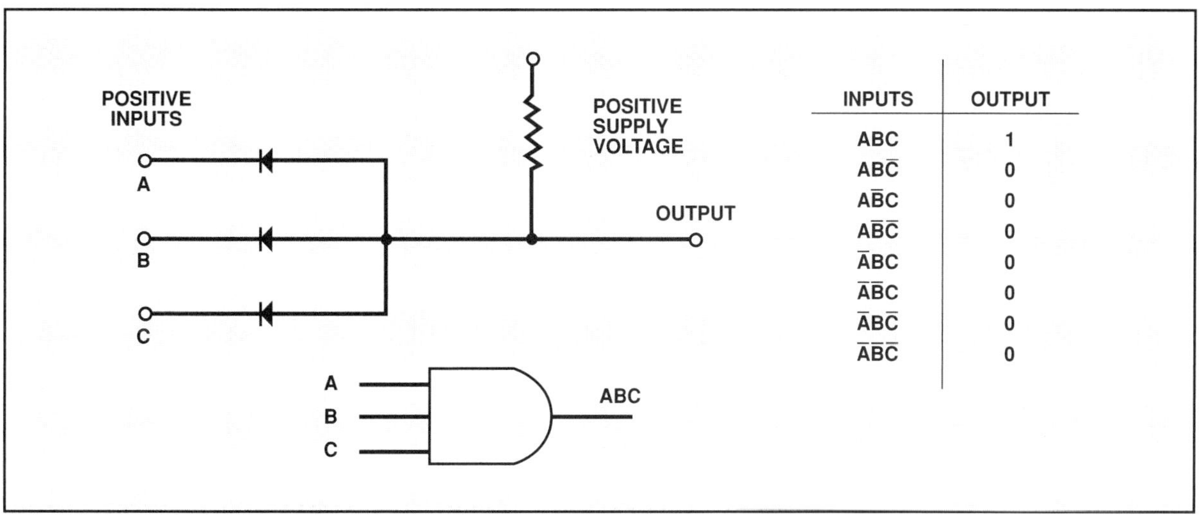

Figure 139 Diode AND circuit

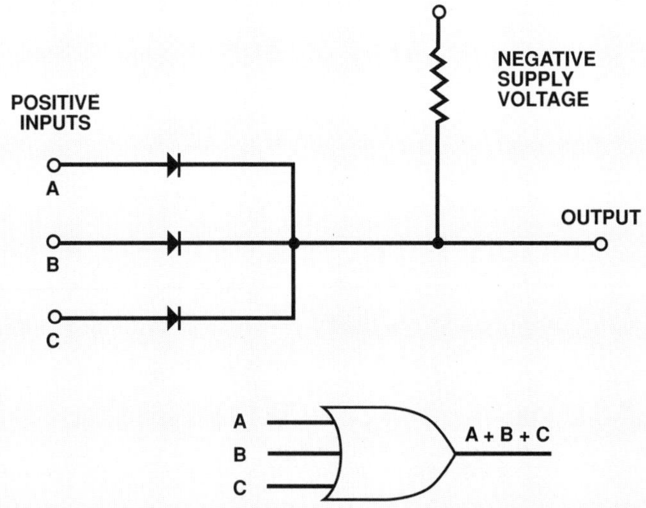

INPUTS	OUTPUT
ABC	1
$AB\overline{C}$	1
$A\overline{B}C$	1
$AB\overline{C}$	1
$A\overline{B}\overline{C}$	1
$\overline{A}BC$	1
$\overline{A}\overline{B}C$	1
$\overline{A}B\overline{C}$	1
$\overline{A}\overline{B}\overline{C}$	0

Figure 140 Diode OR circuit

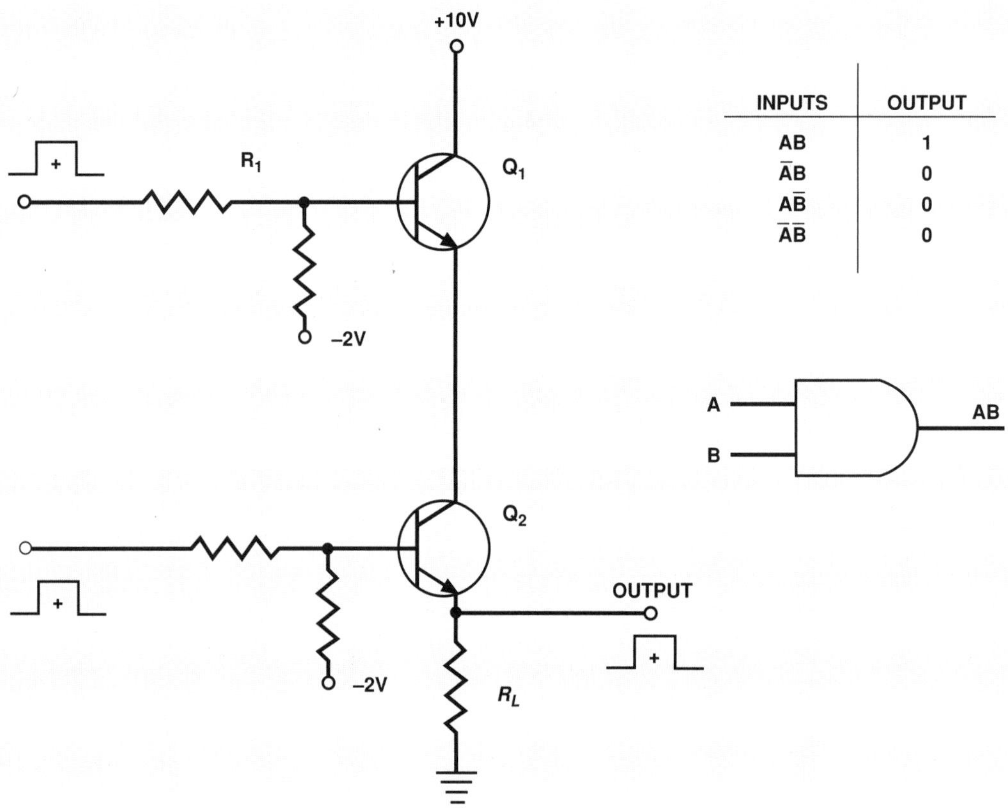

INPUTS	OUTPUT
AB	1
$\overline{A}B$	0
$A\overline{B}$	0
$\overline{A}\overline{B}$	0

Figure 141 Transistor AND circuit

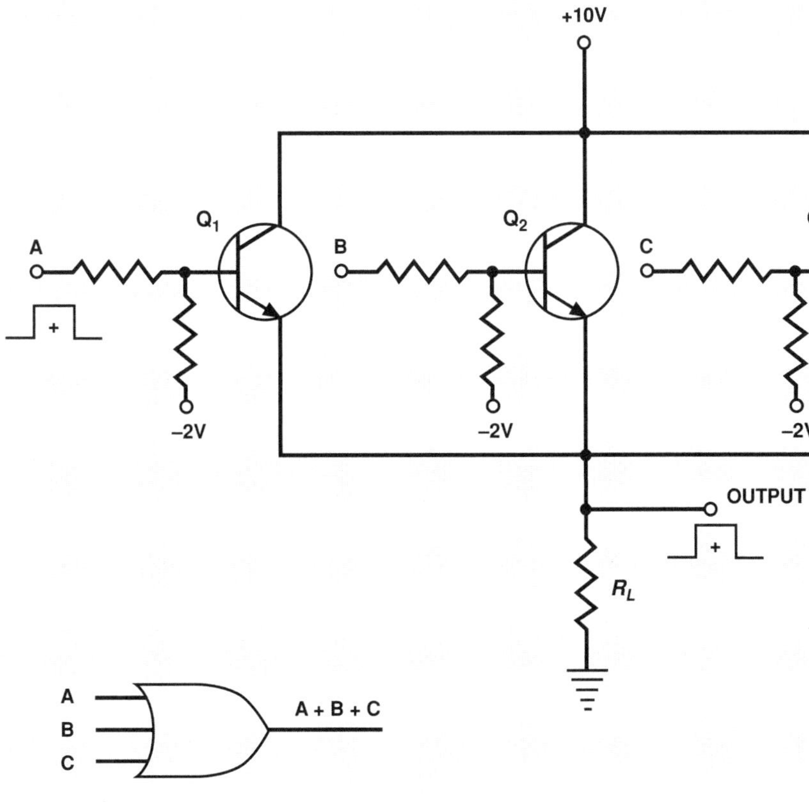

INPUTS	OUTPUT
ABC	1
$AB\overline{C}$	1
$A\overline{B}C$	1
$A\overline{B}\overline{C}$	1
$\overline{A}BC$	1
$\overline{A}\overline{B}C$	1
$\overline{A}B\overline{C}$	1
$\overline{A}\overline{B}\overline{C}$	0

Figure 142 Transistor OR circuit

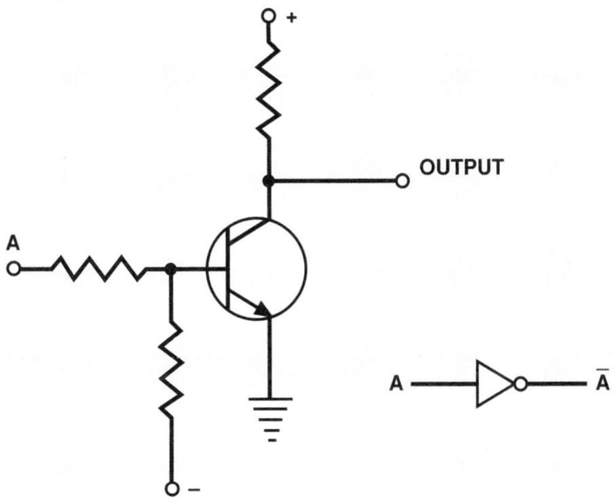

INPUT	OUTPUT
A	0
$\overline{A}$	1

Figure 143 Transistor NOT circuit

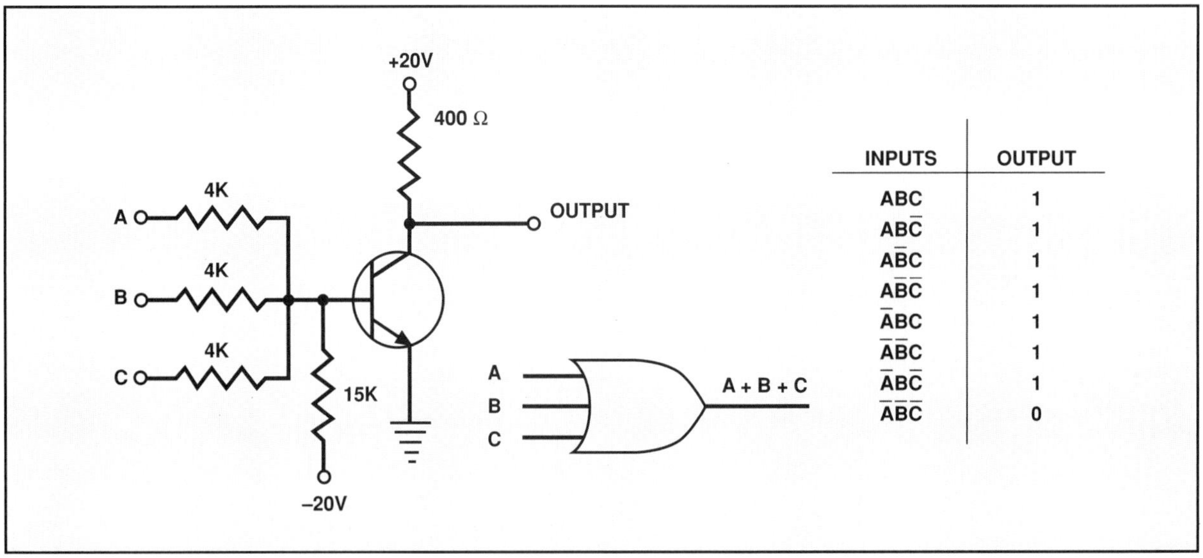

Figure 144 Transistor NOR circuit

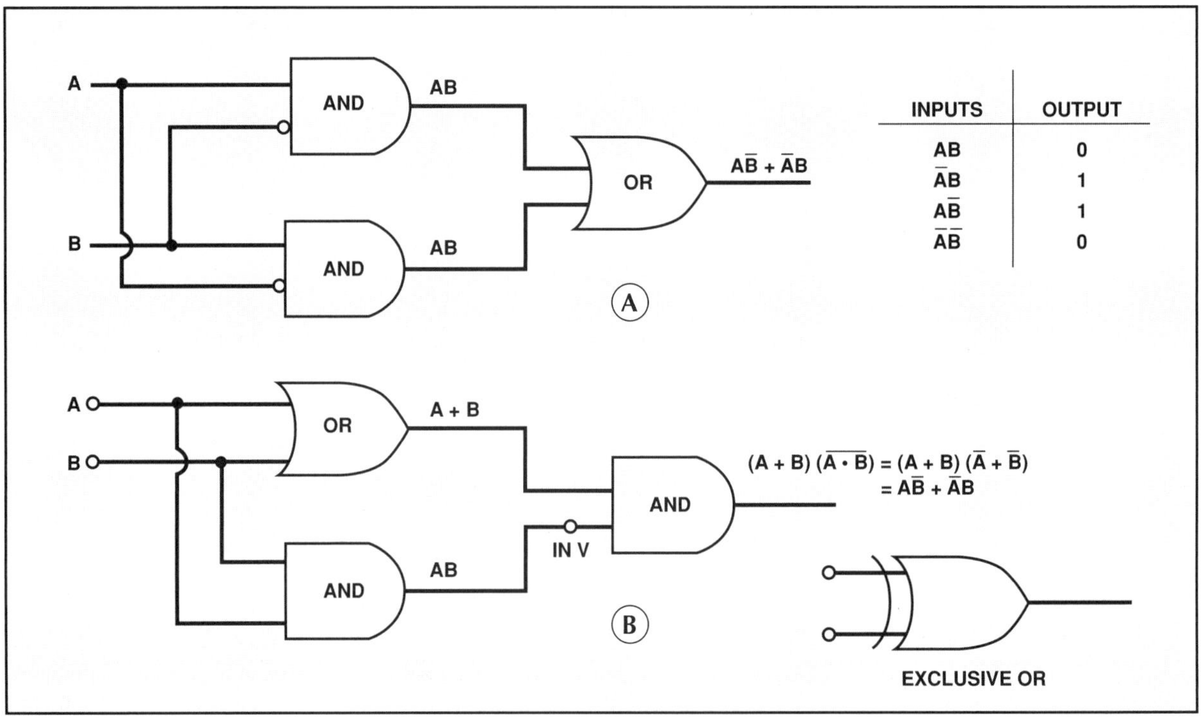

Figure 145 Exclusive OR circuit

logic gates. Each figure also shows the truth table that accompanies each device.

SEQUENTIAL LOGIC AND MEMORY DEVICES

Logic circuits give certain outputs for specified inputs. Because logic circuits function according to physical realties, a small amount of time elapses between the time the inputs reach specified conditions and the time when the output reaches a given logic level. This period of time varies from gate to gate and from logic family to logic family. In some circuits, time is not a factor in the operation of the gate. In such circuits, the negligible amount of time delay before the output reaches its level is not a problem.

However, some logic devices depend on time for operation. Such devices are *sequential logic devices* because most operations take place in a step-by-step manner.

FLIP-FLOP CIRCUITS

The simplest solid-state memory device is the flip-flop circuit, or a bistable multivibrator. Recall that a flip-flop circuit is a circuit in which either of two active devices may remain conducting, with the other nonconducting, until the application of an external pulse. A flip-flop circuit has considerable versatility. Using dual transistors (Q_1 and Q_2 in fig. 146) that alternate in the on-off state, the flip-flop consists of a trigger input, data input, and two complementary outputs.

The switching action of the flip-flop circuit lends itself well to bistable operation, which means that the circuit is capable of assuming either of two stable states. When a flip-flop is triggered, one of the two transistors starts conducting, and the other remains off. With flip-flops, no in-between condition exists except for a very short time when the circuit is switching. To cause a flip-flop to change states, the base of the off transistor is triggered with a pulse that forward biases it and drives it into conduction.

When using npn transistors (fig. 146), a positive trigger pulse is used to forward bias the off transistor, causing it to go into conduction. The transistor that was conducting is almost instantly cut off because of the nature of the bistable circuit. A forward-biasing pulse does not affect the conducting transistor except to drive it deeper into current saturation.

Referring to figure 146, note that an input pulse in input *A* forward biases transistor Q_1 and drives it into conduction. The output of Q_1 becomes zero, which places a low-level signal on output 1. At the same time, the low-level signal from Q_1 is connected to the base of Q_2 through resistor R_1. This low-level signal reverse biases Q_2, causing it to cut off. The collector voltage of Q_2 rises, placing a high-level signal on output 2.

The flip-flop remains in this stable state with Q_1 on and Q_2 off until input *B* is triggered with a pulse. The circuit then reverses its condition and goes into its other stable state. Q_2 goes into conduction, which places a low-level signal on output 2. Q_1 is cut off, which places a high-level signal on output 1. Capacitors C_1 and C_2 serve to speed up the switching action and are sometimes called *speedup capacitors*.

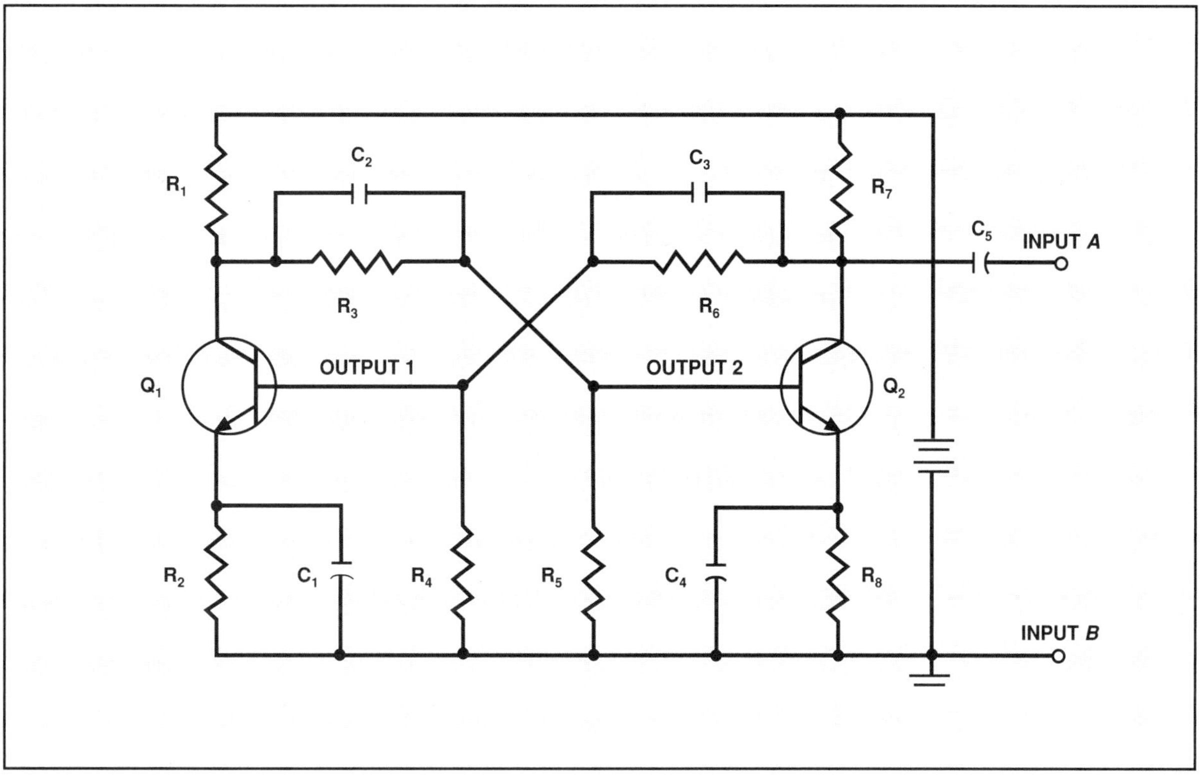

Figure 146 An npn transistor flip-flop circuit

Flip-Flop Logic

The flip-flop is a logic circuit with two possible outputs and from one to three inputs. Each output has two conditions: either a high-level signal represented with a logic 1 and a low-level signal represented with a logic 0 when using positive logic. When an input trigger signal is received, the logic 1 output becomes 0 and the logic 0 output becomes 1.

In the flip-flop circuit shown in figure 147, output 1 is sometimes referred to as the set output, output 2 as the reset output, and the trigger input as the complement input. The set input is used to set output 1 at logic 1 and the reset input serves to reset output 2 at logic 0. A pulse on the trigger input causes the flip-flop to change its condition to the logic state opposite from its current state. A set pulse causes the flip-flop to change states only if output 1 is not already at logic 1; a reset pulse causes the states to be reversed only if output 2 is not at logic 0.

In figure 147, a positive set pulse is directed only to the base of Q_1 because steering diode CR_1 would be reversed biased and would not conduct. This positive charge drives Q_1 into conduction and its output is logic 0. If Q_1 is already in conduction, the positive pulse causes no change.

A positive pulse on the reset input is directed only to Q_2 because diode CR_2 is reversed biased and does not conduct. The positive pulse on Q_2 drives it into conduction, giving a logic 0 output, and the output of Q_1 switches to logic 1. If a positive pulse is now applied to the trigger input, it is directed to both bases of Q_1 and Q_2 because both diodes CR_1 and CR_2 are forward biased and both conduct the signal to the bases.

No matter what conditions Q_1 and Q_2 are in, a trigger pulse will reverse their state. In other words, if Q_1 is conducting and Q_2 is not conducting, a trigger pulse drives Q_1 out of conduction and Q_2 into conduction. The next trigger pulse drives Q_2 out of conduction and Q_1 into conduction. The trigger input is sometimes called the toggle input because of its switching action. It is also called the complementary input.

Table 27 is a truth table for the operation of the flip-flop when pulses are placed on each of the three inputs, set (S), trigger (T), and reset (R). A reset pulse

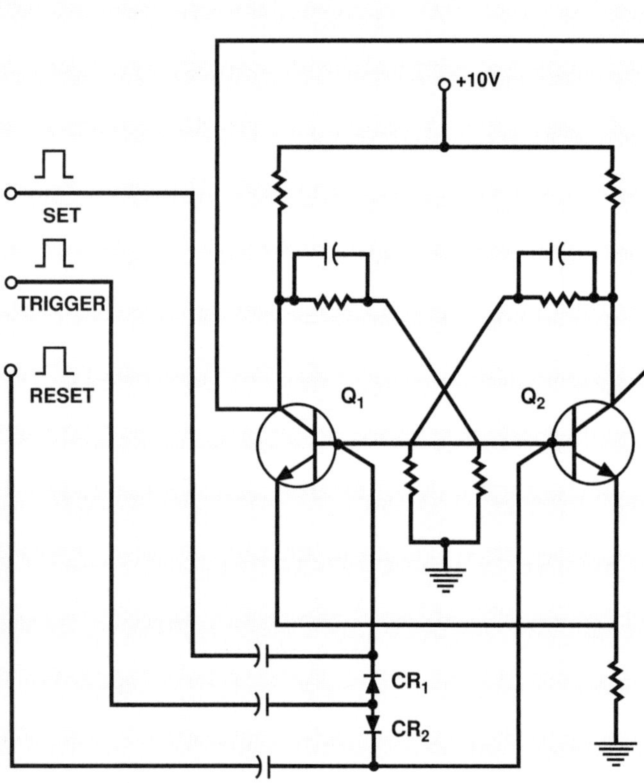

TIME*	S	T	R	1	2	
T_1	0	0	1	1	0	RESET 1
T_2	0	0	0	0	1	
T_1	0	0	1	0	1	RESET 2
T_2	0	0	0	0	1	
T_1	1	0	0	0	1	SET 1
T_2	0	0	0	1	0	
T_1	1	0	0	1	0	SET 2
T_2	0	0	0	1	0	
T_1	0	1	0	1	0	TRIGGER 1
T_2	0	0	0	0	1	
T_1	0	1	0	0	1	TRIGGER 2
T_2	0	0	0	1	0	

(S, T, R = INPUTS; 1, 2 = OUTPUTS)

*T_1 IS THE TIME WHEN THE INPUT PULSE IS APPLIED. T_2 IS A FEW TENTHS OF A MICROSECOND LATER WHEN THE INPUT PULSE HAS BEEN REMOVED AND THE FLIP-FLOP HAS OPERATED.

Figure 147 Flip-flop with 3-trigger inputs

TABLE 27
Flip-Flop Truth Table

Time	Inputs			Outputs		
	S	T	R	1	2	
T_1	0	0	1	1	0	
T_2	0	0	0	0	1	Reset 1
T_1	0	0	1	0	1	
T_2	0	0	1	0	1	Reset 2
T_1	1	0	0	0	1	
T_2	0	0	0	1	0	Set 1
T_1	0	1	0	1	0	
T_2	0	0	0	1	0	Set 2
T_1	0	1	0	1	0	
T_2	0	0	0	0	1	Trigger 1
T_1	0	1	0	0	1	
T_2	0	0	0	1	0	Trigger 2

resets output 2 at logic 1, provided output 2 is at logic 0 when the pulse is applied. If output 2 is already at logic 1, then the condition of the flip-flop remains unchanged. The same is true of the set pulse. If input 1 is at logic 0, a set pulse causes input 1 to change to a logic 1. If not, then a set pulse does not affect the flip-flop condition. A pulse on the trigger input, however, causes the flip-flop to change states no matter what the conditions of the two outputs are. T_1 is the time when the input pulse is applied. T_2 is a few tenths of a microsecond later when the input pulse has been removed and the flip-flop has operated.

Flip-Flop as Frequency Divider

Since it takes two trigger pulses to change the output of Q_1 from a 1 to 0 and back to 1, a division of 2 takes place with frequency. The output signal frequency of Q_1 is only half that of the frequency of the input frequency. Therefore, by connecting two flip-flops in series as a two-stage counter, the flip-flops act as frequency dividers.

Figure 148 shows this action for an input frequency of 1,000 Hz to the trigger input of FF_1 on

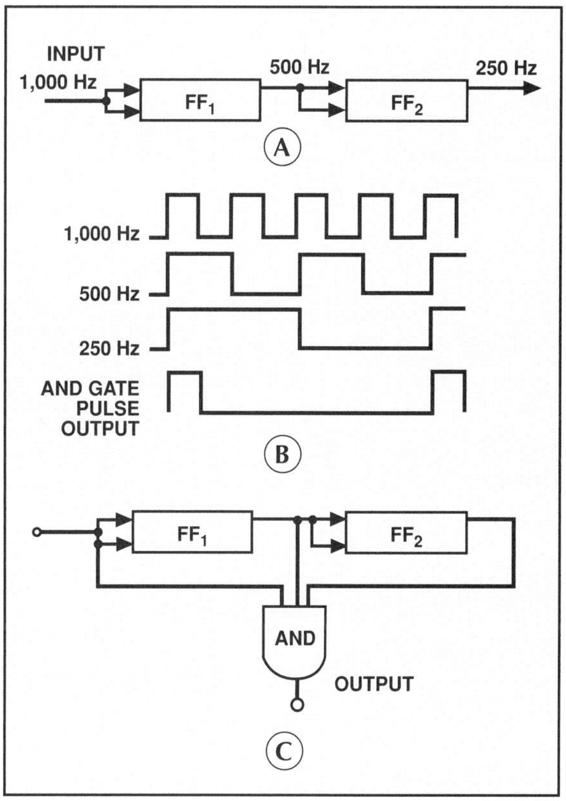

Figure 148 Frequency dividing flip-flops

the left side of the diagram. The input signal is applied to both the set and reset inputs, causing simple complementary switching action and producing one-half of the frequency at the output of FF_1 of 500 Hz. The 500 Hz is in turn applied to the set-reset input of FF_2 to produce 250 Hz at its output.

Further division of frequency can be achieved by cascading other flip-flops in series. Also, another technique is to use an AND circuit with three inputs to give a pulse at the leading edge of every fifth pulse as shown at the bottom of figure 148.

Supplementary Circuits

Other circuits can be developed using flip-flop principles as the basic building block. Such circuits include latching circuits, timing pulse generators, differentiating circuits, and binary counters.

Latching Circuits

A special type of bistable multivibrator (flip-flop circuit) is a *latch circuit* (fig. 149). A latch circuit is the equivalent of a pair of OR and NOT circuits connected in series. The output of the NOT circuits is cross-connected back into the OR circuits for the purpose of latching in the OR circuit once a signal has been applied to the OR circuit input.

This action is similar to the way an auxiliary switch on a magnetic contactor keeps the contactor closed even after the start push button has been released. Once a positive signal is received at one of the input terminals of the latch circuit, a signal appears at the corresponding output terminal and remains there even though the input pulse is no longer present. At this point, the latch circuit is in one of its stable states, which latches, or traps, the information of the inputs *A* or *B* until it is removed.

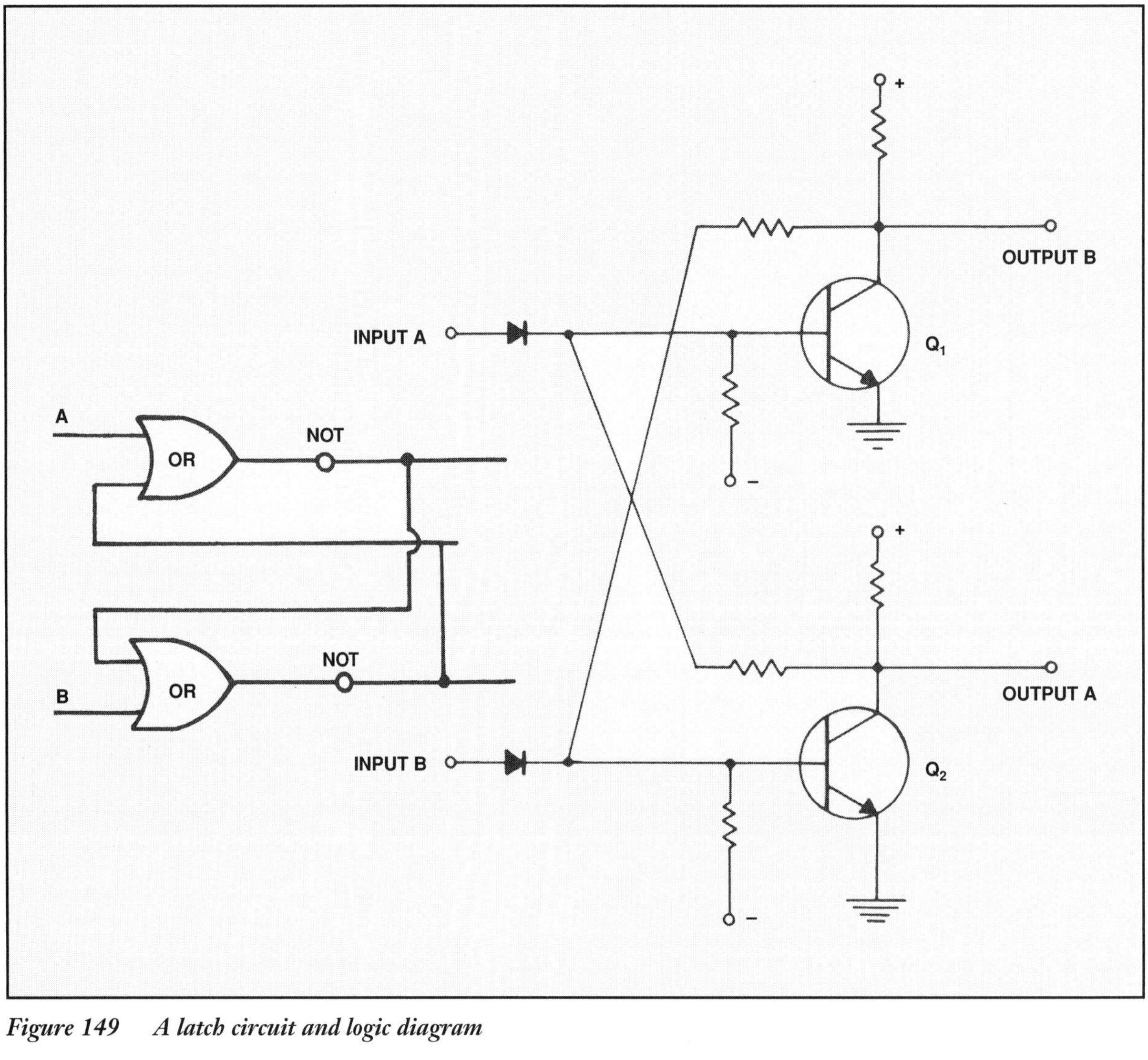

Figure 149 A latch circuit and logic diagram

To get an output signal from the other terminal, the other input terminal must receive a pulse. The output terminals then reverse their conditions: the terminal that was on becomes off and the terminal that was off becomes on.

When a trigger pulse is received at input *A* of figure 149, a high signal is placed on the base of Q_1. Q_1 is a NOT gate and it inverts the input to produce a low signal at output *B*. The zero signal from Q_1 is also fed back to the base of Q_2. Because the latch circuit is triggered with short pulses placed alternately on *A* and *B* inputs, a low signal exists on input *B* at this instant. The low signal on input *B* results in Q_2 not conducting and a high output appears at output *A*. The high-level output from Q_2 is also fed back to the input of Q_1 to latch Q_1 in conduction after the trigger pulse on input *A* has expired. A high signal now exists on output *A* and a low signal on output *B*. Q_1 is latched in conduction and Q_2 is off.

When a trigger pulse is received at input *B*, a high input to Q_2 causes it to conduct and develop a low signal at its output. This action removes the feedback latch from Q_1. Q_1 then develops a high output, which appears at output *B*, and is also sent to the base of Q_2, which latches it into conduction. At this point, a high signal is at output *B* and a low signal is at output *A*. Q_2 is conducting and Q_1 is off. The circuit remains in this state until input *A* is triggered again.

One of the principal uses for the latch circuit is to turn a transmitter or a receiver on and off with precise timing. It can also be used to turn a transmitter on and a transmitter off at the same time.

Timing Pulse Generators

The operation of supervisory control systems and computer systems always involves time. A stable *master oscillator* is used to generate timing pulses. (Recall that an oscillator produces high frequencies without the need for rotational forces. Also, oscillator voltages produce output current that periodically reverses, or oscillates. A master oscillator is an oscillator that establishes the carrier frequency of an amplifier or transmitter's output.) In supervisory control equipment, the master oscillator is usually a carefully designed LC type or crystal-controlled type. The frequency of the oscillator is chosen so that the desired bit frequency to be transmitted is obtained after a certain number of stages of frequency division. In some cases, a means of synchronizing the master oscillator to incoming signals is provided. Because the output of the LC-type oscillator is sinusoidal in nature and because square or rectangular waveforms should be used in the system, a means of converting the sine wave into square waves is needed.

One way to convert the sine wave into square waves is by *clipping* the sine wave several times. (A *clipper*, or limiter, is a circuit that prevents the amplitude of a waveform from exceeding a specified level while preserving the shape of the waveform. Thus, to *clip* the sine wave means to keep it from exceeding a specified level by squaring it off.) A more economical method of transforming a sine wave into a square wave is to use a Schmitt trigger circuit (fig. 150). A Schmitt trigger circuit is a bistable pulse generator in which an output pulse of constant amplitude exists only as long as the input voltage exceeds a certain value.

The Schmitt trigger belongs to the flip-flop, or bistable multivibrator, family of circuits. It is a regenerative circuit because of the common emitter resistor, R, and it changes states very abruptly when the input signal crosses specific DC threshold levels. The input waveforms in figure 150 start at the 0-volt reference line. The output, however, is at a +10-volt level. When the input signal begins to go negative and reaches a critical point on the downward slope, Q_1 snaps out of conduction and Q_2 is forced into full conduction. When the negative half of the input signal returns to zero and begins to go up the positive slope, a critical point is reached again, and Q_1 switches into full conduction. Q_2 is turned off. The output waveform at the collector output of Q_2 is a square wave. Circuits of this type are commercially available as an integrated circuit chip, complete with inputs and outputs.

Differentiating Circuits

A differentiating circuit takes its name from the mathematical operation of differentiation. (Differentiation involves a branch of mathematics called calculus, which is beyond the scope of this manual.) In any case, a differentiating circuit responds to the rate of change with respect to time of the applied voltage. A differentiating circuit is a simple *resistance-capacitance (RC) circuit*, which is an assembly of components that consist of a resistor and capacitor as shown in the middle drawing of figure 151.

An RC circuit has a short time constant because the value of the capacitor is small and the capacitor

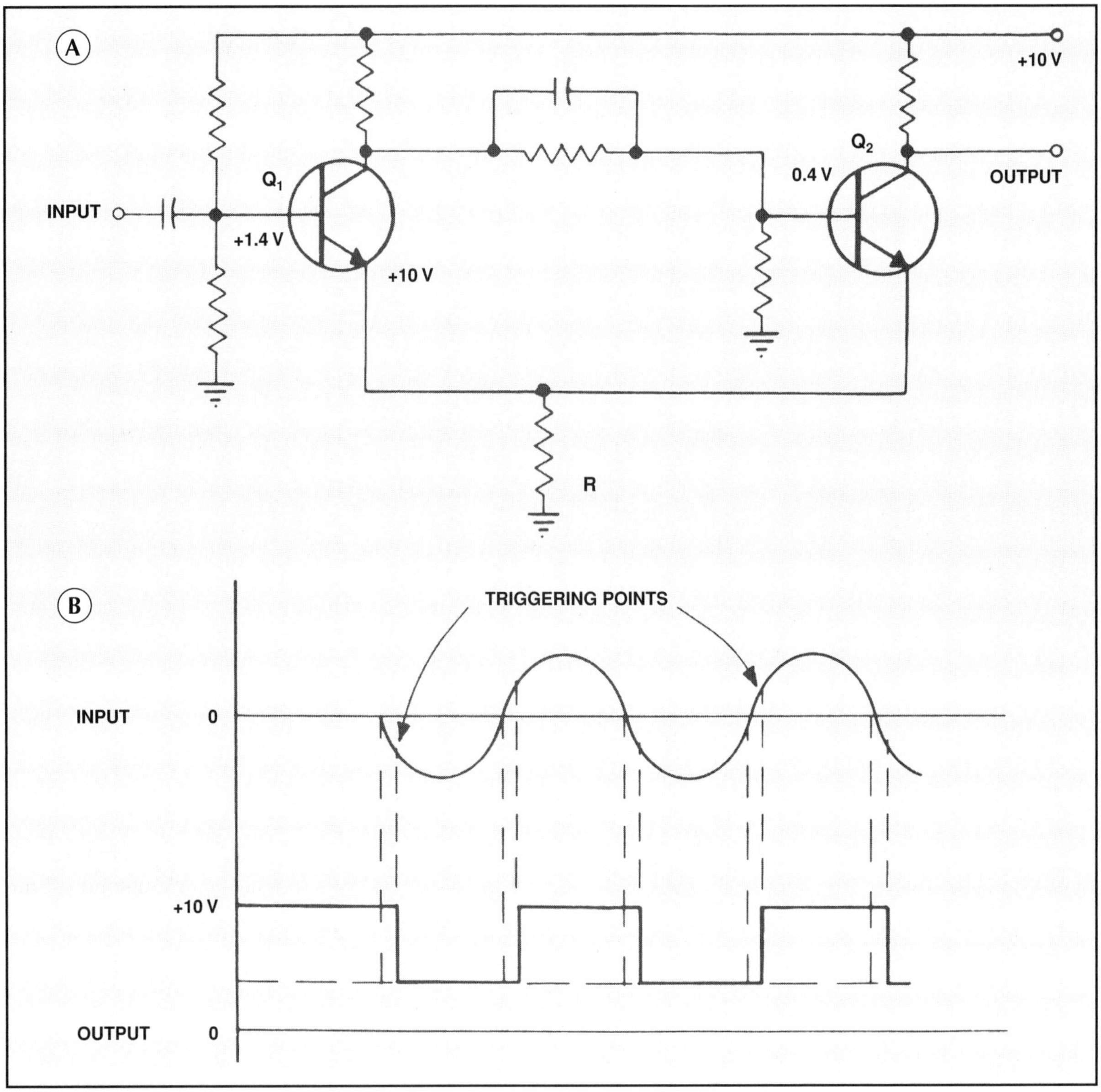

Figure 150 Schmidt trigger circuit and waveform

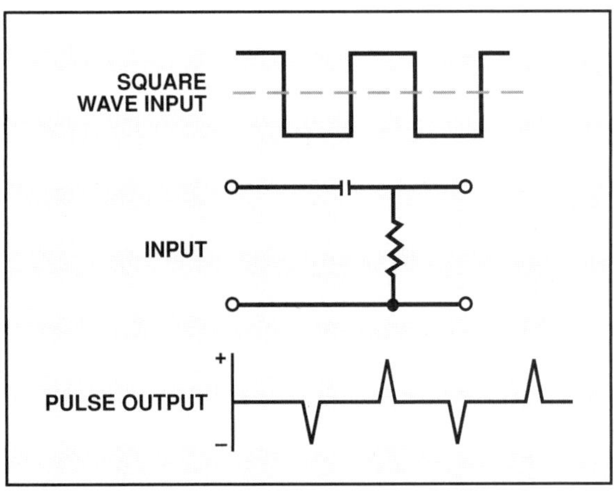

Figure 151 Differentiating circuit and waveform

is able to charge to full applied voltage very quickly. As the square wave input voltage is applied to the circuit's input, the output pulses are created. The output voltage is taken from across the resistor and pulses of voltage are created from the charging and discharging currents of the capacitor.

Output pulses occur at the leading edge and trailing edge of the square wave input. They occur because the rate of change with respect to time is maximum at these two points. The rate of change during the horizontal part of the square wave is zero, resulting in no output. Figure 152 shows that either the leading edge pulse or the trailing edge pulse can be selected for use by means of steering diodes.

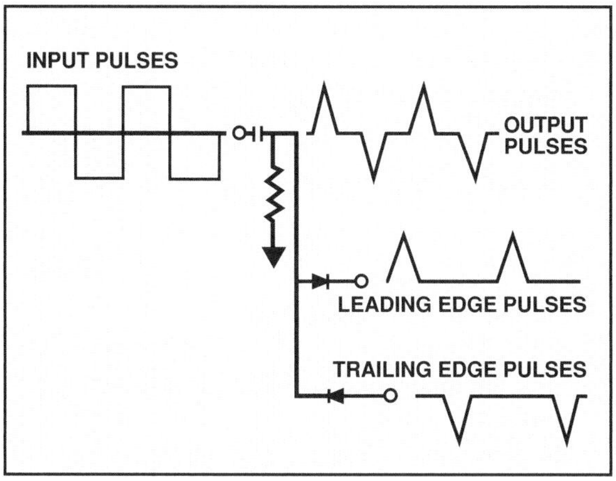

Figure 152 Diode pulse selection technique

LOGIC COUNTER CIRCUITS

Counting pulses and adding binary numbers, and converting to other numbering systems can be accomplished with variations on flip-flop circuits.

Binary Counters

Flip-flops can be arranged so that they can count a sequence of pulses. Figure 153 shows four flip-flops connected to form a binary counter capable of counting to 1111, the decimal equivalent of which is 15. A pulse applied to the reset line returns all flip-flops to zero at their set outputs. The set output of each flip-flop feeds the trigger input of the succeeding flip-flop. The device makes the binary count by noting the state of each set output. When the count pulse goes to its 0 state from a 1 state, a count is made by that particular flip-flop. Flip-flop 1 (FF_1) is the least significant digit, FF_2 is the next significant digit, followed by FF_4 and then FF_8 as the most significant digit in the binary number.

If the set outputs are all at the 0 state and a pulse is applied at *T* of FF_1, FF_1 goes to its 1 state. The pulse does not trigger FF_2, however, because it has the wrong polarity. The binary count at this time is 0001. The second pulse returns FF_1 to its 0 state, and the correct polarity of the pulse triggers FF_2, causing it to go to its 1 state. The binary count is now 0010 or decimal 2.

The third pulse causes FF_1 to go to its 1 state but the output pulse from FF_1 again does not trigger FF_2 because of its polarity. The binary count is now 0011 or decimal 3. The fourth pulse triggers FF_1 and FF_2 back to their 0 states and also triggers FF_4 to its 1 state, making the binary count 0100, or decimal 4. Succeeding pulses trigger the counter until the highest number (1111, or decimal 15) is reached. The sixteenth pulse returns the counter to 0000.

Any number of flip-flops can be connected to increase the counting capability to any desired figure. Counters having thirty-two flip-flops are common and are commercially available on integrated circuit chips.

Decade Counters

Occasionally, a need exists for a counter that can be reset on the tenth pulse. Such a counter operates on binary principles but produces an equivalent decimal output. Data processing equipment commonly use this method. A binary system can be converted to a decade counter by several easy and effective methods.

One method of causing the counter to reset on the tenth pulse involves adding OR and AND gates to the basic flip-flop counter in figure 154. The output of the AND gate is applied to reset inputs for FF_2 and FF_4. The OR gate gets its two inputs from the reset outputs of FF_2 and FF_4, and its output is

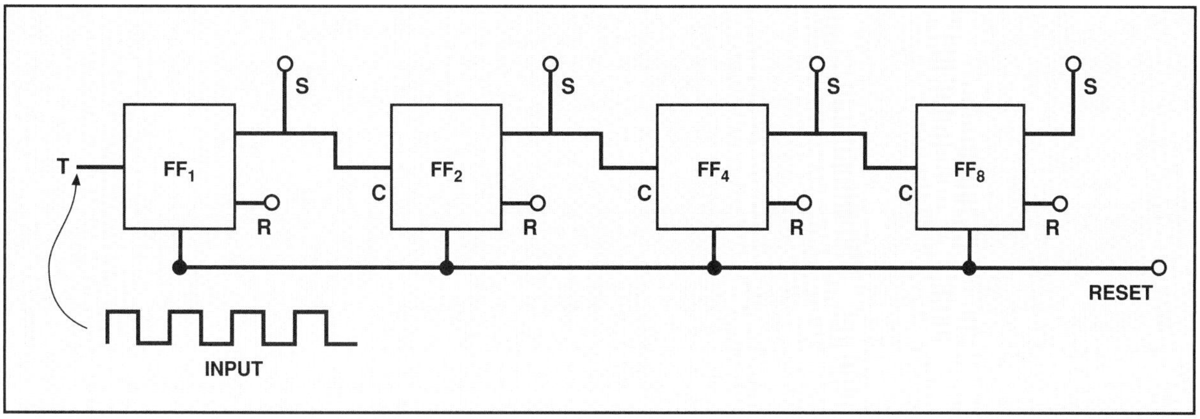

Figure 153 Binary counter using flip-flops and gates

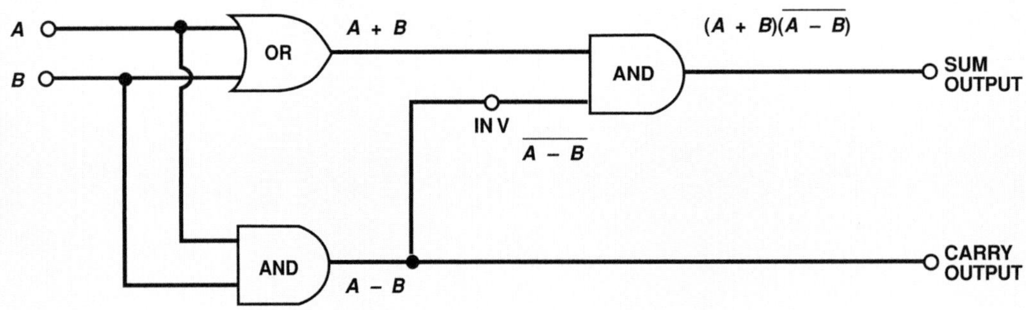

A	B	SUM	CARRY
0	0	0	0
1	0	1	0
0	1	1	0
1	1	0	1

Figure 154 Logic gates forming a half adder

applied to one of the AND gate inputs. The other input for the AND gate is the reset output of FF_8.

Remember that an OR gate produces a logic 1 if either or both of its inputs are pulsed; but, the AND gate must have both inputs pulsed to produce a 1 at its output. When the counter has reached a decimal 9, the binary count is 1001. The tenth pulse restores FF_1 to its 0 state and pulses FF_2 to its 1 state. Without the presence of the AND and OR gates, the action stops because FF_4 does not respond.

The tenth pulse causes FF_2 to apply a 1 to input *B* of the OR gate. This input of 1 triggers the OR gate to produce a 1 at its output. Input *B* of the AND gate already has a 1, so now the output of the OR gate feeds a 1 to input *A*. The AND gate sends a reset impulse to FF_2 and FF_4. This reset pulse does not directly affect FF_4 because it is already at its 0 state. However, the resetting of FF_2 to 0 momentarily triggers FF_4 to its 1 state. Almost instantaneously, the OR gate is able to send another 1 to input *A* of the AND gate, while input *B* of the AND gate is still being pulsed by FF_8. A reset pulse is once again sent to FF_2 and FF_4 by the AND gate. This time, FF_2 is unaffected because it is in its 0 state; but, FF_4 loses its momentary 1 state and returns to 0 as does FF_8. The count now reads 0000.

Another decade counter can be added to the unit in figure 154. The output from FF_8 can be used to trigger this second decade counter with every tenth pulse. By adding additional counters in this manner, a system is achieved that can be quickly interpreted in terms of its decimal numbers.

Binary Adders

Full adders and *half adders* are logic devices that can accurately add binary numbers. Remember that when adding binary numbers, 1 added to 1 results in a sum of 0 for that column with a 1 to carry to the next higher column. A term for this operation is to carry out. A half adder provides this carry-out function; however, it cannot accept a carry-in function from a lower column. A full adder can accept a carry-in 1 and can produce a carry-out 1. (A carry-in is the transfer of the number carried out. In decimal numbers, for example, when the addition of a column of numbers in a row exceeds 9, such as 13, 1 is written down and 3 is carried out to the next higher column of numbers. The transfer of the 3 is carried in to the next column.)

The logic circuitry for a half adder is composed of an OR gate, an AND gate, and an inhibited AND gate (fig. 154). An inverter associated with the second AND gate causes the gate to be called an inhibited AND gate.

Figure 155 shows a full adder and a truth table for it. The full adder is composed of two half adders plus an OR gate. A full adder can accept a carry-in bit and for producing a carry-out bit.

Figure 156 shows a half adder and three full adders arranged to add binary numbers. Two numbers are added in this figure: the decimal numbers 12

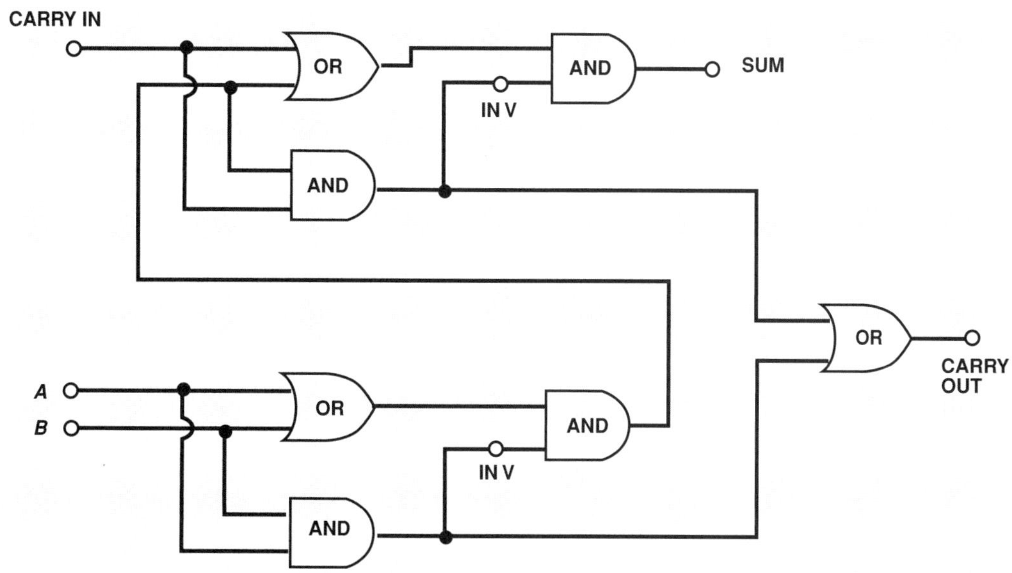

TRUTH TABLE

A	B	C IN	SUM	C OUT
0	0	0	0	0
0	1	0	1	0
1	0	0	1	0
1	1	0	0	1
0	0	1	1	0
0	1	1	0	1
1	0	1	0	1
1	1	1	1	1

Figure 155 A full adder gate arrangement

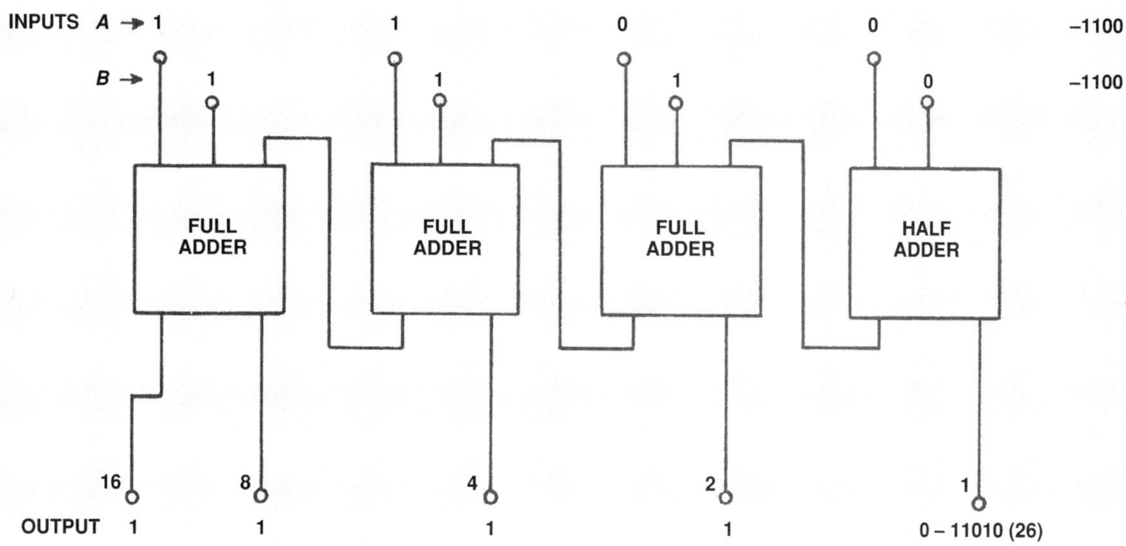

Figure 156 An adder used to add 4-digit binary numbers

and 14, the binary equivalent of which are 1100 and 1110. The resulting addition produces the binary number 11010, the decimal equivalent of which is 26.

ENCODERS AND DECODERS

To convert conventional decimal-number information to binary form for processing and binary numbers to decimal numbers, *encoders* and *decoders* are required. An encoder converts decimal numbers to binary numbers; a decoder converts binary numbers to decimal numbers.

Figure 157 is a schematic of an encoder. The decimal digits 0 through 9 on the right side of the figure at the end of the horizontal lines can be thought of as punch keys. The keys operate switches that send pulses along the appropriate horizontal lines. Diodes connect some of the vertical lines to horizontal lines. The diodes steer the pulse signals to the proper flip-flops (FFs) and prevent them from reaching the input of the wrong FF. This arrangement is a *matrix*, which is a mathematical term for a rectangular arrangement of elements or numbers formed by columns and rows.

Punching the key on the calculator or computer for the number 9 produces a pulse, which goes to the inputs for FF_1 and FF_8. These two flip-flops

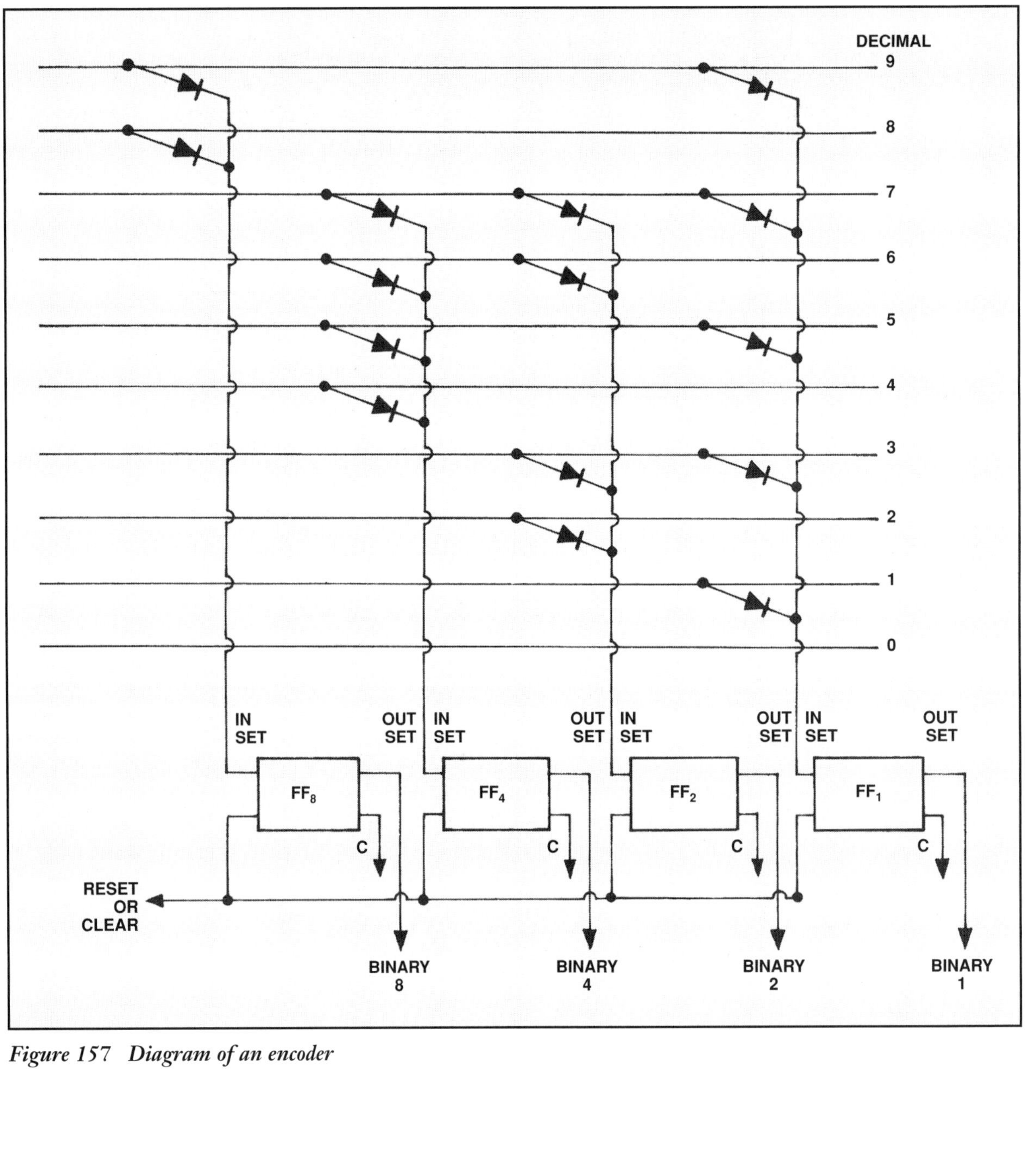

Figure 157 Diagram of an encoder

produce the binary number 1001 or decimal number 9. The encoder's FFs have the appropriate inputs pulsed for each decimal digit.

A bank of four FFs can handle only one digit in one column. For the next decimal digit that appears in the next higher column, another bank of four FFs must be used. Banks of four FFs can be added to achieve the required number of decimal columns. For example, a decimal number of twelve digits requires an encoder with 48 FFs and 45 steering diodes. Such encoders can be produced at reasonable cost using integrated circuits.

As previously mentioned, decoders perform the opposite function of encoders; they convert binary numbers to decimal numbers. Figure 158 shows a decoder. It receives its input from both the regular

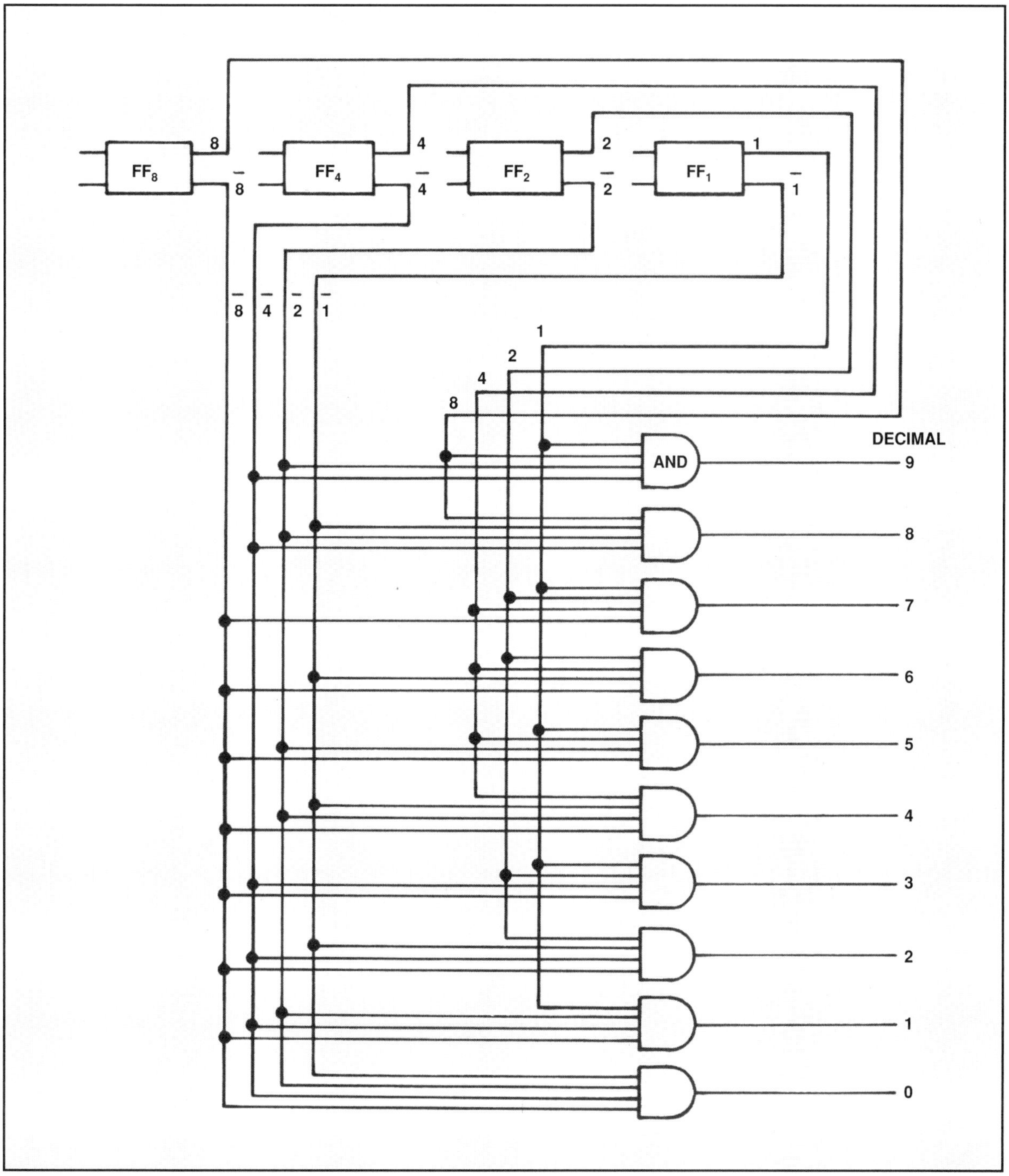

Figure 158 A decoder for convening binary numbers to decimal digits

and complementary outputs of a group of four FFs. In figure 158, bars over the numbers indicate the complementary outputs.

The decoder circuit consists mainly of four input AND gates. One gate is required for each decimal digit of 0 through 9; so, 10 AND gates are needed. The outputs of the AND gates provide the decimal signal. The logic of the AND gate requires that all inputs must be at 1 for a 1 output. If the FF bank is storing a binary 9, or 1001, the regular outputs for FF_1 and FF_8 and the complementary outputs for FF_2 and FF_4 are 1s.

With this information in mind, the circuit can be traced to show that a 1 exists at each of the four inputs to the AND gate for decimal digit 9. Tracing the circuit for other possibilities shows that the output for the 0 digit AND gate is a 1 only when all regular outputs of the FFs are 0, making all complementary outputs 1. As with the encoder, an additional encoder is needed for each column of decimal digits, but integrated circuits make these complex devices easy to produce.

STORAGE OF BINARY INFORMATION

The storage, or memorizing, of logic 1s and 0s for the purpose of retaining instructions, data, or other needs in a digital system can be achieved in several ways. The flip-flop can be considered to be a storage device since it can retain a 1 or 0 indefinitely unless power is lost to the circuit. In the past, magnetic devices, such as magnetic core memories, stored data. These memory systems relied on the magnetic saturation of a core. Data could be stored for an unlimited amount of time as long as power was applied. Unfortunately, if power was removed from the core, loss of memory occurred. Because of this characteristic, magnetic core memories were called destructive or volatile memory devices.

Other types of magnetic memory devices include magnetic-drum and magnetic-tape machines. These devices store data permanently since removal of electrical power does not change the stored bits in their memory. However, modern computing systems use magnetic disks and electronic devices.

Electronic Memory

Erasable programmable read-only memories (*EPROMs*) are integrated circuits with inputs in which the manufacturer programs the circuit with permanent information. Once programmed, these devices are not reusable for other data storage and can only be accessed for their information. In other words, they are *read-only memory* (*ROM*) devices.

Electronic memory that allows a user to program it, or write to it, and then access (read from) it, erase it, and reprogram it is *random-access memory* (*RAM*). Memory devices that can be programmed and erased with electrical pulses are further called electrically erasable programmable read-only memory (EEPROM).

Permanent, or nonvolatile, memory that continues to hold its memory when power to it is turned off includes switches, magnetic and photo disks, tapes, and memory chips. In some cases, a small backup battery can make volatile memory nonvolatile by maintaining power to it when power is turned off. Users must replace the battery periodically—perhaps every 1 to 5 years. Otherwise, memory will be lost when the power goes off and the battery is dead.

REVIEW QUESTIONS

1. Convert the decimal numbers 15 and 33 to binary numbers and add them together.
2. What is the logic output of a NOT gate if the logic input is a 0?
3. What is the logic output of a three-input AND gate if the logic inputs are 1, 1, 0?
4. Describe an application of a Schmitt trigger circuit.

12 Electrical Test Instruments

A large part of the world's population is served by a group of persons who perform special services related to the repair and maintenance of machines. These persons are mechanics, electricians, and technicians who use special instruments to locate points of trouble. And, once they determine the problem, they often use special tools to repair malfunctioning equipment and machines. For example, auto mechanics use special tools such as pressure, vacuum, temperature, and feeler gauges to measure and adjust engines. Mechanics also use a variety of electronic instruments when dealing with automobiles. Special scopes allow them to study the performance of ignition systems, *dwell meters* allow them to set timing (on older cars), and rpm meters (tachometers) allow them to set engine idle speeds. They also use voltmeters, ammeters, and ohmmeters to locate trouble in electrical systems and to help them make engine adjustments.

However, electronics technicians are likely to use the largest array of test and measuring instruments. They are engaged in the maintenance and repair of industrial instrumentation systems that employ electronic controls and they must know about several testing, measuring, and calibrating devices.

An instrumentation technician or mechanic not only uses ordinary pressure gauges and mechanical measuring devices, but also many test instruments that measure and observe the various aspects of electricity. These aspects include voltage, current, resistance, frequency, and wave shape.

D'ARSONVAL MOVEMENTS

An electricity detection device that has been in existence for over 200 years is the electroscope (fig. 159). An electroscope consists of a metal rod with a ball on the end of it, a strip of gold leaf, a metal cylinder, and a rubber stopper. When the rod and leaf take on the same electric charge—either positive or negative—the like charges drive the free end of the thin and limber leaf away from the rod, just as the same poles of a magnet repel each other.

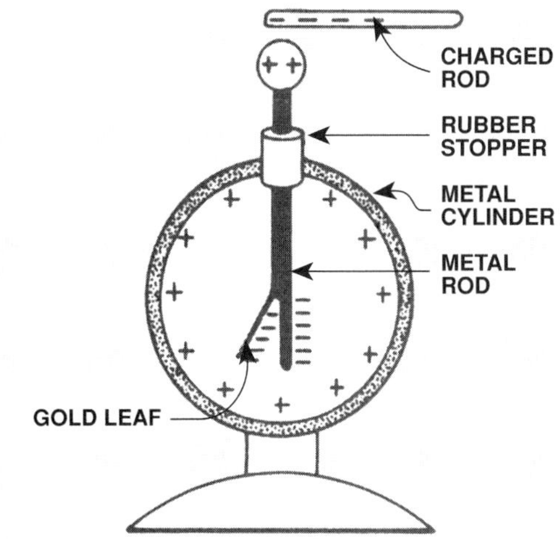

Figure 159 Simple electroscope

In figure 159, a negatively-charged rod is held near the metal ball. Electrons are repelled to the lower end of the metal rod and onto the gold leaf. Electrons in the rod and leaf repel one another, pushing the leaf away from the rod. As scientists such as Volta, Galvani, and Coulomb, all of whom worked in the early 1800s, recognized, the leaf's movement confirms the presence of electricity. Later, other scientists recognized the close relationship between electricity and magnetism and began using electric current produced by chemical and mechanical means. From these beginnings, electrical researchers rapidly developed measuring and testing devices using magnetic fields.

Most modern testing and measuring instruments use a meter. Several kinds of meters are available, but most have a scale of values printed on a background that makes them easy to read. A pointer, when energized by the meter's leads being attached to the circuit or component to be tested, moves to some point on the scale. A user then reads the value opposite the pointer's tip. Meters display values of voltage, current, resistance, and other variables, such as rpm and *dwell angle*.

Most meters use a mechanism called a *d'Arsonval movement*. A meter with a d'Arsonval movement forms the basis of most electrical measuring instruments. These instruments include voltmeters, ammeters, and ohmmeters, which are electrical measuring instruments. Other meters include dwell angle meters, rpm meters, temperature indicators, pH meters, and light meters.

A d'Arsonval movement (fig. 160) consists of two major components: (1) a powerful permanent magnet and (2) a moving coil assembly that drives a pointer. The pointer sweeps across a dial that can be marked with numerical values expressing current, voltage, resistance, or whatever is appropriate to the application of the meter. The movement is named after one of its inventors, Jacque-Arsène d'Arsonval (pronounced dar-son-val), a French physiologist who, in the mid-1800s, studied electrical effects on living organisms.

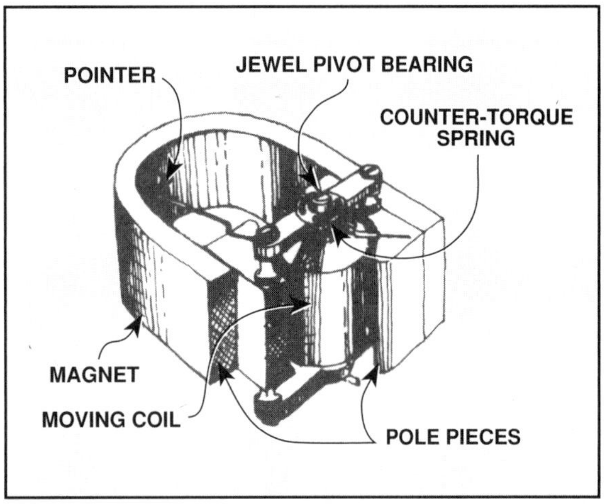

Figure 160 Example of a d'Arsonval meter movement

The permanent magnet assembly not only includes the magnet, but also pole pieces shaped to surround the moving coil. A cylindrical core of magnetic material fits within the coil. Besides the pointer, the moving coil assembly contains spiral *hairsprings* and jeweled bearings at either end. The springs create torque to resist motion of the coil and conduct current to the coil.

With direct current flowing through the coil in the proper direction, the magnetic field of the permanent magnet and the magnetic field set up by the current-carrying coil attempt to align. The hairsprings resist the torque caused by the interaction of the two magnetic fields. The respective torques of the spring and magnetic fields are balanced within the range of the instrument, and the pointer indicates a numerical quantity on the scale.

Generally, the coil in a d'Arsonval meter movement has fairly low resistance—anywhere from a few to hundreds of ohms. The typical movement without shunt or series resistances measures current in the milliampere or microampere range. Through the use of shunt and series resistances, the movement can be adapted to measure virtually all practical ranges of current and voltage. While the d'Arsonval movement is not the only meter movement—more modern ones are available—it will likely be used for years to come.

Analog Voltmeters

A meter using a d'Arsonval movement can easily be adapted to measure any reasonable range of voltage by adding series resistances to the meter circuit. As previously noted, current flowing in the meter movement's coil causes it to turn the pointer. To bring about current flow, it is necessary to apply voltage to the coil. Typically, the voltage appearing across the coil is quite small. Because the coil's resistance is low, the field strength of the permanent magnet is sizable. Further, the friction of the system is negligible, so only a small amount of voltage produces enough current and torque to overcome the spring tension.

For any d'Arsonval movement, a specific value of current causes full-scale deflection of the pointer. Many modern meter movements are capable of full-scale deflection with a current flow of 25 to 100 microamperes. Such movements work well for voltmeters because their sensitivity avoids excessive loading of the circuit being tested. A load on a high-resistance circuit causes a voltage drop proportional to the current being drawn from the circuit. The voltage drop not only results in a false voltage reading, but also can interfere with the operation of the circuit being tested.

Meters augmented with solid-state amplifier circuits achieve sensitivity ranges that extend to very high values—typically 10- or 11-megohm input resistance, regardless of scale. These meters are good choices for measuring sensitive circuits that are easily upset by extraneous loading.

The circuit for a simple single-scale voltmeter consists of a meter movement and a series resistance (fig.161). The meter can be made more versatile by adding a combination of several resistances and a

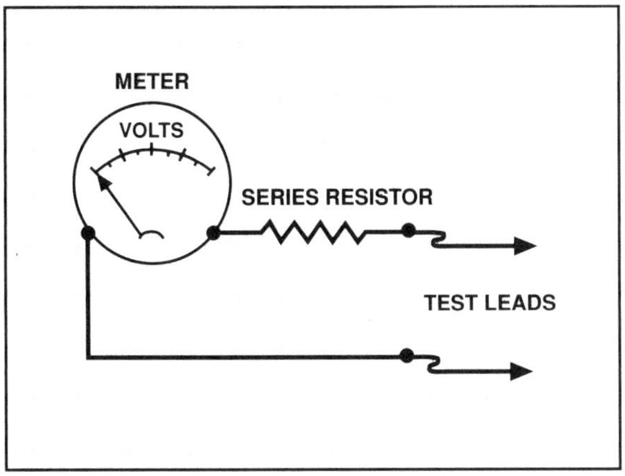

Figure 161 A simple single-scale voltmeter

range switch (fig. 162). The range switch introduces series resistances into the circuit, beginning with a minimum value for the lowest scale and proportionately higher values for each succeeding scale. The number of scales on a voltmeter and their arrangement is dictated by the service for which the meter is intended. The schematic of the voltmeter in figure 162 also has a polarity reversing switch, which is used when testing motor circuits. It allows the correct polarity to be selected to match that of the motor.

A voltmeter used for automotive electrical service usually has two or three scales, including 3- and 16-volt scales, because these ranges are the most useful for testing automobile circuits. On the other hand, a voltmeter intended for testing solid-state electronics usually has scales for 0.3, 1, 3, 10, 100, and 300 volts. It also has a polarity reversing switch so the polarity of the test leads can be easily changed (see fig. 162).

Direct-Current Analog Ammeters

A meter movement used only in an ammeter does not need high sensitivity. Typically, only about 1 milliampere of current through the coil circuit is needed to deflect the pointer to the highest value on the meter's scale. Resistance of the coil circuit is about 50 ohms. Shunting the coil circuit with a 50-ohm resistor enables the meter to indicate 2 milliamperes for full scale, because the current divides equally between the meter and shunt (fig. 163).

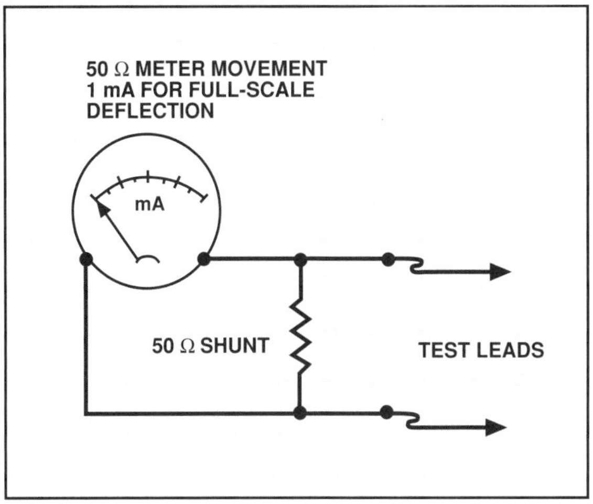

Figure 163 Analog current meter circuit

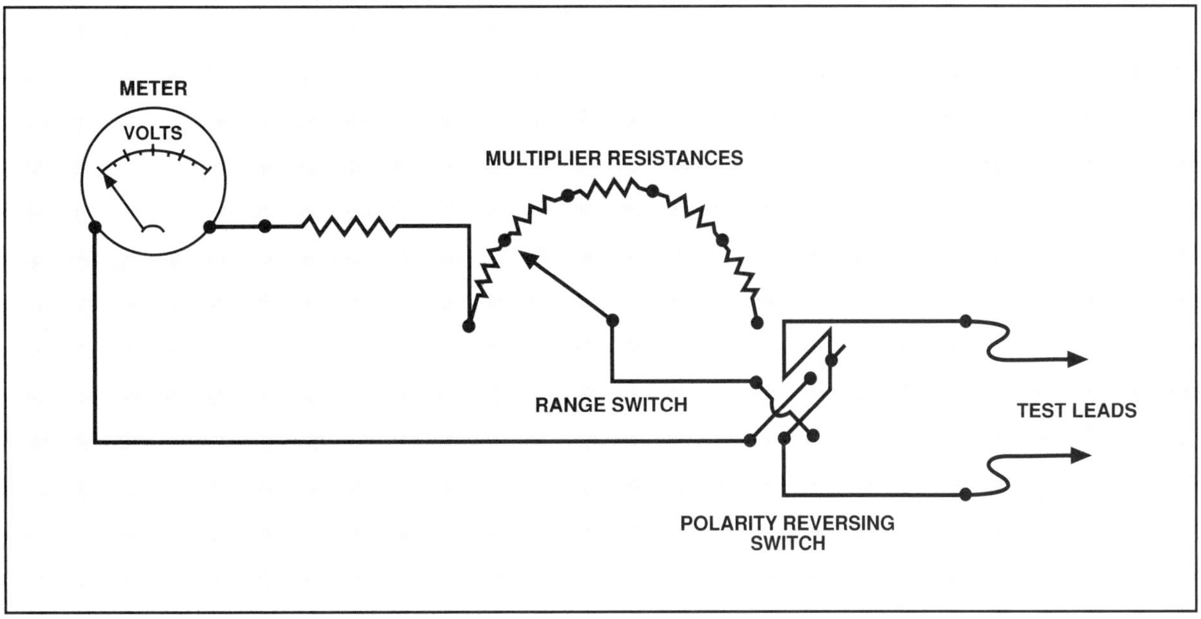

Figure 162 An analog voltmeter with selector switch for choosing any of five voltage ranges

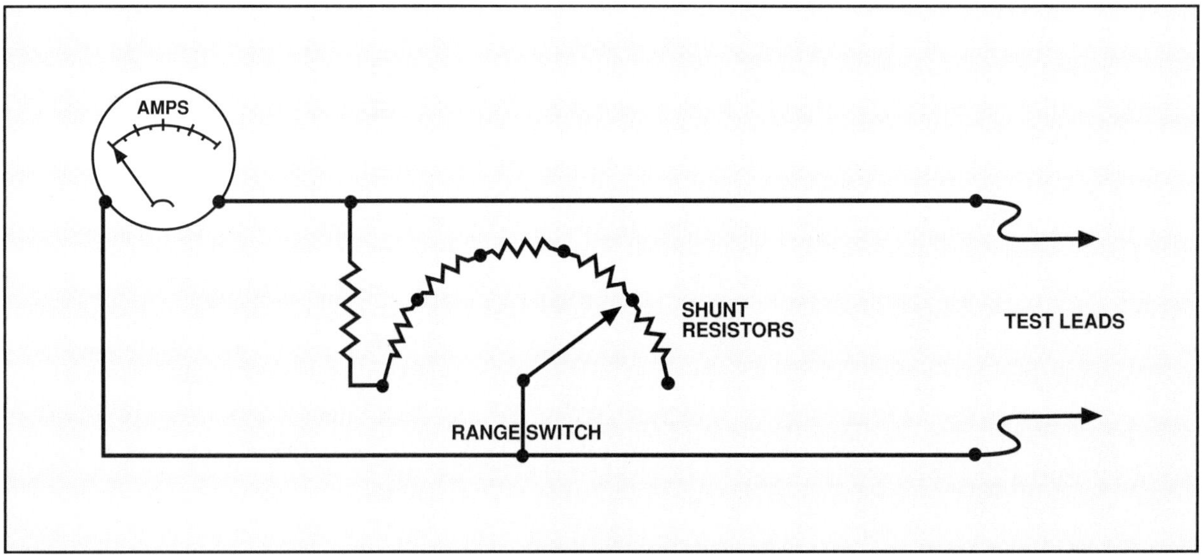

Figure 164 An ammeter arrangement with five ranges

Using a shunt resistance of 0.005 ohm enables the meter to indicate 10 amperes full scale or, to be more exact, 10.001 amperes. The current flow through the meter and shunt is inversely proportional to their resistance values. The following equations show this relationship.

$$\frac{I_s}{I_m} = \frac{R_m}{R_s} \qquad \text{(Eq. 50)}$$

where

I_s = shunt current, amperes
I_m = meter current, amperes
R_m = meter resistance, ohms
R_s = shunt resistance, ohms.

$$I_xR_s = I_mR_m \qquad \text{(Eq. 51)}$$

where

I_s = shunt current, amperes
I_m = meter current, amperes
R_m = meter resistance, ohms
R_s = shunt resistance, ohms.

In the circuit shown in figure 163, 2 milliamperes must flow in the test leads to obtain full-scale deflection of the 1-milliampere meter movement.

Panel or switchboard ammeters generally have a single scale, although exceptions exist. Portable ammeters typically have several scales, each of which can be selected with a range switch (fig. 164).

Analog Ohmmeters

A sensitive direct-current meter can be used to measure resistance of a circuit or component. One simple type of ohmmeter uses a voltmeter and a battery of sufficient voltage to provide full-scale deflection when the meter leads are shorted together (fig. 165). The *ohms-adjust potentiometer* is set to indicate full scale with the meter leads shorted. The scale can be calibrated to read in ohms or it can be calibrated to allow a curve of volts versus ohms to be drawn after numerous resistors of known value have been measured by placing them across the meter leads.

The values on the calibrated scale are very close together at the left end of the indicator's travel and relatively far apart at the right end of the scale. Such a scale means that a small change in resistance value is determined easily at the right end, but even sizable changes are difficult to detect at the left end. The switch (labeled S_W in fig. 165) is turned off when the meter is not being used to prevent the battery from draining. The switch is part of the potentiometer assembly.

Another more common type of ohmmeter measures resistance by placing the unknown resistance in parallel with one or more multiplier resistors (fig. 166). This arrangement is found in many *multimeters*, which is a meter capable of measuring volts, ohms, and milliamperes. Several multiplier resistors are connected in series and placed across a sensitive meter movement. One resistor is a variable resistor that serves as a zero-adjust control. Before measurements are made, the user shorts the meter leads together and moves the zero adjust until the pointer reads zero. The user then places the test

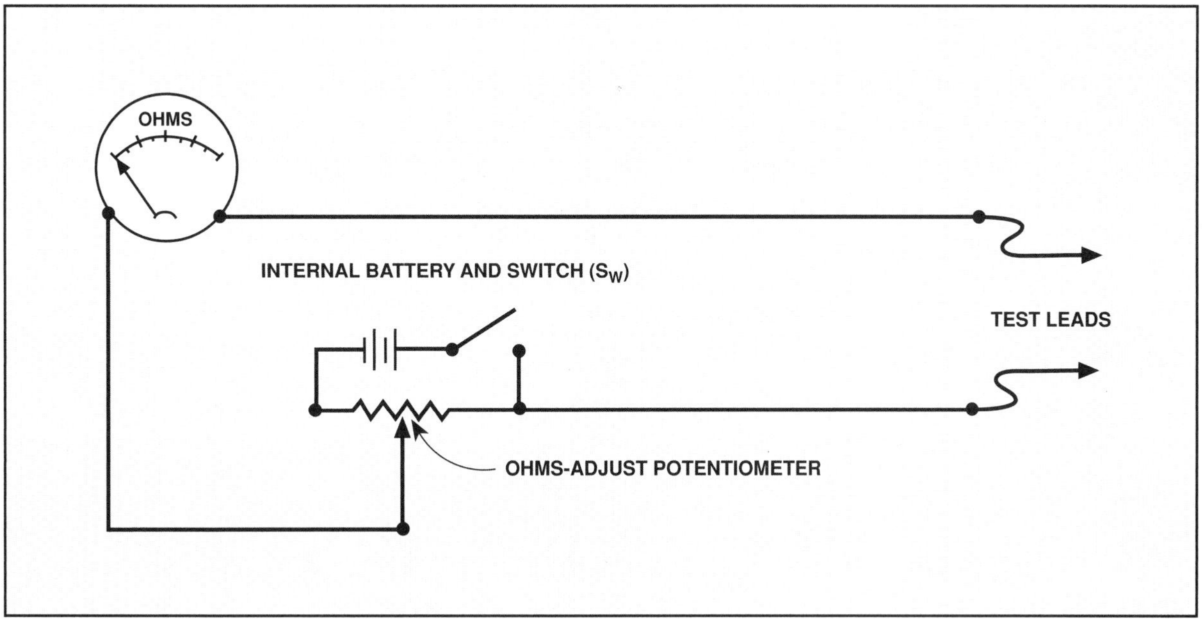

Figure 165 A simple form of ohmmeter

leads across the leads or terminals of the component being tested, and reads the resistance value on the scale. It is advisable for the user to select a range that causes the pointer to settle near mid-scale when the unknown resistance is measured. The operator turns switch S_w to its on position when the range selector is moved to any one of the ohmmeter positions, R_1, R_2, R_3, or R_4.

Measuring Alternating Current and Voltage

A voltmeter, direct-current ammeter, and ohmmeter cannot measure alternating current and voltage. Connect a meter of this type to a source of 60-Hz alternating current and the only result is a rapidly vibrating pointer near the zero point. However, installing solid-state rectifiers into a meter circuit enables its d'Arsonval movement to respond accurately

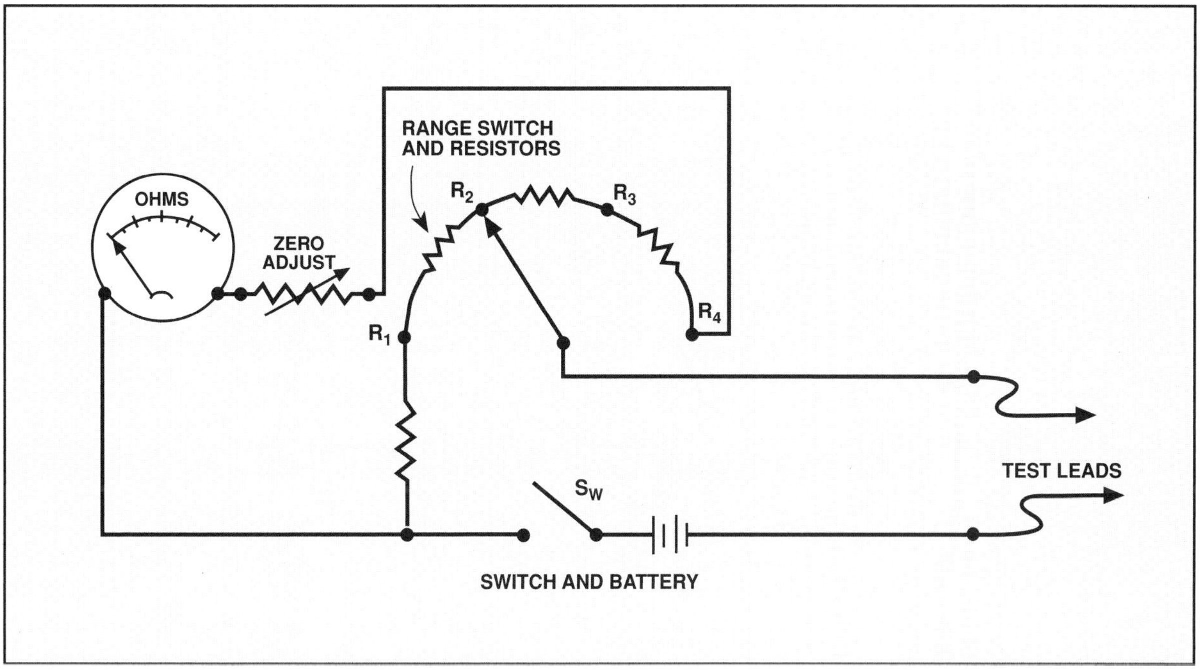

Figure 166 A multi-range ohmmeter

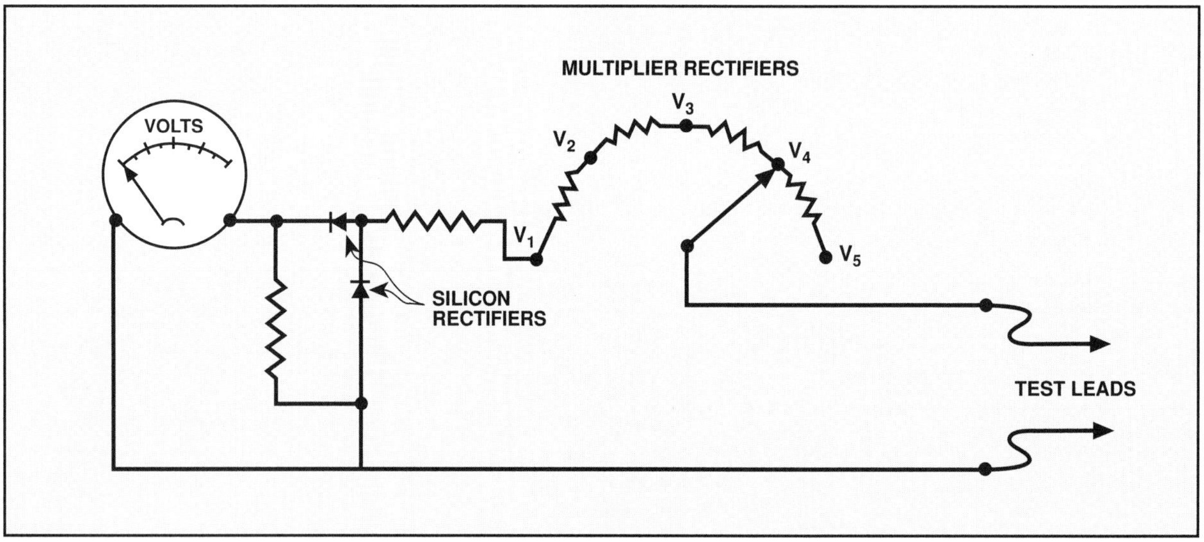

Figure 167 AC voltmeter with rectifiers and range switch

to alternating current and voltage. A meter circuit suitable for measuring AC voltage uses two silicon diodes to rectify the current before it is applied to the meter coil (fig. 167).

Figure 168 shows a sine wave produced by 60-Hz current that has passed through full-wave rectification. A typical meter cannot respond fully to the rapid pulsations. Proper design and calibration enable the meter to indicate the effective value of the voltage at about 0.707 of the peak value for a sine wave current. This factor, 0.707, is also known as the root-mean-square (RMS) value.

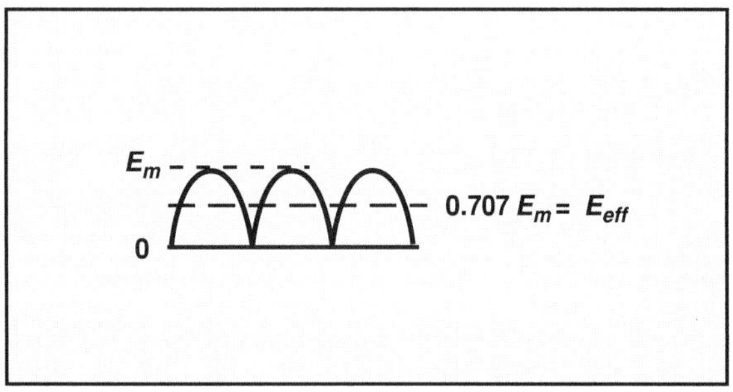

Figure 168 Sine wave after full-wave rectification

Figure 169 shows a sine wave that undergoes half-wave rectification. Half-wave rectification causes a typical DC voltmeter to indicate an average value of the half-wave pulses. The average value is about 0.32 of the peak value. Proper design and calibration produces a meter that indicates the effective value of the AC voltage.

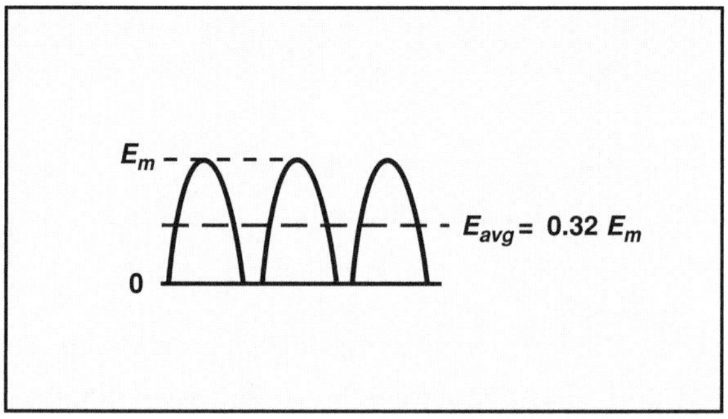

Figure 169 Sine wave after half-wave rectification

To measure alternating-current flow with a d'Arsonval meter movement, manufacturers also install rectifiers; however, the arrangement differs from that used to measure AC voltage. The setup in figure 170 includes a current transformer. The primary in a current transformer is a single turn of heavy conductor material. The secondary winding consists of 100 or more turns of comparatively fine wire. A voltage and corresponding current is induced in the secondary that is proportional to the current flowing in the primary. A full-wave rectifier changes the induced secondary AC voltage to DC. DC then flows to the meter coil. The range switch provides various ranges for the meter.

A special form of AC voltmeter-ammeter, called a *clamp-on meter*, uses the principles of operation shown in figure 170. It has a magnetic core in the form of hinged jaws that can be clamped around a current-carrying wire to measure AC voltage or amperage. Such meters offer great convenience,

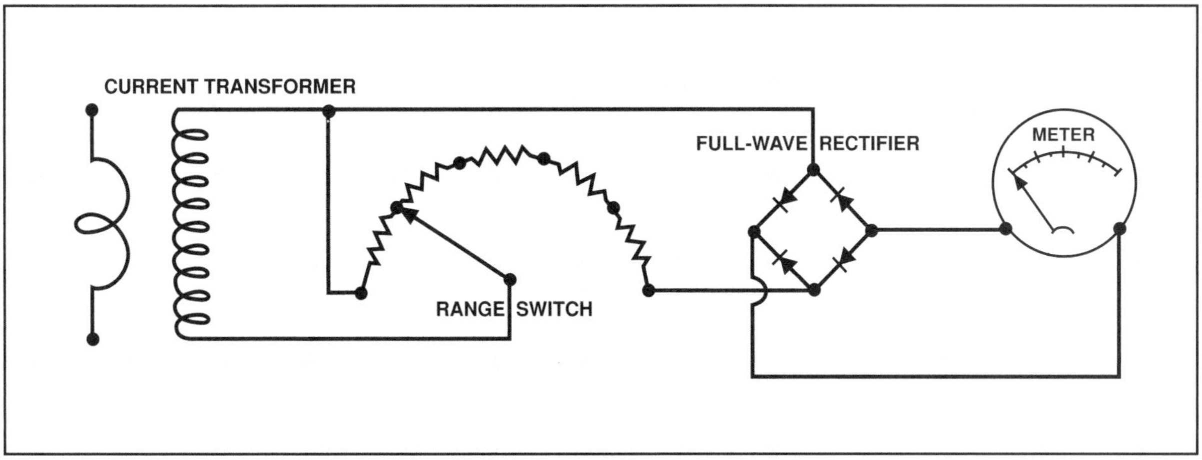

Figure 170 Arrangement for measuring alternating current

particularly when making measurements on power circuits, since they do not require that circuitry be disturbed for the installation of shunts. This fact also enhances the safety aspects of the meters.

MULTIMETERS

Multimeters, as the name suggests, combine the functions of several meters into one unit. These meters require that the user select the desired parameters to be measured with a selector switch and to choose the proper leads. Improper selection can cause injury or damage, so always exercise caution when using multimeters.

Analog Volt-Ohm-Milliammeters

The *volt-ohm-milliammeter* (VOM) combines the capabilities of the instruments discussed to this point. Most VOMs cannot measure alternating current, but they can measure AC and DC voltage, direct current, and resistance. Figure 171 shows a popular VOM. It has a number of scales and a selector switch for choosing the appropriate range and scale.

When using a VOM, it is important to choose the appropriate scale with the selector switch and to properly place the leads in the VOM before applying them to the circuit or component being tested (fig. 172). For example, if a user applies voltage to

Figure 171 Analog multimeter

Figure 172 Testing field-effect transistor (FET) circuits with a VOM

the test leads when the selector switch is set for measuring resistance, the voltage will likely blow a fuse in the instrument.

Digital Multimeters

Because digital readout displays and logic devices using integrated circuits are inexpensive and widely available, digital multimeters are now the industry's standard. The detailed circuitry of a typical digital meter is enormously complex; but, thanks to the modern technology of solid-state electronics that permits the combining of thousands of circuit components into a single chip, such meters can be manufactured at relatively low prices.

Lightweight digital readout multimeters that use either internal batteries or 60-Hz 120 VAC as a source of power, are widely available. Figure 173 is a photo of a typical digital readout multimeter. Figure 174 shows the external parts of the meter. This instrument can measure AC and DC voltage, ohms, and AC and DC current with remarkable accuracy.

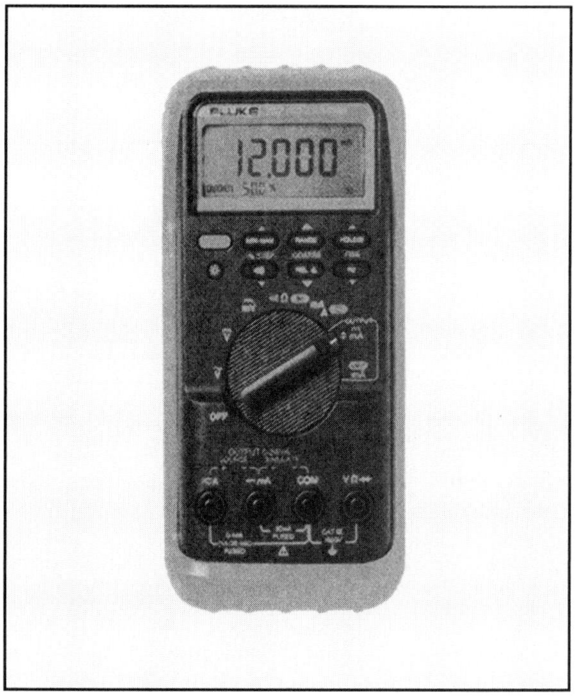

Figure 173 Digital multimeter

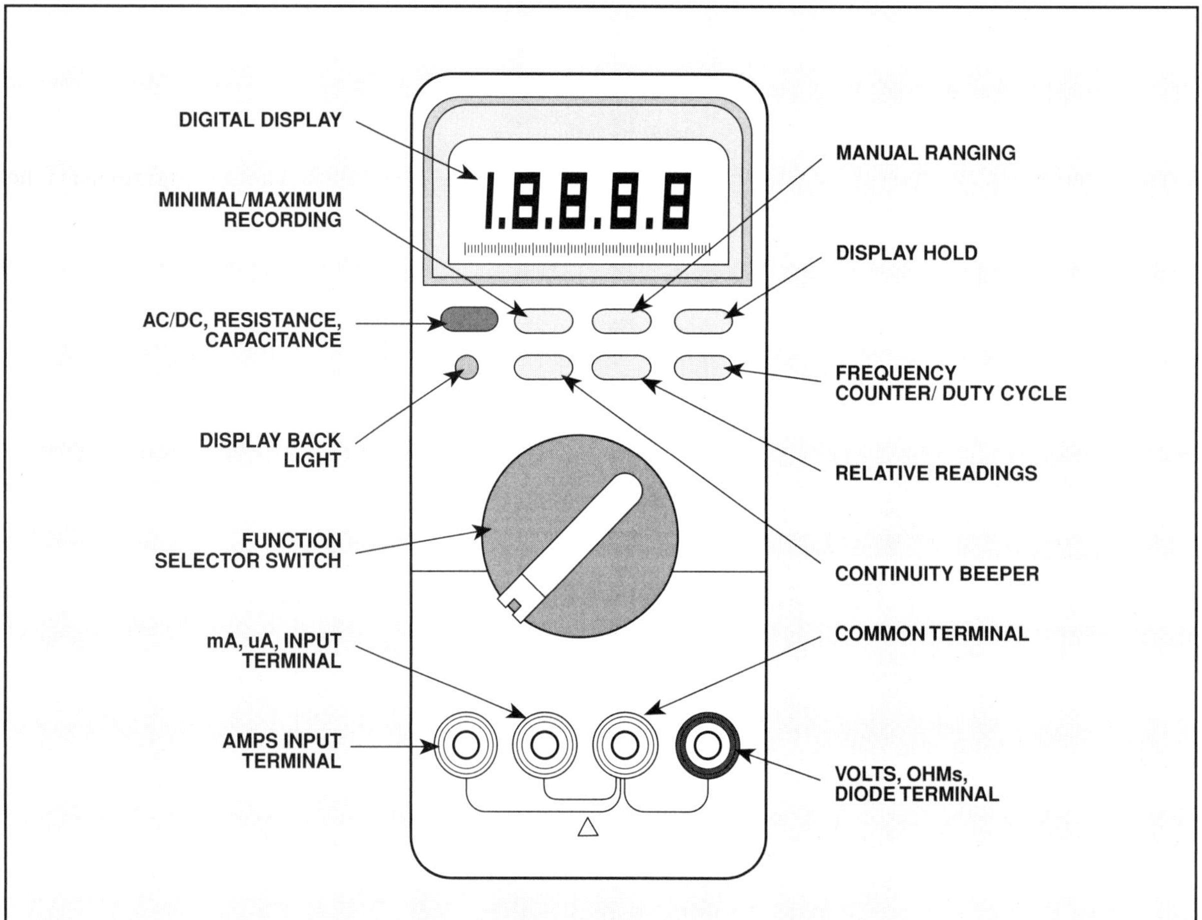

Figure 174 Digital multimeter displays

The display is usually arranged to measure the electrical values to four or five significant figures. In this regard, the digital meter has an advantage over the conventional analog meter, similar to the way a handheld calculator has an advantage over a slide rule.

The most common digital multimeter is a 3½-digit instrument. It is called a 3½-digit instrument because it can display a maximum reading of 1999. The left digit space can only display 0 or 1, which is loosely thought of as being only half the other three digits. (It is not, of course, but it is a short way to indicate the instrument's ability.) Two-and-one-half and 4½-digit meters are also available. Their maximum readouts are 199 and 19999, respectively. The decimal may be manually set at any point in the readout—for example, 1.99, 0.9999, 225.6, or 1.05. Or the meter can be set to automatically place the decimal point.

Digital meters are available that not only automatically set the decimal point, but also automatically zero and automatically sense the polarity needed for the circuit. While digital meters should be handled with reasonable care, they are more rugged than meters with d'Arsonval movements. They are also less sensitive to overranging but can be ruined just as any meter if abused or improperly used.

OSCILLOSCOPES

A measuring device that operates differently from the meters previously discussed is the *oscilloscope*. An oscilloscope uses a cathode-ray tube to make visible on a fluorescent screen the instantaneous values and waveforms of electrical quantities that rapidly vary as a function of time or other quantity. An analog oscilloscope enables an operator to measure voltage by observing how much the voltage deflects a beam of electrons. A digital oscilloscope selects *pixels* for display. The oscilloscope gets its name from the fact that it displays a visible pattern of the voltages that exist in an irregular, or oscillating, circuit. It not only measures the voltage of an AC circuit, but also displays its wave shape.

Oscilloscopes became valuable test and measuring instruments in the 1930s. At the time, electronics technology had progressed to the point where the observation of waveforms was useful in making adjustments to special circuits. Moreover, technicians discovered that analyzing waveforms was a valuable way to troubleshoot circuits. Although early oscilloscopes were not simple, they had only limited capability when compared to the versatile instruments of today (fig. 175).

The Cathode-Ray Tube

The heart of older style oscilloscopes is the cathode-ray tube (CRT). It is a complex arrangement of parts inside an evacuated bottle-like glass structure (fig. 176). A filament (not shown in the figure) heats the oxide-coated cathode, which emits a rich supply of electrons. The electrons are attracted to the first and second anodes that carry relatively high positive

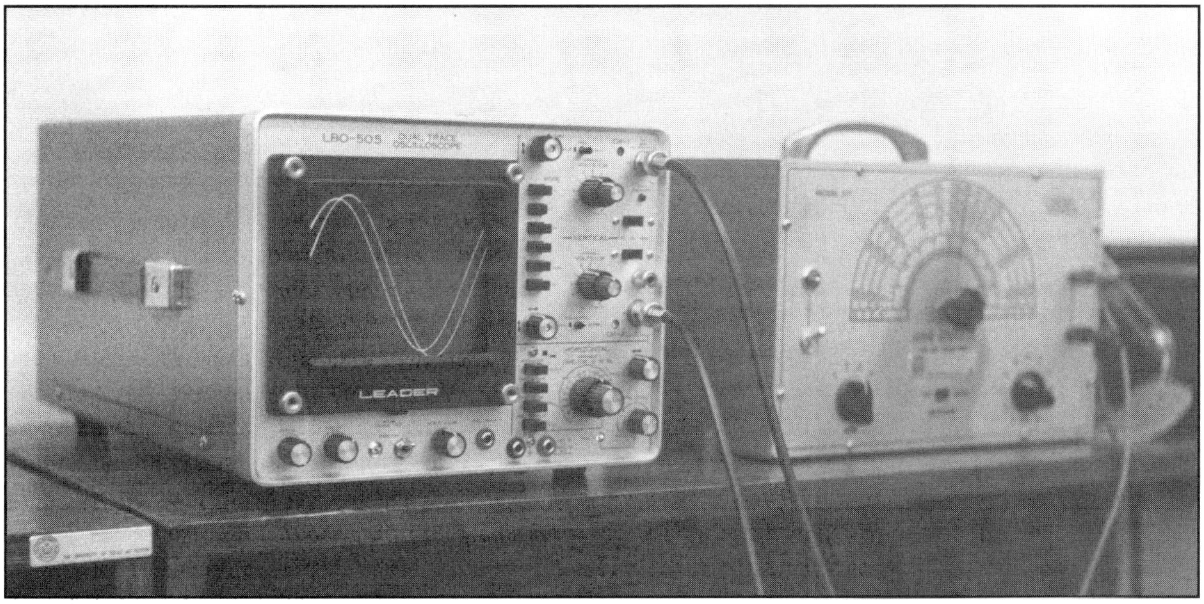

Figure 175 Dual-channel oscilloscope

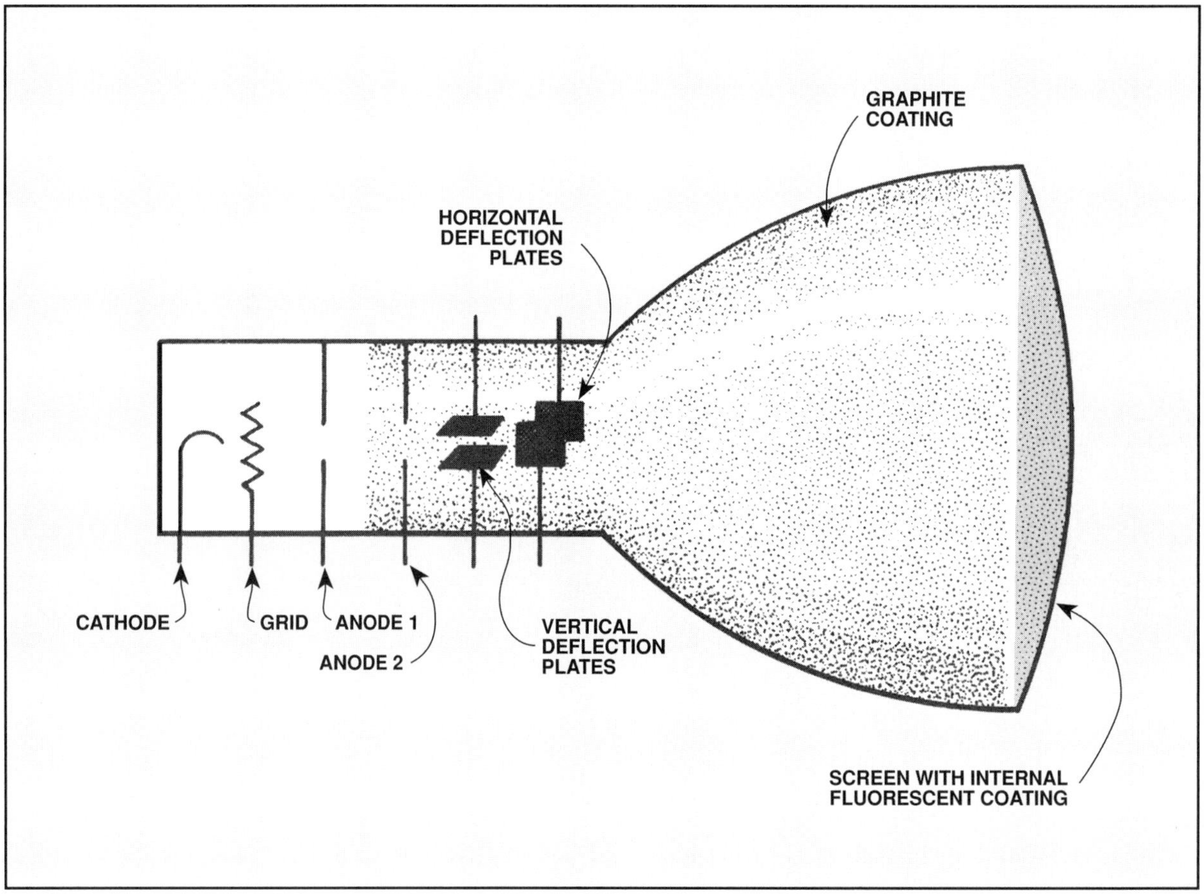

Figure 176 Principal parts of a typical cathode ray tube

potentials. A beam of electrons stream from the cathode and go through small apertures in the grid and past anodes 1 and 2. The electrons have such high velocity that when they reach anode 2 most of them continue through the aperture and eventually strike the fluorescent material that coats the inside of the screen.

It is the nature of the fluorescent material to emit light when fast-moving electrons strike it. The color of the emitted light is determined by the composition of the material. The fluorescent material also has a characteristic quality of persistence, which means a bright dot continues to glow after the electron beam has ceased to excite that particular spot. This afterglow is usually of short duration in an oscilloscope.

The electrodes affect the electron beam as it passes through them on its way to the viewing screen. Further, the potential that is applied to the grid and anodes determines the effect the electrodes have on the electron beam. A negative potential with respect to the cathode applied to the grid suppresses the electron stream. The amount of suppression is in proportion to the value of the negative potential. Varying the amount of suppression controls the brilliance of the screen's illumination.

Anode 1 has a relatively positive potential applied to it, and anode 2 has an even higher positive potential. The two anodes with their potentials act to accelerate the electron beam; in addition, they serve to focus the stream of electrons into a narrow beam. Anode 2 is normally supplied with a fixed potential, while anode 1 has a smaller potential that can be varied over a narrow range to adjust focus of the beam to its optimum fineness.

After the electron beam emerges from anode 2, it passes between two pairs of deflection plates. If these plates are at zero potential, the electron beam is not affected in its path to the fluorescent screen. It strikes the screen somewhere near the center and causes a bright spot. The electron beam's path can be bent by placing a potential on the deflection plates. If one of the vertical deflecting plates is made positive and the other negative, the beam bends

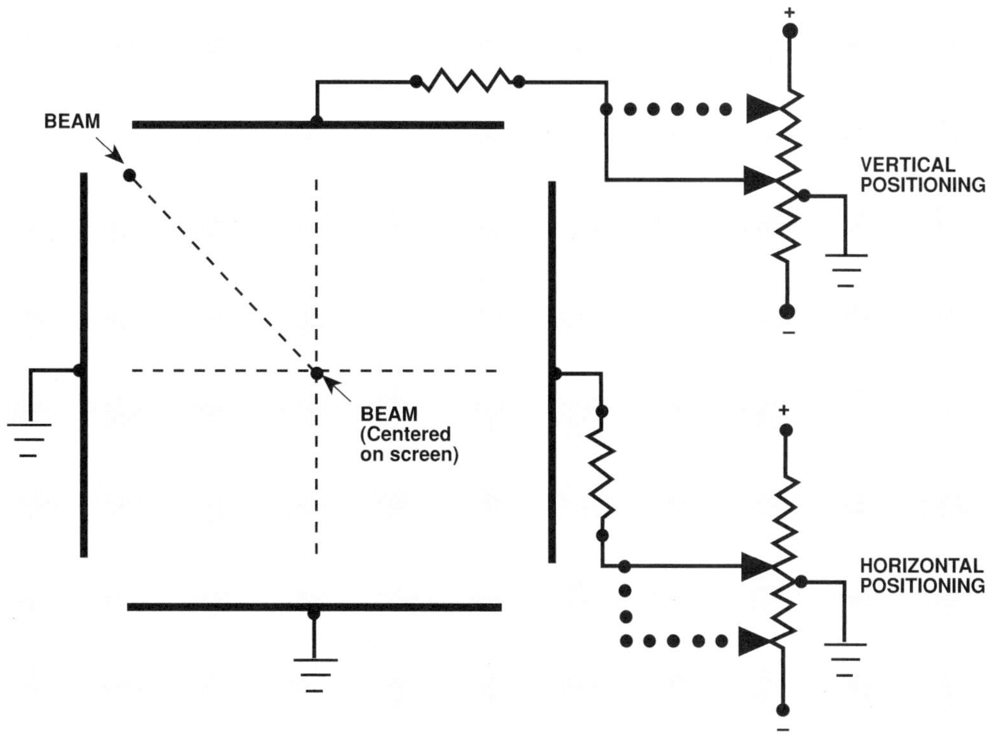

Figure 177 Positioning controls for an oscilloscope

toward the positive plate. The amount the beam is bent is proportional to the difference in potential between the two plates.

The same effect occurs if similar potentials are applied to the horizontal deflecting plates. In one instance, the beam is deflected in a vertical direction, and in another instance, it is deflected in a horizontal direction.

By applying various potential values to the deflecting plates, the electron beam can be moved to any point on the fluorescent screen. Thus, oscilloscopes have a circuit for centering the electron beam on the screen and for positioning it to a desired point (fig. 177). Such a circuit provides for changing the relative potentials and the polarities of each set of deflection plates. Adjustment of a positioning potentiometer can change the polarity of one plate from positive to negative with respect to the other plate. Setting the controls near their extremes, as shown by the arrows with dotted lines, drives the beam to the upper left corner of the screen.

Cathode-ray tubes are internally coated with a graphite film that serves as a return circuit for electrons. The graphite film, which is also called *aquadag*, is not in direct contact with the fluorescent material of the screen but gathers electrons through a secondary emission process. The electrons are returned to the power supply of the oscilloscope. Removal of the electrons from the screen is necessary to avoid an intense accumulation of negative charges on the screen. Such a charge acts as a retarding force to the electron beam.

Auxiliary Circuits and Components

Auxiliary circuits and components are combined with the cathode-ray tube (fig. 178) to make it a useful instrument for studying waveforms, observing phase relationships, and measuring voltage and frequency.

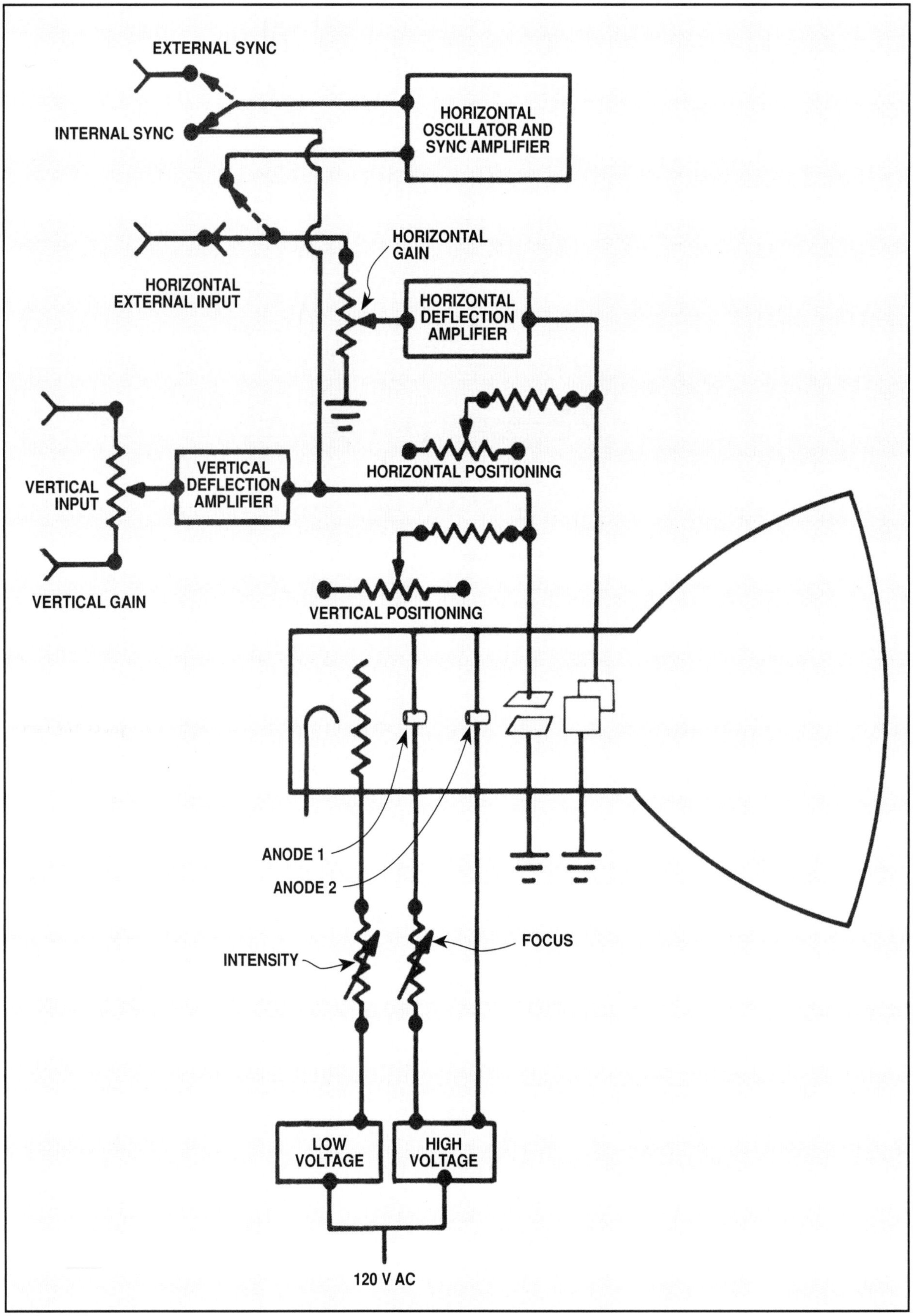

Figure 178 Minimum number of special circuits in an oscilloscope

A 120-VAC, 60 Hz power source operates most oscilloscopes. Anode 2 of the CRT requires a voltage in the range of 1,500 to 2,500 volts. Anode 1 requires less voltage. The low-voltage section of the power supply provides relatively low potentials for the remaining circuits. Two front-panel controls on a conventional oscilloscope are associated with the power supplies: a focus control adjusts the voltage on anode 1 to obtain the sharpest image; another control adjusts the bias on the grid electrode to control the intensity, or brilliance, of the image. (Manufacturers sometimes combine the oscilloscope's on-off power switch with the intensity control.)

Each set of deflection plates is connected to an amplifier. In the simplified diagram shown in figure 178, one of each pair of plates is at ground potential, and the amplifiers are capable of driving the other two plates to potentials well above and below (+ and -) ground potential. The horizontal deflection amplifier can be driven from an outside, or external, source or by an internal *sawtooth sweep generator*. A sawtooth generator is a device that creates, or generates, an output voltage that has a waveform in the shape of a sawtooth. Such a generator produces *sweep voltage* for cathode-ray tubes. Sweep voltage is periodically varying voltage applied to the deflection plates of a cathode-ray tube.

Horizontal-Sweep Frequency Oscillators

Conventional oscilloscopes have an internal *sawtooth oscillator* associated with the horizontal deflection system. (A sawtooth oscillator is an electronic circuit that converts energy from a direct-current source to a periodically varying electric output; its waveform has the shape of a sawtooth.) In addition to this internal system, manufacturers usually provide connections for an external device to drive the horizontal system. One purpose of the horizontal oscillator is to provide a time base. The beam sweeps across the screen at a constant velocity—that is, the beam neither speeds up nor slows down. This uniform movement enables the oscilloscope to accurately display the pattern of the signal that is applied to the vertical deflection system. Most signals are dynamic in nature, and vary in voltage according to a time-related function. For example, a 60-Hz wave goes through a complete cycle of voltage changes in 1/60 of a second.

The sawtooth wave is so named because it resembles the teeth of a saw (fig. 179). The wave consists of two straight lines per cycle: one line, which gradually slopes upward, represents rising voltage; the other line, which is nearly vertical, represents an abrupt drop to the starting point for a new cycle. The upward sloping, horizontal line represents voltage that drives the electron beam across the screen of an oscilloscope. It must be straight (and not curved) to drive the beam at a uniform rate and serve as an accurate time base. The slanted portion of the wave causes the electron beam to move from left to right; the nearly vertical portion is the retrace part of the wave that moves the beam from right to left.

Note in figure 179 that at time t_0, the sawtooth wave begins below an arbitrary zero line, which is indicated as E_0. The zero line (E_0) represents a set of voltage values that is applied to the horizontal deflection plates. These voltage values place the electron beam at the center of the screen. For voltage values below the E_0 line, the beam is deflected left of center. For voltages above the E_0 line, the beam is right of center.

The wave eventually reaches a maximum value above the line, noted by E_m, then falls abruptly to the value it had at t_0, which is designated as E_{-m}. The sudden drop from maximum to minimum occurs in a very short time, and the beam sweeps to the left side of the screen in this brief instant. This brief sweep to the left is called a *retrace*. Sometimes, this retrace can confuse the screen's pattern; so, a

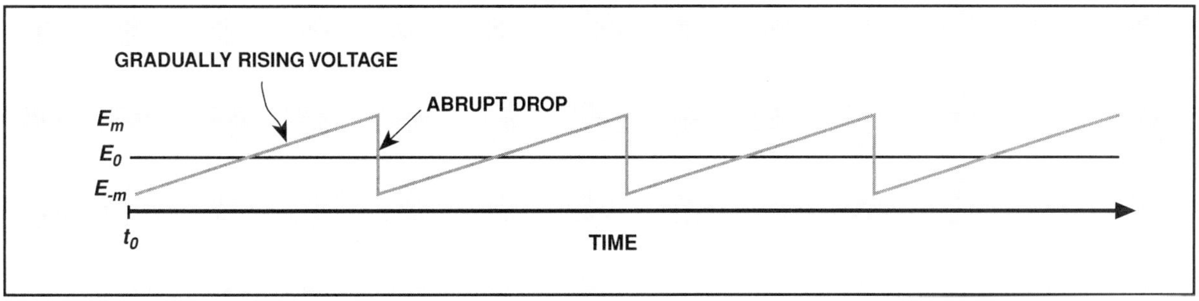

Figure 179 A sawtooth wave of the sort used to drive the horizontal deflection amplifier

blanking pulse is usually employed to shut off the beam during retrace. (A blanking pulse is a positive or negative square-wave variation in electrical power—a pulse—that switches off a part of the oscilloscope's screen for a predetermined length of time.) The blanking pulse is applied to the grid electrode, which cuts off the electron beam during retrace.

Oscilloscopes generally use a *multivibrator oscillator* for horizontal sweep. (A multivibrator oscillator has two transistors; resistors and capacitors couple the output of each transistor to the input of the other. Such coupling obtains in-phase feedback voltage.) Multivibrator oscillators have good *linearity*, which means that they produce a uniform time base. (Linearity is the relationship between two quantities that occurs when a change in one of the quantities produces a directly proportional change in the other quantity.) Multivibrator oscillators can produce an enormous range of sweep frequencies—from a few Hz to 25 MHz or more—which is a range modern oscilloscopes need. The range of frequencies is covered in five or six steps using a coarse sweep frequency selector switch. A potentiometer provides for fine adjustment of the frequency.

Synchronization Amplifiers and Controls

Assume that the horizontal-sweep oscillator is driving an oscilloscope's electron beam at a particular frequency. Also assume that the vertical input to the oscilloscope has a frequency near that of the horizontal sweep or a multiple of it. The fine-adjust of the sweep frequency brings the two inputs into step. However, even the least bit of variation in either the vertical or the horizontal frequency causes the pattern to drift right or left on the screen. To lock them together, a *synchronization amplifier* and a *synchronization control*—usually shortened to sync amplifier and sync control—are used. Locking the vertical and horizontal frequencies creates a steady pattern that can be studied under stable conditions.

The sync amplifier receives a small portion of the vertical input signal and feeds it to the horizontal-sweep oscillator as a pulse that triggers the oscillator in step with the vertical input signal. The sync control varies the amount of pulse strength that reaches the oscillator, providing just enough pulse to stabilize the pattern. Conventional oscilloscopes provide for external synchronization signals. Such signals are usually associated with other signals that drive the horizontal deflection amplifier from an external source.

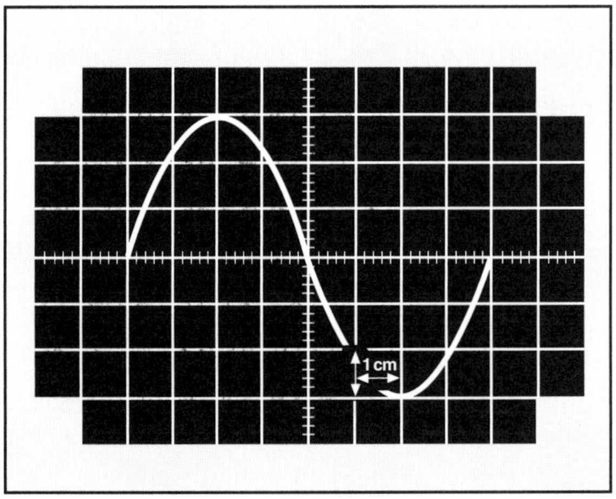

Figure 180 Pattern of sine wave on oscilloscope screen

Measurements with the Oscilloscope

Oscilloscopes commonly have a round-faced CRT, although some are fitted with rectangular screens having a 3-to-4 *aspect ratio*. (The aspect ratio is the ratio of the screen's height to its width. For example, if a rectangular screen is 3 in. high, its width is 4 in. Another example is an automobile ignition analyzer's screen, which is 9 by 12 in.) An oscilloscope's CRT usually has a plastic overlay, which is lightly colored and contains vertical and horizontal lines to form a grid. Typically, each grid is 1 cm square. A horizontal line and a vertical line cross at the center of the screen. These lines are typically divided into increments of 2 mm (fig. 180). These grid patterns are required to make measurements. In figure 180, the sine wave peaks up at 6 cm and peaks down at 6 cm. Peak heights can be used to determine peak-to-peak voltage in a circuit being measured.

Measurements that an oscilloscope can make include—

- AC and DC voltage,
- frequency,
- phase shift, and
- time of a period.

Measuring Voltage

While older oscilloscopes cannot measure direct-current voltage, most modern units can. To measure DC voltage, the oscilloscope must have *direct coupling* from one stage to the next in its internal circuitry. Direct coupling is the joining, or coupling, of two circuits with a nonfrequency-sensitive device, such as a wire, resistor, or battery, so that both direct and alternating current can flow through the coupling path. (Besides direct coupling, *capacitive coupling* and *inductive coupling* are also available; however, such coupling circuits in an oscillator do not allow it to measure DC current. In capacitive coupling, energy is transferred from circuit to circuit through a capacitor. In inductive coupling, energy is transferred from circuit to circuit through the inductance provided by a transformer.)

To measure voltage, an oscilloscope must first have a calibrating voltage sent to it. Put another way, it is necessary to know how much the electron beam will be deflected in a vertical direction for a given voltage value. Some oscilloscopes have a terminal from which a fixed voltage can be obtained (1 volt, for example). This voltage may be taken from a secondary winding on the low-voltage power transformer, or it may be obtained from a closely regulated DC source within the instrument. The latter method provides a more accurate calibrating voltage, but it is applicable only to oscilloscopes capable of measuring DC volts. In any case, the known voltage is applied to the vertical deflection input, and the desired amount of deflection is adjusted through use of the vertical gain control. The instrument is then ready to measure peak-to-peak AC signals or DC voltage, as appropriate.

Some oscilloscopes have a selector switch that features a calibrating step, or position. Additional switch steps provide various multiples of the volts-per-cm range on the grid. Other oscilloscopes rely on occasional checks that can be performed by applying a known voltage value to the vertical input, and then adjusting an internal potentiometer to recalibrate the unit.

Once the oscilloscope is set for voltage measurements, the unknown voltage is applied to the vertical input and the result interpreted from the amount of the electron beam's vertical deflection. Applied DC voltage deflects the horizontal line to a level either above or below the centerline, depending on polarity of the applied voltage. AC voltage deflects the beam above and below the centerline. Also, the oscilloscope displays the input's wave shape if the horizontal oscillator is set properly. The peak-to-peak value of the AC voltage is determined by counting the number of grid squares from top to bottom of the wave. In figure 181, the sine wave rises 2.5 cm above and 2.5 cm below the reference line, which totals 5 cm from peak to peak. If the sensitivity of the oscilloscope is 3 volts per cm, then

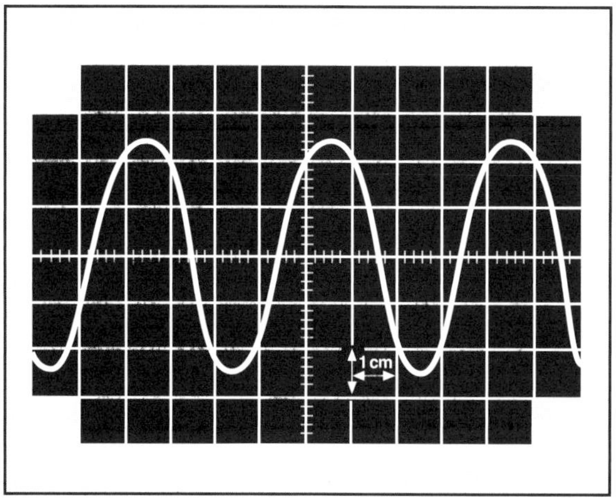

Figure 181 Sine wave of 15 volts value

the sine wave indicates a value of 15 volts, peak to peak. This voltage is determined by multiplying the peak amount times the sensitivity setting for the oscilloscope. Thus, $5 \times 3 = 15$ volts.

In many instances, it is necessary to use an *attenuating probe* between the voltage source and the vertical input to the oscilloscope. An attenuating probe is a small device (a probe) that is put into contact with or inserted into a circuit to make measurements of the circuit and that reduces (attenuates) the strength of the value (such as voltage) being measured in the circuit. Such a probe may have a 10-to-1 ratio, and this must be taken into account when calculating the voltage.

Measuring Frequency

An oscilloscope is not the best instrument with which to measure frequency. Reasonably close approximations can be obtained by matching the vertical input signal's frequency with that of the horizontal-sweep oscillator. Accurate measurements entail the use of calibrated signal generators to produce the horizontal-sweep frequency. It is then a matter of adjusting the horizontal frequency to match the frequency of the vertical input.

Measuring Phase Shift

It is sometimes desirable to detect and measure the phase shift that occurs in a circuit between its input and output. Phase-shift measurements can be done with a conventional oscilloscope, although the use of a *dual-trace oscilloscope* is preferred. (A dual-trace oscilloscope can compare two waveforms on the face of its single cathode-ray tube.) If the input signal of a circuit is applied to the oscilloscope's horizontal-sweep section, and if the output signal of the circuit is applied to the vertical section, the resulting display can show whether a phase shift is present. A straight diagonal line indicates no phase shift, while an oval display indicates a phase shift. For a small amount of shift, the oval is thin; as more shift occurs, the oval opens up and becomes quite pronounced at a 90-degree phase shift.

Measuring the Time of a Period

Using an oscilloscope to measure the time of a period is not often required in practical electronics. However, in research and some manufacturing processes, it may be necessary. In studying waveforms, for example, it may be desirable to determine the rise time of a pulse. Figure 182 shows a sawtooth pulse that occurs when the time of a period is measured. From this pattern, the rise time of the sawtooth pulse can be calculated. Accuracy in measuring time requires knowing exactly the frequency of the horizontal sweep. The period in seconds is the reciprocal of the frequency in Hz. Thus, a 60-Hz signal has a period of $\frac{1}{60}$ of a second.

In figure 182, the sweep is 1,000 Hz, and the gain of the horizontal amplifier is adjusted to limit the sweep to a width of 10 cm (5 cm on each side of the vertical centerline). In this case, the beam sweeps across 10 cm in $\frac{1}{1,000}$ of a second (1 millisecond). The top of the pulse is positioned to lie on the vertical centerline. Note that the bottom of the pulse on the left side is exactly 1 cm to the left of the centerline. The rise time is then the time it takes the beam to scan 1 cm, or one-tenth of the total scan distance. Thus, $\frac{1}{1,000} \times \frac{1}{10} \times \frac{1}{10,000}$ second, or 100 microseconds. Many moderately expensive oscilloscopes contain internal circuitry and controls that enable a technician to measure period time with considerable ease.

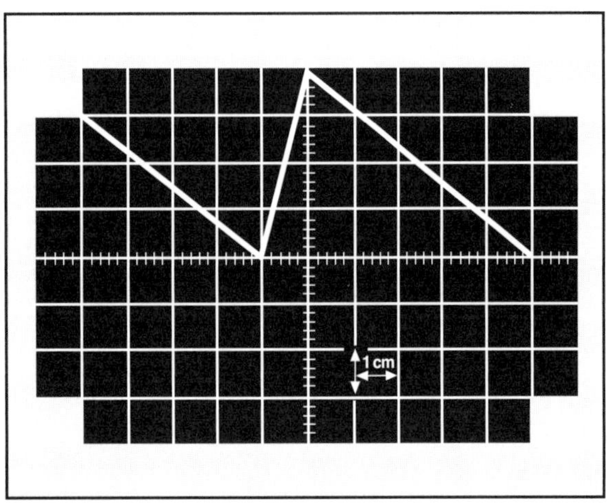

Figure 182 Sawtooth pulse

Dual-Trace Oscilloscopes

With added circuitry and controls, an oscilloscope can simultaneously display two separate traces, or waveforms (see fig. 175). A dual-trace oscilloscope is required to study the behavior of two signals and their time relation to each other. While the dual display appears to be simultaneous, in reality the cathode-ray tube's circuits are switched rapidly back and forth between the two channels, bringing about an illusion of two steady patterns.

Separate circuits operate the two channels in a dual-trace oscilloscope. The separate controls position the baseline for each channel and control the vertical input gain. Dual-trace oscilloscopes have advantages over single-trace units for performing several functions. For example, the phase shift in a circuit is easily and clearly observed and measured using a dual-trace oscilloscope.

Oscilloscope Probes

Earlier, this manual mentioned the use of a 10-to-1 probe when measuring voltage with an oscilloscope. Such an attenuation probe is more commonly called a *lo-cap* (for low capacitance) *probe*. This probe allows the oscilloscope to be attached to a sensitive circuit with a very small loading effect. Typical characteristics of lo-cap probes are 5 to 10 megohms and 10 to 20 picofarads.

Another probe used with oscilloscopes is a straight-type probe. It consists of a shielded single-conductor made of stranded wire. It is used for measuring nonsensitive circuits of low frequency, such as those found in the audio range.

A third probe is a *detector*, or *demodulator*, *probe*. It is used in radar and television alignment work. The probe *demodulates* a signal composed of a high-frequency carrier and low-frequency modulation. Only the effects of the low-frequency signal are displayed on the screen when using this probe.

Combination Scope-Meters

Manufacturers now produce a measuring device that incorporates a multimeter and a dual-trace oscilloscope (fig. 183). It is rugged, has excellent performance, and is versatile. It contains internal batteries; so, it can be used in remote locations without an AC supply. It has multiple capabilities and has all

Figure 183 Dual-trace scope-meter

but made the table-top oscilloscope obsolete. As a digital multimeter, it can measure—

- AC and DC voltage, including root-mean-square (rms) values, peak values, peak-to-peak values, average values, and effective values;
- frequency, period, and duty cycles;
- diode function, resistance, and capacitance;
- DC and AC current;
- signal hold and changes;
- temperature (with a separate probe); and
- decibels (power or voltage ratio).

As an oscilloscope, it can present, function as, or measure—

- single or dual traces;
- AC, DC, and invert functions;
- amplitude and timing adjustments;
- plus or minus trigger slopes;
- frequency responses up to 20 MHz;
- rise times up to 17.5 seconds;
- range selections from 1 to 40 MHz;
- external signal triggering; and
- plot trends and changes on the screen.

Figure 184 shows a typical screen display of a dual-trace scope-meter. This type of instrument also

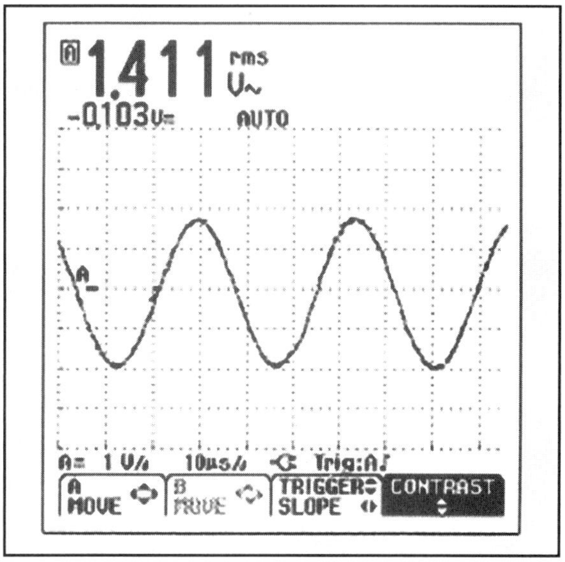

Figure 184 Typical screen display of dual-trace scope-meter

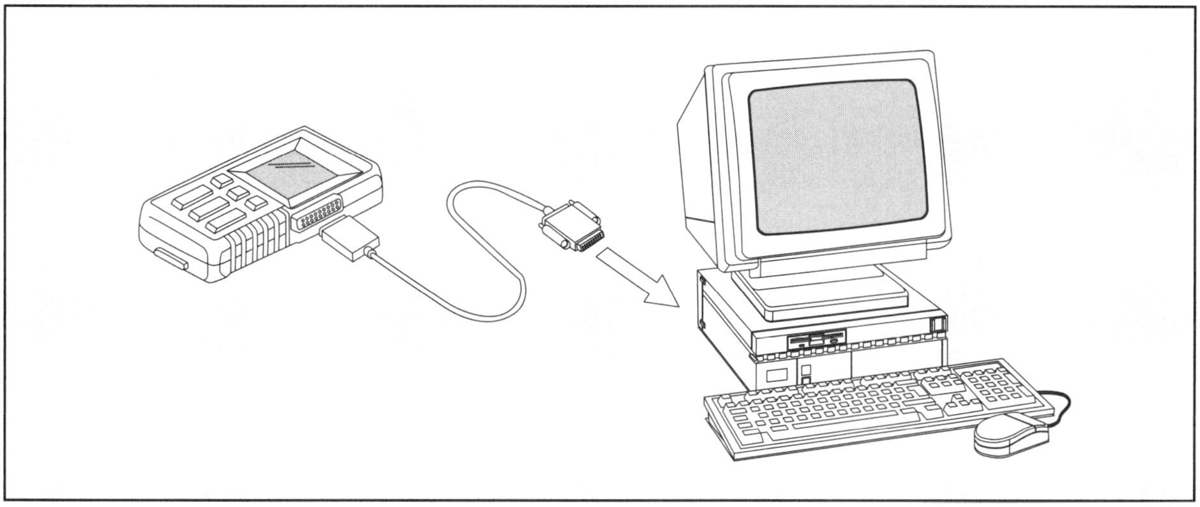

Figure 185 Connecting scope-meter to computer

has computer-compatible capabilities by providing a *Recommended Standard (RS)-232* connector for downloading oscilloscope display information to a printer, computer memory, or floppy disk (fig. 185). (RS-232 is an Electronic Industries Alliance—EIA—standard that describes the requirements for the interface between data processing and data communications equipment. It is widely used to connect computers to peripheral devices.)

ADDITIONAL TEST AND MEASURING INSTRUMENTS

Current Measuring Meters

One of the safest methods used to measure high-power AC or DC currents is to use a clamp-on meter that measures the magnetic field produced by the current in the cable (fig. 186). The meter displays current in amperes; an operator can select current ranges from low to high. Contact with a live circuit is not required, nor are changes in wiring.

Figure 186 Digital current meter

Frequency Meters (Cycle Counters)

A *frequency meter*, or cycle counter, is a device that measures the frequency of a signal by counting the number of cycles in the signal during a specified time interval. Modern frequency meters are solid-state devices that are readily available. However, early versions were bulky and used several dual-triode vacuum tubes in flip-flop circuits. Neon bulbs provided a crude readout means.

Modern frequency meters use LEDs or *liquid crystal displays* (LCDs), and some are capable of reading frequencies higher than 1 GHz. A liquid crystal display (LCD) is an electronic digital display that consists of two sheets of glass separated by a sealed-in, normally transparent, liquid crystal material. The outer surface of each glass sheet has a transparent conductive coating etched into character-forming segments that have leads going to the edges of the display. Voltage applied between front and back electrode coatings disrupts the orderly arrangement of molecules, which darkens the liquid enough to form visible characters although no light is generated.

Today's frequency meters are often used to provide a direct readout of the operating frequency of radio transmitting and receiving equipment. Instrumentation systems in the oil and gas production industry also use frequency meters where data is transmitted from remote locations to a central receiving point.

Signal Generators

Signal generators are generally classified as radio- or audio-frequency types. Audio-frequency signal generators may produce sine waves or square waves. Radio-frequency signal generators may produce a continuous wave, or a *frequency-modulated* (FM) signal. (Frequency modulation is the varying, or modulating, of a wave in which the instantaneous frequency of the modulated wave differs from the carrier frequency by an amount proportional to the instantaneous value of the modulating wave.)

Signal generators are limited to use in instrumentation work, and are usually used to troubleshoot microwave or other radio transmitting and receiving equipment.

A special form of signal generator is a pulse generator, used in troubleshooting and adjusting electronic controllers. A pulse generator produces repetitive pulses or signal-initated pulses. Most pulse generators have a way to adjust the width of the pulse as well as its amplitude and frequency. Pulse generators typically serve to produce timing and trigger pulses needed in testing and adjusting electronic instrumentation systems.

TEST INSTRUMENT OVERVIEW

Once technicians have achieved a good understanding of the basic operating principles of test instruments, they should also study the specific instruments they will use in maintaining the control system for which they are responsible. Keep in mind that a good source of information is the instruction manual that accompanies the test and measuring equipment. Another source is the operating, maintenance, and troubleshooting manuals that are pertinent to the control system.

Should manuals be missing, a technician should obtain new ones from the manufacturer of the test equipment or from the installer of the control system. Obtaining new or additional manuals usually requires a formal request from supervisory personnel and may entail a small cost. In any case, it is worth the effort and cost to obtain them.

REVIEW QUESTIONS

1. Common voltmeters, ammeters, and ohmmeters use a mechanism called a ___________ meter movement.
2. To adapt a meter movement to measure voltage, it is usually necessary to use a resistor connected in _____________ with the meter coil.
3. The sensitivity of a voltmeter can be stated in terms of _____________ per volt or input ___________ for all scales.
4. Ammeters can be prepared by placing _____________ resistance across the meter movement coil.
5. Alternating-current voltage can be measured with a d'Arsonval meter movement that has been adapted through use of solid-state _______________.
6. Within its range, a 3½-digit digital multimeter can display a voltage reading to _____________ or ___________significant figures.
7. An oscilloscope can measure AC and DC voltage. It also will display the ___________of an AC voltage.
8. The intensity of the stream of electrons in a cathode-ray tube (CRT) can be controlled by the voltage applied to the _______ of the CRT.
9. The electron beam in a cathode ray tube can be centered on the screen by use of the ______________________ controls.
10. In an oscilloscope, the principal purpose of the horizontal-sweep frequency generator is to provide a _____________ _____________.
11. A sweep frequency generator produces a sawtooth wave. The steep, almost vertical, portion of the sawtooth wave represents the _____________ portion of the cycle.
12. The sync amplifier and sync control of an oscilloscope provide a means to help the sweep oscillator stay in step with the _____________ input signal.
13. To accurately measure voltage with an oscilloscope, it is necessary to know how much the electron beam will be _________________ for a given value of voltage.
14. When using a 10-to-1 lo-cap probe between voltage source and oscilloscope, it is necessary to multiply the value of the displayed voltage by a factor of _____________.
15. The input signal to an AC circuit is applied to the horizontal-sweep section of an oscilloscope, and the output of the circuit is applied to the vertical section. The display is an oval pattern that indicates a ___________ ___________ is occurring within the AC circuit components.
16. A dual-channel oscilloscope has an advantage over a single-channel unit for studying _____________ _____________ between points in an AC circuit.
17. Persons involved in maintaining electronic equipment and test instruments should make sure they have proper ________________ to guide them in their work.

13 Troubleshooting Techniques and Safety

Electronic control systems used in the petroleum industry are usually complex groupings of integrated circuits (ICs) that represent thousands of discrete components. Circuits are complicated; indeed, even the simplest systems can be very involved. To understand the function and characteristics of every component or module is virtually impossible. At the same time, however, modern systems possess many excellent features and they are relatively easy to troubleshoot—certainly easier than the equipment of two or three decades ago.

Consider, for example, a solid-state operational amplifier (op amp), called a 741 (fig. 187). It replaces a 1950s vintage assembly that contained at least a dozen vacuum tubes and scores of other components. To determine whether a 741 op amp is operating properly, a technician only has to take measurements at two or three points. On the other hand, troubleshooting the 1950s model required considerable skill and an extensive amount of time.

Besides integrated circuits, modular construction has also made modern-day systems easy to troubleshoot and service. Making components in modules dates back to the vacuum-tube era, but in those days the modules were bulky and expensive (fig. 188). Today, however, a technician can quickly and easily extract a malfunctioning module that may contain tens of thousands of discrete components from a cardholder or socket and replace it with a module known to be operational (fig. 189).

RECORDKEEPING

It is very important to keep accurate troubleshooting records. Often, the operation and maintenance manuals that accompany a particular piece of equipment contain forms or special pages for recordkeeping. If they do not, then the technician responsible for the equipment's upkeep should start a record. Indeed, most companies have detailed

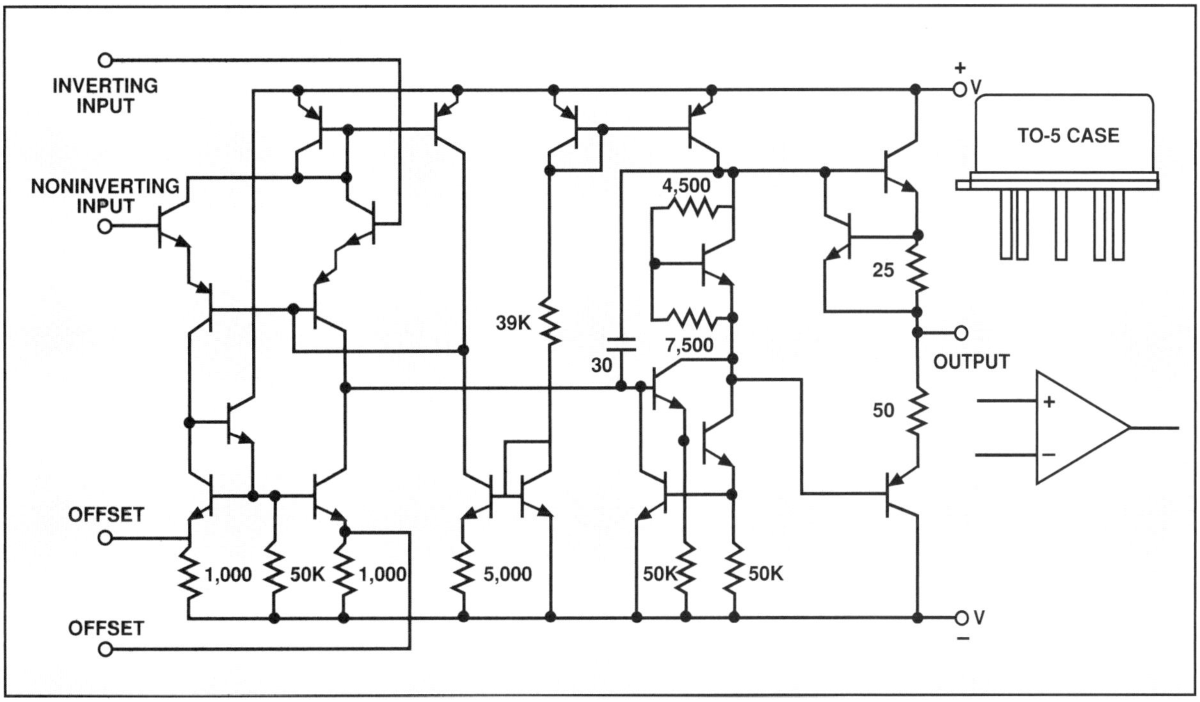

Figure 187 Circuit diagram of a type 741 op amp with a typical package format

Figure 188 A 1950s-era modular device. Note the vacuum tubes and large relays.

Figure 189 Printed circuit card containing integrated circuits and numerous discrete components

operating procedures that include requirements for maintenance and repair records of all equipment. In any case, a typical troubleshooting record should contain—

1. date of the failure or malfunction;
2. symptoms of the trouble (record the problem specifically—for example, trouble-panel light on, blown fuse, system not responding to signals, or actions operator took when problem occurred);
3. where the trouble was found and how—that is, was the trouble determined by measurements or by deduction;
4. how the problem was solved—for example, by replacing a fuse or a module;
5. disposition of the faulty component or module (repaired locally, sent elsewhere, or junked);
6. a list of the new components or modules used or ordered (for replacement of spare parts); and
7. additional remarks that might be of value at a later date.

FAMILIARITY WITH EQUIPMENT

In the matter of troubleshooting, perhaps the best asset for success is being familiar with the equipment. The best troubleshooter is one who knows how to operate the equipment and knows what it is supposed to do. The person normally assigned to operate equipment, however, is not necessarily capable of full-scale troubleshooting. However, the operator is always a good source of information about the system's failure and is typically the first person to detect and report the difficulty.

OPERATION AND MAINTENANCE MANUALS

Technicians should not troubleshoot a malfunctioning system without the instructions contained in the manufacturer's literature, unless they are thoroughly familiar with the system. Figure 190 depicts two large and extensive manuals that cover the operation, maintenance, and troubleshooting of one company's control equipment.

Reliable manufacturers spare no effort to produce excellent operation and maintenance manuals to accompany their products. Large corporations that contract to buy control systems typically require adherence to detailed specifications. A part of the specifications package usually calls for the supplier to provide acceptable manuals covering operation, maintenance, and parts lists.

A good maintenance manual includes a section on troubleshooting. Troubleshooting guides may

Figure 190 Manufacturer's instruction manuals

take several forms. A common one is a three-column arrangement (fig. 191). The first column lists the symptom; the second column lists the probable causes; and the third column suggests corrective measures. The example in figure 191 is a troubleshooting chart for an oscilloscope. Such charts are effective if they list enough symptoms and include virtually every cause of the symptoms.

Figure 192 is another troubleshooting guide. The procedure column lists test points (TPs) to which a voltmeter is to be connected; the next column gives the normal range of voltage; and the right column directs attention to the components that are likely to be faulty if the voltage readings are out of normal range. Figure 193 points out the test points listed in figure 192 on a diagram of the equipment's circuit board. The procedure for locating trouble spots can be a slow process, but it usually leads to the trouble source.

Figure 194 is another effective troubleshooting guide called a flow diagram. The diagram in figure 194 troubleshoots a generator control. It resembles a maze. Consequently, flow diagrams require some practice before their use. A flow diagram involves a step-by-step procedure coupled with a yes–no or go–no-go response technique. A particular step is taken, for example, and the response is noted as yes or no (normal or not normal). The next step involves branching one way or another according to the response.

Usually, maintenance manuals contain circuit diagrams, which can also be helpful in troubleshooting. Most diagrams include appropriate values for all the discrete components and, where applicable, have waveform information at various test points.

TEST INSTRUMENTS

The most popular troubleshooting instrument is the digital multimeter. It is moderately rugged, portable, and self-contained. It can measure resistance, current, and a wide range of voltage values without seriously loading a circuit.

The standard VOM is another instrument that should be included in a troubleshooting kit. Also, small portable instruments that combine the multimeter with oscilloscope functions are available. Other test instruments commonly used in troubleshooting include *signal generators*, *digital multimeters*, and oscilloscopes. Additional instruments include *wattmeters*, frequency counters, and clamp-on meters. (A signal generator delivers a sinusoidal output at an

TROUBLE CHART		
Symptom	**Probable Cause**	**Corrective Measure**
Can obtain dot, but no horizontal or vertical trace.	Scope selector switch S6 (at rear of PS 148) in improper position.	Set scope selector switch to normal position.
Trace tilted.	CRT positioned incorrectly.	Remove scope from case, loosen screws holding CRT, and carefully rotate CRT so that trace is not tilted.
Trace bounces up and down at a slow rate.	Open peaking coil L2, L3, L4, or L5 or leaky capacitor L6.	Use ohmmeter to determine which coil is open. Replace capacitor if defective.
Intermittent or erratic vertical output.	Short at input jack. Gimmick (C1) touching chassis.	Check for short at input jack or see if gimmick is leaning against chassis.
Poor dot size or shape.	MU shield on CRT positioned improperly. Power transformer tilted incorrectly.	Set scope selector switch to external direct to eliminate pickup from amplifiers. If dot shape does not improve, move MU shield and/or change transformer tilt for best dot.
Unit completely dead.	1. Check OFF Standby-ON switch, circuit breaker.	1. Switch all the way to right. Push circuit breaker into reset.
Unit completely dead.	2. V7 & V8	2. Replace if necessary
Unit completely dead.	3. R75, R76, & R71	3. Replace if necessary
Unit completely dead.	4. C22, C36, & C37	4. Replace if necessary
Insufficient intensity or brightness.	V7, V8, V9	Replace V7, V8, or V9 if necessary.
Insufficient vertical output.	V1, V2, or V3, R77, C40A, C41C	Replace if necessary.

Figure 191 ***A troubleshooting chart***

accurately calibrated frequency ranging anywhere from audible to microwave frequencies. A digital multimeter displays its measurements digitally rather than as analog readouts. A wattmeter is an instrument that measures the magnitude of the power in an electric circuit. It may read watts, kilowatts, or megawatts.)

Any troubleshooter must be familiar with the test instruments. The most complicated instrument is an oscilloscope, mainly because of its versatility. An oscilloscope is not only useful for studying waveforms, but also it measures voltage accurately, in spite of the fact that voltage frequency may range from 0 (direct current) to 15 MHz.

Where troubleshooting takes place often determines the instruments used. Usually, troubleshooting in the field limits the kinds of instruments. On the other hand, searching for circuit faults in a shop or laboratory allows the use of virtually any specialized instrument.

A STEP-BY-STEP TROUBLESHOOTING TABLE			
Step	**Procedure**	**Normal Indication**	**Action**
1	Connect voltmeter + to TP 12, Figure 32, and – to TP 11.	24VDC ± 1.2	Normal: Go to Step 2. Abnormal: Check D6 circuit.
2	Connect voltmeter + to TP 13.	24VDC ± 2.4	Normal: Go to Step 3. Abnormal: Check Q3, D4, C3 circuit.
3	Connect voltmeter + to TP 14 and – to TP 10.	16VDC ± 0.8	Normal: Go to Step 4. Abnormal: Check D7 circuit.
4	Connect voltmeter + to TP 15.	15VDC ± 1.5	Normal: Go to Step 5. Abnormal: Check Q4, D5, C4 circuit.
5	Connect voltmeter + to TP 16	6.2VDC ± 0.3	Normal: Go to Step 6. Abnormal: Check D12 circuit.
6	Connect voltmeter + to TP 17	0.3VDC ± 0.015 for 0 to 1V or 0.25 to 1.25VDC inputs or 1.45VDC ± 0.07 for 1 to 5VDC inputs	Normal: Go to Step 7. Abnormal: Check A10 circuit.
7	Connect voltmeter + to TP 18	-16VDC ± 0.8	Normal: Go to Step 8. Abnormal: Check D8 circuit.
8	Connect voltmeter + to TP 19	-15VDC ± 1.5	Normal: Go to Step 9. Abnormal: Check Q5, D5, C5 circuit.

Figure 192 A step-by-step troubleshooting table

TROUBLESHOOTING STEPS

Troubleshooting should be carried out in a logical fashion. A logical method usually prescribes a natural or obvious approach. A logical approach follows a step-by-step procedure. Each step either pinpoints the trouble as being in a particular section or eliminates the section as the trouble source.

Determine and Interpret

Trouble in electronic circuitry always produces one or more signs of malfunction. Modern large systems typically provide extensive warning signal circuits that produce visual or audible evidence of system trouble. Such signal systems may indicate that a valve has failed to close or open properly, that pressure or temperature has gotten out of bounds, or that power has been lost to a part of the system. Such straightforward signs of trouble usually lead the troubleshooter directly to a faulty component. On the other hand, the failure of an obscure or small component on a printed circuit board can truly test a technician's skills.

Whether obvious or obscure, troubleshooting begins with a study of the problem's symptoms. A troubleshooter should determine the symptoms either through direct observation or from a knowledgeable operator of the equipment. The latter is particularly important if the faulty equipment is located some distance away, because it allows the troubleshooter to bring the needed manuals and equipment to the job.

Troubleshooting is partly an intuitive endeavor. For example—

1. if a system is totally inoperative, it is likely that a failure has occurred near the power end of the system.

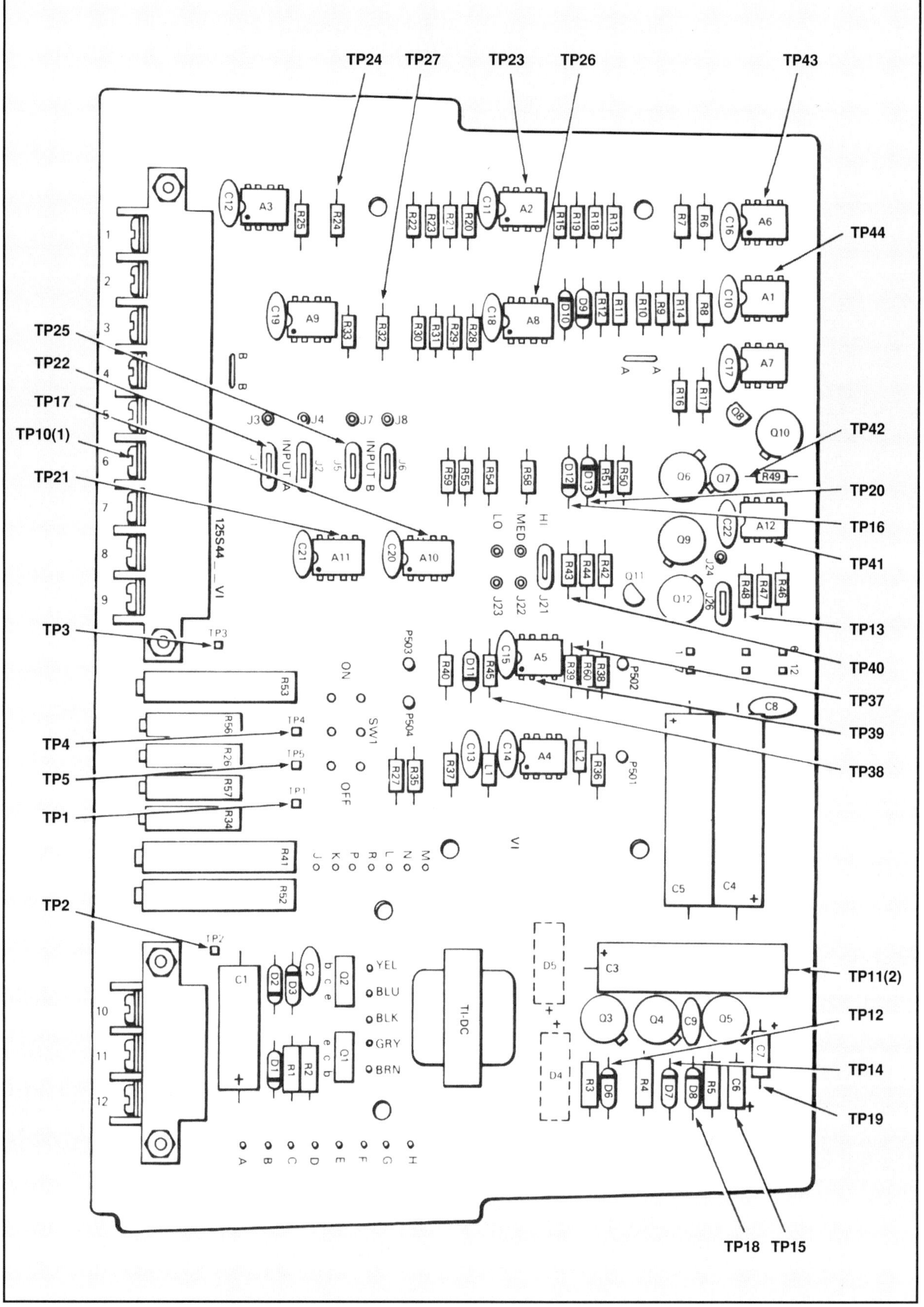

Figure 193 Diagram of test points

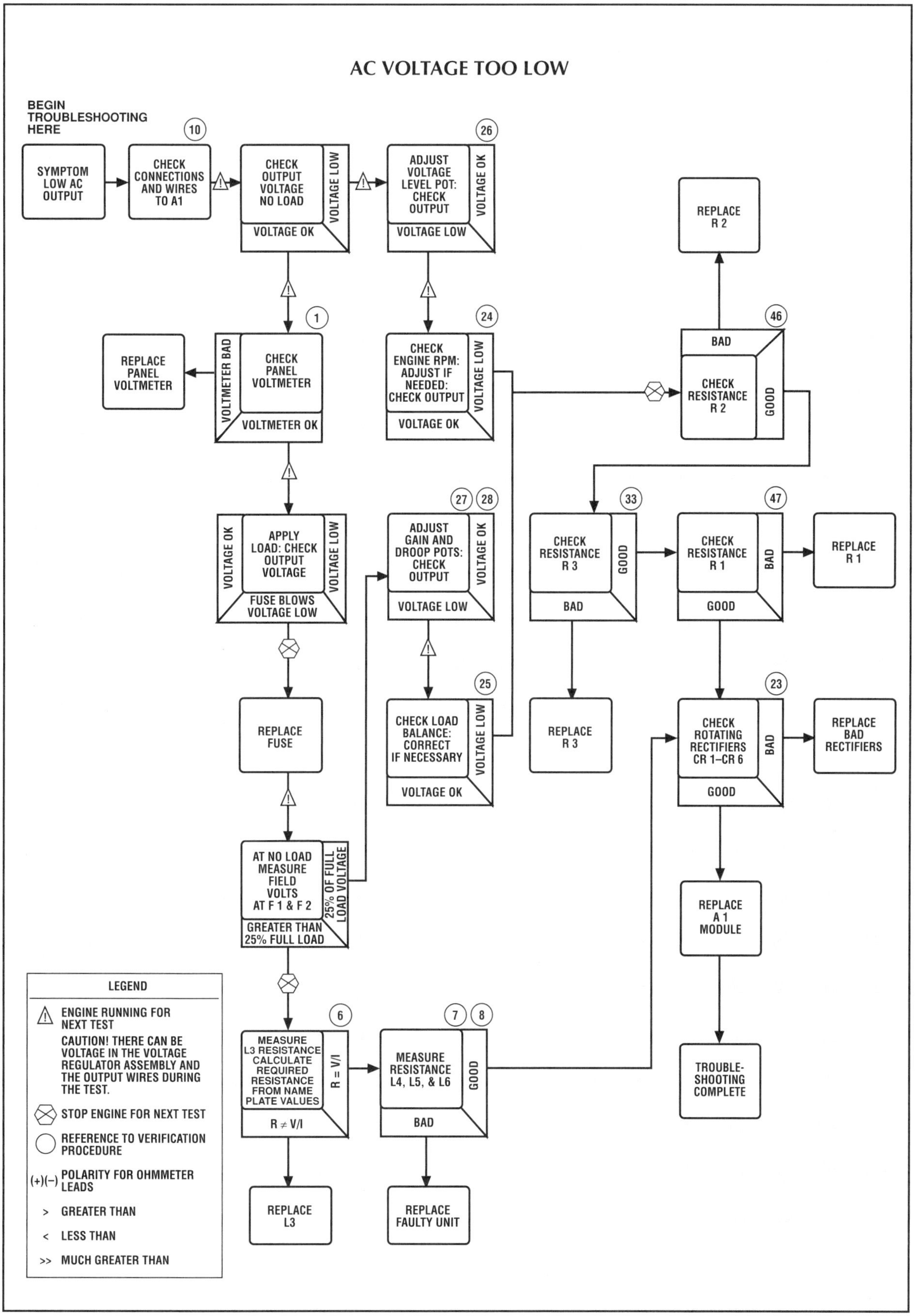

Figure 194 Flow diagram for troubleshooting generator controls

2. if some parts of a system are operating satisfactorily, these parts offer an opportunity to proceed by a process of elimination—that is, if one part works, then the fault is probably in some other part.

A good troubleshooter notes the symptoms and refers to the manufacturer or supplier's manuals to see if the symptoms are listed in the troubleshooting guide.

Assemble the Tools and Instruments

If the trouble lies at a remote location, familiarity with the system really pays off. Otherwise, the technician may take tools and equipment to the location that are not related to the system. If a spare parts kit for the faulty equipment is available, it should also be carried to the site. Moreover, any modular components that can be substituted in the faulty system should also be taken to the site. Also, the technician should consider whether the failed system is inside or outdoors, far from suitable power sources.

Keep a pad and pencil handy to record observations involving voltage and resistance values and waveform comments. A pad and pencil should be a part of the instrument and tool kit.

Make a Visual Inspection

Once at the location, make a visual inspection before using instruments to trace trouble. Turn off power to the assembly to be inspected. If discrete components are part of the assembly, damage might be obvious. Sometimes, faulty components can be smelled. Resistors and transformers are good candidates for these observations. Then, look for broken connections and poor solder joints on printed circuit boards (a rarity in commercial equipment). Finally, make sure that all components and assemblies that plug in are in place and tight.

Apply Module or Card Substitution Technique

Where applicable, the substitution technique is the quickest way of returning the malfunctioning equipment to normal operation. But before applying the technique, be careful to avoid electric shock and make every effort not to cause further damage. Most solid-state devices operate on low DC voltage, but video display units are high-voltage, low-current users. Also, the primary source of power is usually 120–230 VAC, 60 Hz, which is very dangerous.

At any rate, if the substitution technique is appropriate—that is, if spare modules or circuit cards are available and the manufacturer's troubleshooting guides or charts indicate that its use is satisfactory, then—

1. turn off power before substituting modules or printed circuit cards.
2. be aware that a module or card may have failed because of a fault in another part of the system; consequently, the substitute module or card might also be damaged when it is placed in use.

Assuming that substitution of one of the cards or modules succeeds in solving the difficulty, closely observe the equipment's operation for a short time. Watch carefully for any sign that the trouble will return. If satisfied that the problem is solved—

1. retain the faulty module for repair, if feasible, or ensure that a new spare is made available without delay.
2. make a permanent record of the symptoms of trouble and the steps followed to correct the matter, noting part numbers and system sections involved.

CIRCUIT TESTING

Sometimes substitution does not clear up the difficulty, either because it is not applicable or because the trouble is not in modules or cards that can be substituted. In such cases, test and measuring instruments must be employed. Always follow the recommendations for troubleshooting found in the manufacturer or installer's manuals or literature.

Locate the Difficulty

Using information available from circuit diagrams, such as waveforms and values of resistance and voltage, measure and observe their shapes and values with the appropriate instruments. Where deviation is observed, consider which malfunctioning component may be causing the problem. Measure the suspected component if practicable.

Once a faulty component is located, attempt to diagnose why it failed. Troubleshooters should ask—

1. was it simply a component that had an inherent weakness and failed under normal conditions?
2. was it overloaded by another malfunctioning component? Except for resistors, it is

difficult to tell whether a unit failed from overload. Even a resistor can burn up in normal service if its value changes. For example, assume that a 2-watt, 1,000-ohm resistor operates with 40 volts across it. In this case, normal power across the resistor is 1.5 watts. Suppose, however, that the resistor changes value to 500 ohms. The current doubles to 80 milliamperes, and power dissipation rises to I^2R, or $0.0064 \times 500 = 3.2$ watts, which is well above the resistor's 2-watt power rating.

Repair and Replace Components

Once the cause of circuit failure has been located, the major part of troubleshooting has been accomplished. It is usually easier to repair a fault than to find it. Good repair practices require skills that are attained mainly through experience and practice. However, general guidelines are available. For example—

1. use proper tools to work on electronic assemblies; improper tools and work habits can cause damage.
2. printed circuit cards and compact modular assemblies are rugged, but misuse can inflict damage to mounted components and the foil-like circuit wiring—therefore, only use desoldering devices and small soldering irons that have the proper heat range.
3. avoid damaging solid-state components by using too much heat or by exposing them to static charges; static charges can ruin MOSFET components.
4. if a module or a card is substituted to repair a system, troubleshoot the faulty card and restore it to operational quality if possible.
5. enter a list of repairs and comments in a logbook associated with the equipment and note what was done with faulty components; order new modules, cards, or other parts to ensure that replacement parts remain available.

TROUBLESHOOTING INTERMITTENT MALFUNCTIONS

Occasionally, an operator reports a failure in a system; then, when the technician arrives on the scene, the failure has cleared up of its own accord. Intermittent failures are among the worst sort of trouble, especially when they recur often, because they can be difficult to solve. Intermittent malfunctions in control equipment are similar to a watch that will not run a full day without stopping a few times.

Components changing value, shorting out, or opening up because of heat or other environmental conditions can cause intermittent troubles. In discrete component assemblies, the use of special sprays to cool the components or devices to accelerate heating may speed detection of faulty components. However, do not use sprays on printed circuits or modules containing integrated circuits.

Where possible, substituting cards or modules is a good approach to solving intermittent problems. Generally, the failure symptoms point to the part of the system producing the trouble. Occasionally, wholesale replacement of large assemblies is the only solution when the problem cannot otherwise be determined.

FIELD TROUBLESHOOTING

Most modern automated oil leases have remote signal systems that pinpoint trouble in the field. Automated systems usually automatically shut down systems if conditions occur that threaten the environment or equipment. Typical shutdown devices include those that close in flowing or pumping wells when the lease tanks are full. Another type of shutdown device cuts off power to a pumping well when a sucker rod string breaks.

Other trouble indicated at a central control point include failure of flow valves to open or close, malfunctioning of a *sediment and water* (*S&W*) monitor, loss of power to electric motors, and erratic performance of various *transducers* in the field: those that transmit pressure, flow, and temperature values, for example. (A transducer is a device actuated by power from one system and supplying power to another system, usually in a different form.)

REMOTE TERMINAL UNITS

A greatly automated lease, one capable of transmitting information to and receiving commands from a distant point of control, contains a *remote terminal unit* (*RTU*). An RTU is the central point on the lease that receives command signals from the distant control point and relays the signals to the wells, pumps, tank batteries, and custody transfer units in the field.

The RTU also receives feedback from these points and relays the feedback to the distant control point. Wire circuits normally connect the RTU to units in the field. However, the RTU may communicate with the distant control point with wire circuits or by microwave radio.

For troubleshooters, the RTU is the place to start for diagnosing trouble in the field. They can pinpoint most difficulties from the RTU with a VOM. When a valve fails to respond, for example, troubleshooters can determine if its control relay is getting the proper signal from the RTU. Moreover, they can determine whether the transducers are sending a signal to the RTU.

To be successful, troubleshooters should be thoroughly familiar with the system. For example, they should—

1. know the overall layout of the field; they should know the field's electric power source and the location of the transformers, circuit breakers, and other equipment. Also, they should know the location of the wells, tank batteries, vapor recovery units, custody transfer units, valves, flow lines, and the RTU.
2. be aware of the type of transducers used to monitor pressure, temperature, and flow; they should know their manufacturer and signals they produce.
3. know how the flow valves operate; they should know whether they are electrically or electropneumatically operated.
4. be familiar with the shutdown devices on the producing wells—for example, they should know if strain gauges indicate parted rods or if liquid level floats in the tanks indicate tank levels.
5. know the type of signal the S&W monitor delivers to the RTU; they should know the indications of normal operation as well as the indications of failure of the temperature or capacity probe circuit.
6. know the components, modules, and the terminal strips in the RTU.

This information can be gathered from several sources, including the installer's drawings and literature. The information probably will not be available as a single form unless someone collects it from various sources.

Technicians should become familiar with a lease and learn its system before trouble develops. For example, they should determine the number and location of the terminal strips that connect incoming lines from various lease units to the RTU. It is also helpful to know a VOM's indications when it is applied to certain terminals.

FIELD TROUBLESHOOTING INSTRUMENTS

A multimeter that operates on internal batteries serves for most troubleshooting in the field. If a technician is expected to troubleshoot AC motor circuits, a clamp-on ammeter-voltmeter is a useful addition. Also, a small and portable oscilloscope is useful for troubleshooting the RTU, particularly if trouble develops in parts of its *telemetry* section. An oscilloscope can also be used in some instances to pinpoint trouble in most S&W monitor installations. The signal pattern can be studied at the RTU.

SHOP TROUBLESHOOTING

A shop designed to troubleshoot electronic instrumentation systems can quickly locate malfunctioning elements. Sophisticated test and measuring instruments, along with literature and diagrams, can make troubleshooting much easier than in the field.

Instruments

Most well equipped shops have—

- digital and analog multimeters,
- single and dual-trace oscilloscopes,
- signal generators,
- frequency counters,
- transistor testers, and
- capacitor testers.

A shop may also have special equipment intended for particular instrumentation elements, such as special power supplies, electronic analogs, and pulse generators.

Features and Organization

In many cases, a foreman directs a shop that serves a company with extensive instrumentation and control systems. Usually, a crew works under the direction of the foreman. Their responsibilities may extend beyond the repair of instruments. They may also stock spare parts and order replacements. Moreover, the shop may keep equipment records and repair histories.

Once equipment reaches the shop for repair, it may or may not remain the responsibility of the field troubleshooter, depending on the shop's policy. The shop is the logical place to stock circuit cards and modules to replace those that have been found faulty in the field. Regardless of who does the troubleshooting on the faulty pieces, the shop offers a good environment for the work. Usually, the shop allows technicians plenty of time to follow the manufacturer's troubleshooting guides.

Further, large elements of a control system can be handled in shops having only moderate facilities. Elements such as a *multitrace recorder* combined with a *three-mode controller* can be repaired and calibrated through the use of standards having high accuracy. (A multitrace recorder is an instrument that senses, or traces, and records two or more signals generated by a component to which the recorder is attached. A three-mode controller is an instrument that continuously measures the value of signals generated by three variable quantities or conditions and acts to control any deviation from a desired preset value.)

ELECTRICAL SAFETY

While electricity improves operations and reduces the work of personnel, it can also cause serious injury if not handled properly. Since electricity cannot be seen, personnel sometimes get a false feeling of security when working around it. However, electricity can be as dangerous as a rotating flywheel and cause serious injury or death. Therefore, technicians should develop a cautious attitude toward electricity and consider it a threat.

Electrical Hazards

Injury from electricity can come in several forms. Hazards include electrical shock, electrical flashes from arcs, and explosions from the release of energy and the flying shrapnel and material.

Electrical Shock

Electrical shock occurs when persons place their body in a live electrical circuit. For example, shock can occur when standing on a surface that is an electrical ground and touching a live circuit with another part of the body. Another way is to place one hand on an electrical circuit potential and the other hand on a different electrical potential.

Keep in mind that electricity always flows along the path of least resistance. Also, current, and not voltage, damages the human body. While voltage is required to produce electrical current, the magnitude of the current is also related to the conductivity of the body. A dirty and sweaty body offers lower resistance to either current flow or lethal shock than a clean and dry body.

When someone is shocked, the most important thing is to get the victim away from the source of the shock. However, never subject yourself to the same source of electricity. Rather, safely disconnect the power first. Once power has been definitely turned off, summon medical help immediately. Calm the victim and make him or her as comfortable as possible. If the victim's heart has stopped beating, start cardiopulmonary resuscitation (CPR) and continue it until medical help arrives.

Shock Levels

Different current levels cause the body to respond to shock in different ways. For example, static electricity is generated when a person moves his or her shoes on a dry surface. Touching a metal doorknob or other conductive body can discharge the electricity. Current then flows off the person's skin with a loud zap. The current in the discharge does not flow through the body; instead, it flows but only on the body's surface. The sensation and the effect are minor.

Under controlled conditions, very high voltage can be applied to the human body without damaging it as long as the current levels are below certain levels. For example, a person can be subjected, without harm, to thousands of volts of electricity as long as the amperage is small. Lots of volts and little amperage only raise hair on the head and do no harm. In such cases, electrical generators or transformers supply the voltage. They have very high internal impedance, which limits the amount of current flow to a low level.

The next level of shock can occur at voltage levels normally associated with 120-volt AC appliances. While voltages at this level can be lethal, many people have had the experience of being shocked when accidentally coming in contact with a live circuit at this level. The sensation is one of tingling throughout the body and an inability to control muscles. In most cases, the body's reaction causes a person to release the live circuit.

Generally, severe shock occurs at voltages above 120 volts AC and at current flows above 10 milliamperes. High currents cause loss of muscle control, which nullifies a person's ability to release the electrical object and remove the body from the hazard. Sustained exposure to the electrical source can cause burns, heart fibrillation, and heart stoppage.

Table 28 shows current levels and the body's response to them. Keep in mind that Ohm's law states that current is voltage divided by resistance ($I = V/R$). Thus, the lower a body's resistance, the higher the current is for a given voltage. Current levels in Table 28 are in milliamperes (ma)—that is, one-thousandth of an ampere.

TABLE 28
Body Response to Various Current Levels

Current Levels	Body Response
0.4 ma	Slight sensation on the hand
1.1 ma	Threshold of shock perception
1 to 8 ma	Unpainful shock; muscle control not lost
8 to 15 ma	Hazardous and uncomfortable shock; can cause cardiac arrest if shock lasts long enough
16 ma	Painful shock; usually, the let-go threshold; probability of cardiac arrest increases at this point upward
16 to 20 ma	Painful shock; victim cannot let go of energized object; muscle control lost
20 to 50 ma	Painful shock and severe muscle contraction
50 to 100 ma	Heart and nerve damage; severe burns and strong possibility of death
100+ ma	Probable death; in most persons, instantaneous death results

Hazardous Conditions

Electrical hazards occur under many conditions. It is therefore vital to be constantly alert to the possibility of shock. Some points to keep in mind follow.

1. Water and electricity do not mix. Avoid working on wet or damp surfaces because water can provide a low resistance path between a live electrical source, you, and a ground. Never touch or operate equipment while standing or in contact with water. Do not operate electrical appliances while taking a shower or bath. Exercise extreme care when washing down areas around motors and generators.
2. Portable equipment, such as an electric drill, can pose threats if the insulation is frayed or damaged or the electrical cord is spliced incorrectly. Also, do not drill into a surface without knowing whether live electrical circuits are behind it.
3. A metallic surface may appear to be a mechanical support; however, it may also be an electrical part. For example, bus bars, connecting lugs, tie joints, and male plugs appear to be merely mechanical but in reality are also electrical. Never touch equipment whose function is unknown.
4. Avoid disconnecting electrical equipment while it is operating. Unplugging live and operating circuits causes arcing and sparks, which may injure your eye or hand. Also, it may damage the receptacle and plug.
5. When working on electrical equipment, be sure all power is off and the disconnected equipment is tagged with a sign and locked out with your lock to prevent accidental application of power by other personnel. Further, if you see a locked and tagged out circuit, do not unlock it for any reason. Always assume that locked and tagged out equipment is being worked on.
6. Never apply power to circuits or equipment unless it has been inspected for proper operation and you have checked that no other personnel are in any danger.

Arc Flashes

When air ionizes in the presence of a voltage, current flows through the air to produce a white or bluish arc. An electric arc produces intense heat and radiation that can cause severe skin burns, eye damage, and other injuries. Moreover, an arc can start a fire if it occurs near combustible material.

The intensity of an arc is influenced by several factors, including—

- the magnitude of the current,
- the voltage across the electrode gap,
- the number of phases involved, and

- whether the arc occurs in the open or in an enclosure.

Besides the intensity of the arc, the injury incurred from an arc is also influenced by—

- the arc's duration,
- the distance between the arc and the person,
- the type and fit of clothing worn,
- the percentage of the body burned, and
- age and health factors.

Scientists have analyzed arcs and determined the incident energy they pack. This analysis can guide personnel who work on, around, or with electrical equipment in selecting the proper personal protective equipment (PPE) for flash protection. Researchers present the energy and intensity of arcs in calories per square centimetre (cal/cm^2). Factors that influence arc intensity include the amount of fault current in the system, size of arc's gap, supply voltage, number of phases, distance from arc's centerline to the worker, and whether the arc is in air or in an open box.

Table 29 shows the results of software calculations that determined the flash energy from an arc. Figures in the table demonstrate the energy and thermal exposure acting on a person subjected to an arc under certain conditions. In Table 29, the conditions are that (1) the arc lasts 10 cycles of the AC, (2) the arc's length is 12 in. (30 cm), (3) the worker is 12 in. (30 cm) from the arc, and (4) the equipment is powered by 4,160 VAC. The resulting energy is in cal/cm^2.

These figures can be used to select the appropriate PPE for those working near equipment that may arc, to identify areas and distances that are dangerous for observers, and to plan electrical maintenance and repair with confidence.

Flammable Gas Hazards

Another hazard exists when electrical equipment is used in the vicinity of combustible gases. Combustible gases include hydrocarbon gases that occur around oil and gas facilities. Combustible gases ignite into an explosive flame at a given temperature. The temperature at which they ignite depends on whether the gases evolve from flammable liquids or volatile solids, or whether the material itself is a gas. The National Electric Code (NEC), Article 500, lists ignition temperatures for which equipment has been approved when used in the presence of gases and combustible materials.

TABLE 29
Arc Flash Energy

Fault current (amps)	Single-phase arc (in open air)	Single-phase arc (enclosed in box) factor = 2	Three-phase arc (enclosed in box) factor = 3.4
2,000	2.5 cal/cm^2	4.9 cal/cm^2	8.3 cal/cm^2
4,000	5.0 cal/cm^2	9.9 cal/cm^2	16.8 cal/cm^2
6,000	7.5 cal/cm^2	15.0 cal/cm^2	25.5 cal/cm^2
8,000	10.1 cal/cm^2	20.2 cal/cm^2	34.3 cal/cm^2
10,000	12.8 cal/cm^2	25.5 cal/cm^2	43.4 cal/cm^2
15,000	19.6 cal/cm^2	39.3 cal/cm^2	66.8 cal/cm^2
20,000	26.9 cal/cm^2	53.8 cal/cm^2	91.5 cal/cm^2
40,000	60.4 cal/cm^2	120.8 cal/cm^2	205.4 cal/cm^2
65,000	115.8 cal/cm^2	231.6 cal/cm^2	393.7 cal/cm^2

Because of the ever-present danger of sparks or arcs being developed around electrical equipment, care and consideration must be taken when the equipment is installed in or near atmospheres that contain hydrocarbons. The American Petroleum Institute (API) publishes *Recommended Practice for Classification of Locations for Electrical Installations at Petroleum Facilities*, RP 500, that classifies certain areas within production facilities for potential hazards. (RP 500 is available from the American Petroleum Institute; the web address is www.api.org.) RP 500 was derived from NEC, Article 500. In response to the recommended practices, manufacturers have designed equipment and companies have established procedures for placing electrical systems in hazardous areas.

To summarize API RP 500—

1. If an explosive gas mixture exists or may exist under normal conditions of operation or maintenance, the area is designated as a class I, division 1 area. Therefore,

explosion-proof, enclosed equipment built to the National Electrical Manufacturers Association's (NEMA's) standard 7, or equipment rated for class I, division 1, must be used.

2. If an explosive gas mixture exists or may exist only in the event of equipment failure and such failure does not cause a simultaneous electrical failure, then the area is designated as class I, division 2. Equipment must be suitable for class I, division 2, areas but not necessarily explosion proof. In some cases, totally enclosed, fan-cooled equipment, such as AC induction motors, may be used.
3. If an area has no reasonable chance of a gas leak, and is more than 10 ft (3 m) from process equipment, it is unclassified. In chemical plants, gas plants, and refineries, general purpose or rain-tight equipment is normally acceptable unless procedures dictate otherwise.

Explosion-Proof Equipment

One of the most popular methods used to operate electrical equipment in class 1, division 1 and class 1, division 2 areas is to place the electrical components in an explosion-proof enclosure. The enclosure is constructed of heavy-gauge cast aluminum or other metal that can sustain an internal explosion resulting from combustible gas and electrical sparks. Manufacturers machine the covers to a smooth finish, which helps keep explosive flames from escaping from the enclosure when it is bolted or screwed together.

To complete the installation, an explosion-proof enclosure should have seals placed in the conduit that contains wiring or metal-clad cable should be used to prevent an explosion from going to other parts of the electrical system.

Intrinsic Safety

Equipment that is called intrinsically safe has certain assumptions associated with it. In general, the label, intrinsically safe, suggests a component or equipment that is incapable of producing sufficient energy to ignite combustible mixtures under normal or abnormal conditions. This assumption must be made knowing the conditions in which the component will be used, the type of gas present, the maximum ambient temperatures present, and the energy available from an intrinsically safe device.

When installing intrinsically safe equipment, always refer to standards such as the NEC's Article 500-1, and to the manufacturer's specifications and recommendations. Intrinsically safe equipment bears a tag stating that it is intrinsically safe and the conditions in which it is safe to be used.

Intrinsically safe also means that the component or equipment is a part of an intrinsically safe system. The equipment is designed within controlled barriers to make it intrinsically safe. Removal of these barriers eliminates its intrinsically safe rating. Intrinsically safe systems must be designed and maintained in accordance with the NEC's Article 500-1.

Purging

One method used in the petroleum industry to prevent explosions and make an electrical installation safe is to purge equipment with continuous fresh and clean air, which removes combustible gas that may be present before operation. By purging the equipment with adequate airflow and pressure, the need to house equipment in explosion-proof enclosures is eliminated.

Certain precautions must be followed in such installations, however, as outlined in NEC Article 501-8 and in the *National Fire Protection Agency's* (*NFPA's*) regulation 496. Precautions include:

1. The amount of air pressure provided for purging must maintain a positive pressure, be totally enclosed, and provide a complete purging cycle to remove 10 volumes of air before the equipment is operated.
2. The source of the fresh air must be assured so that contamination from other sources of gas is eliminated and the exhaust must be diverted to a safe location.
3. A sensor must be provided to detect loss of airflow, activate an alarm, and safely turn off the equipment until purging is resumed.
4. Detection of potential hazards that may occur during normal operation of the equipment, such as overheating or overloads, must be provided and a safe shutdown implemented.

Typically, production operations do not use purging; however, certain drilling applications, such as control enclosures on the rig floor and DC motors used to power rigs, do use purge systems.

Grounding

Several reasons exist for grounding electrical equipment. One is to protect the operator of electrical equipment from being subjected to high, and potentially deadly, voltages when a live wire accidentally touches its enclosure. Another is to provide a path for fault currents (short circuits) so that protective devices such as fuses and circuit breakers can operate and isolate the fault condition. Still another is to establish a drain source for any electrical noise that may interfere with sensitive electronic equipment.

Grounding establishes a zero voltage reference for a voltage supply since the operator's body is usually at ground potential. Any voltage source that is connected to ground is then at zero voltage because it is shorted through the low-resistance ground connection.

Grounding also can operate protective circuit devices such as fuses, circuit breakers, and ground fault detectors. If a circuit is properly grounded, all other voltages are referenced to it. If one of the circuits that normally operates above ground accidentally comes in contact with ground because of insulation breakdown, accidental contact, and the like, very high current flows, which causes the protective fuse or circuit breaker to operate.

If a circuit is not referenced to ground, contact with ground potential may not cause the protective device to operate. In such a case, if an accidental ground was maintained without detection, accidental shock or equipment damage could occur.

Companies have established criteria for grounding based on NEC, Article 250. In addition to these requirements, standardized grounding industry practices include:

1. All metal parts of equipment containing electrical components that do not carry electricity shall be grounded to prevent any rise in voltage on such parts in relation to earth or structure near such equipment. The resistance between equipment enclosures and the ground circuit should be less than 1 ohm.
2. The grounding conductor should be low enough in resistance and have adequate current-carrying capacity to safely handle any expected ground fault currents. Acceptable conductors include—
   a. copper conductors no smaller than that required by NEC's Table 250-95.
   b. rigid metal conduit.
   c. the metal sheath of *metal-clad (MC) cables*. In large-sized cables where the metal sheath alone is not adequate, a ground wire is included in the cable.
   d. on new installations, all power feeders (480, 208, and 120 volts) must have an internal equipment-grounding conductor in addition to all other ground paths.
3. On steel platforms or rigs, the structure should be considered ground.
4. On concrete platforms, a grounding conductor or conductors adequately sized for maximum faults and connected to an earth ground are required.
5. NEC allows the resistance to ground of the grounding electrode to be as high as 25 ohms depending on the circuit's kVA size and power conductors. However, for improved personnel protection, the resistance to ground should be 1 ohm or less.
6. The grounding conductor shall not be a load current-carrying conductor.
7. Grounding conductors and connections shall be approved and made of corrosion-resistant materials or adequately coated to protect against corrosion.
8. Grounding circuits shall be installed to minimize the possibility of physical damages. For mechanical reasons, it is recommended that no ground conductor outside a *raceway* be smaller than No. 6 AWG.

Grounding Methods

In all equipment installations, whether portable or fixed, at least one earth ground should be established. NEC, Article 250-83, specifies that the ground be attached to a fixed metal frame when the frame is effectively grounded to an earth ground.

Bonding is the next most effective means of providing a ground for equipment reference and personnel protection. The housings or structures of electrical equipment that do not carry electricity should be bonded together through welding or with a substantial cable or wire braid as specified in NEC, Article 250-94.

Even in portable systems, such as skid-mounted equipment, the skids and equipment housings should be bonded together with copper braid or conductors to assure a ground bond. In some permanent or semipermanent installations, welding the structures to each other and establishing a low-resistance connection can accomplish bonding.

Regardless of how the bond or grounding is established, a technician should always test the bond or ground with an ohmmeter after the work has been completed to assure that the desired continuity and resistance levels have been met.

Dissimilar Metals

Where possible, similar metals should be used for electrical equipment connections. When dissimilar metals are brought in contact with each other, oxides can form. These oxides can create a high-resistance contact that generates heat, which damages equipment. If dissimilar metals are used, such as copper and aluminum, technicians should apply a chemical on all aluminum lugs to keep aluminum oxide from forming. A better solution is to use cadmium-plated copper lugs rather than aluminum lugs. In any installation where dissimilar metals are used, frequent inspection and surveys using infrared heat detectors are recommended.

Circuit Protective Devices

Equipment manufacturers and designers provide several electrical and mechanical devices to protect equipment users. Among the mechanical devices are insulating equipment that prevents current flow from the voltage source to ground, barriers that isolate one voltage level from another, and enclosures that prevent the accidental touching of live electrical parts.

Electrical devices provide automatic detection of equipment fault conditions. These include fuses, circuit breakers, and overload devices. Also, devices that sense an unwanted ground condition (caused, for example, by wiring insulation breakdown) can sound an alarm, indicate the problem, or interrupt the electrical circuit.

Ground Fault Detectors

In many systems, *ground fault detectors* are used to detect grounds and provide a visual alarm in equipment whose electrical output cannot be readily grounded. These systems are referred to as floating because they are not directly connected to a ground reference. Whenever a ground occurs in such systems, evidence may not exist that indicates that the ground has occurred because of insulation breakdown or accidental connection. In such a case, devices called ground fault detectors and indicators are used. These devices incorporate sensors, such as transformers or resistors, in a common ground circuit. In some cases, the power system continues to operate without disconnection because of operating considerations.

When a ground in the main system occurs, ground currents flow in the detector's sensors, and the detector illuminates a lamp or activates an audible alarm. A technician should promptly locate the ground faults and repair the problem to prevent injury to personnel or damage to equipment.

Ground Detecting Systems

A typical ground fault sensor used in low-voltage installations is the high-resistance type. A high-resistance sensor (typically 200 ohms) is placed in the neutral ground circuit of the 480-volt, three-phase, generating system, or a *wye-connected* transformer, to detect a ground fault with the aid of a voltage sensitive relay. It is usually located in a generator *switchgear* cabinet and operates an alarm.

Because ground fault currents are limited to 1.38 amperes with a 200-ohm neutral ground resistor, the power system may continue to operate; however, the fault should be located and corrected as soon as practical. A ground fault voltmeter indicates 277 volts when a ground fault exists. High-impedance faults on a voltmeter show up as voltages between 0 and 277 volts.

Similar ground detecting methods are used in high-voltage systems (those over 1 kV), but the resistance used for detecting the ground fault is considered to be either a low-resistance type or high-resistance type.

Low-resistance ground fault detection in high-voltage systems consists of a resistor connected to the neutral of the generator or transformer. When a ground fault occurs, a relay or similar device senses the fault, current flows through the resistor, and trips the main circuit breaker to off. In the case of generators, a low-resistance ground fault system also de-excites (cuts off exciting current to) the generator to prevent further voltage from being developed into the ground fault.

High-resistance ground fault detection in high-voltage systems uses a bus-connected, three-phase transformer with a wye connection primary and broken *delta connection* secondary. (A delta connection is a combination of three components connected in series to form a triangle similar to the Greek letter delta.) The wye primary uses a resistor connected to its neutral and is further connected to ground. The secondary broken delta uses a resistor-relay connected to the open portion of the secondary that is operated when the primary has a ground fault occur on one of its phases. When the relay detects a ground fault, it causes generator circuit breakers to trip, which isolates the fault and de-excites the generators.

Ground Fault Circuit Interrupters (GFCIs)

In areas where personnel are frequently exposed to water, such as in bathrooms, kitchens, or at outdoor receptacles, devices called *ground fault circuit interrupters* (GFCIs) are installed. A GFCI is a fast-acting circuit breaker that senses very small ground fault currents and which breaks the circuit when it detects such faults. When a GFCI detects a ground fault current greater than 5 mA, it opens the circuit (typically 120 VAC power) to protect personnel. GFCIs can also be installed in main circuit breaker panels to protect a complete branch circuit.

Arc Fault Circuit Interrupters (AFCIs)

Recent fire investigations have revealed that insulation failure and accidental contact can create arcing between circuit conductors, which does not produce enough current to trip protective circuit breakers or fuses. Such intermittent arcing can cause a fire and substantial damage. GFCIs cannot sense arcing because the current does not go to ground; rather, the current arcs between load carrying conductors or inside a partially broken conductor.

In the U.S., residential electrical fires cause 700 deaths, 3,000 injuries, and $700 million in property damage each year. Electrical faults cause several of these fires. Arcing causes electrical fires that start without warning. Sputtering electric arc faults between damaged and deteriorating wiring kindle the fires by generating heat in nearby combustible material.

An *arc fault circuit interrupter* (*AFCI*) has a coil and electronic circuits within a circuit breaker. When an AFCI detects arcing, the circuit breaker trips to isolate the circuit from additional voltages and arcs. The NEC introduced the AFCI in 1999, but with few details or instructions for its implementation. The 2002 edition of the NEC recognized the AFCI and incorporated it in Section 210.12. Section 210.12 states that AFCIs must be installed in all residential dwellings built after January 1, 2002.

SAFETY SUMMARY

It is very important to understand the dangers associated with electricity. Injury can occur when it is least expected. A shock can cause serious internal injuries and death. An electrical flash can produce burns that vary from mild to life threatening. To avoid shock, turn the power off whenever work is being performed on or near electrical equipment. If the power cannot be turned off for maintenance or repair, wear proper personal protective equipment (PPE) to prevent shock and burns from arc flashes. To select the proper PPE, refer to OSHA regulations and NFPA 70E.

REVIEW QUESTIONS

1. The modular concept of modern solid-state electronic systems provides for ____________ troubleshooting practices.
2. The date of the failure or malfunction of equipment and details of its correction should be a matter of entry in ____________________.
3. Before attempting to troubleshoot a new or unfamiliar piece of equipment, what source of information should be consulted?
4. What is the most popular form of troubleshooting instrument?
5. Oscilloscopes are more complex than most other instruments, partly because they have more extensive ___________________ than most instruments.
6. A step-by-step procedure is designed to pinpoint trouble in a particular section or free that section from suspicion through a process of ________________.
7. Once symptoms of a malfunction are noted, the troubleshooter should refer to the ________________.
8. Where it is applicable, the ______________ is an expeditious way to restore equipment to normal operation.
9. A module or card that has failed is removed, and a good replacement card is substituted. If it fails immediately, where does the fault probably lie?
10. If it is not feasible to repair a faulty component or module without delay, what step should be taken?

11. It is usually an easier task to ______________ a fault than to find it.
12. In discrete component assemblies, it is sometimes permissible to use special sprays to ______________________ a suspected component. Such sprays should not be used on printed circuits or those containing ______________.
13. The ______________________ is the central point on a lease where command signals are received from a distant control point and relayed to various units on the lease.
14. A foremost requirement for troubleshooting an automated lease is ____________________ with the system.
15. If a technician is expected to troubleshoot AC power circuits, _____________ is a handy addition to the tool kit.
16. Troubleshooting in a shop is usually easier because of the better availability of __________ __________ and __________ __________.
17. To avoid shock or arc flash injury, personnel working near, on, or with electrical equipment should wear _______________.

Glossary

A

A *abbr*: ampere.

AC *abbr*: alternating current.

addend *n*: one of a collection of numbers to be added.

AFCI *abbr*: arc fault circuit interrupter.

air *n*: a colorless, odorless, tasteless, gaseous mixture, mainly nitrogen (approximately 78 percent) and oxygen (approximately 21 percent) with lesser amounts of argon, carbon dioxide, hydrogen, neon, helium, and other gases.

air-cored coil *n*: in a coil of wire carrying electric current, the central interior portion of the coil in which only air exists.

alnico *n*: a metal alloy of aluminum, nickel, and one or more of the elements cobalt, copper, and titanium. This alloy is used to make strong permanent magnets.

alnico magnet *n*: a permanent magnet made from the alloy alnico. See *alnico*.

alternating current (AC) *n*: current that flows in one direction in a circuit and then reverses flow. Reversal of current flow happens rapidly—for example, in many AC circuits in the United States, reversal happens 60 times in 1 second.

alternation *n*: in an alternating current generator, the rising and falling of induced positive and negative voltage as the coil rotates from 0 to 180 degrees to produce positive voltage and from 180 to 360 degrees to produce negative voltage. The rising and falling of positive voltage from 0 to 180 degrees is one alternation. The rising and falling of negative voltage from 180 to 360 degrees is also one alternation. One positive and 1 negative alternation is 1 cycle.

alternator *n*: an electric generator that produces alternating current.

AM *abbr*: amplitude modulation.

American National Standards Institute (ANSI) *n*: serves as a clearinghouse for nationally coordinated voluntary standards for fields ranging from information technology to building construction. Address: 11 W. 42nd Street, 13th floor; New York, NY 10036; (212) 642-4900; www.ansi.org.

American standard code for information interchange (ASCII) *n*: in computer science, a standard for assigning numerical values to the set of letters in the Roman alphabet and to typographic characters. For example, in ASCII code, the binary number 1001011 is assigned the letter K, and the binary number 1011110 is assigned the typographic character ^.

American War Standard (AWS) *n*: outdated specification for supplies and equipment purchased by the military in World War II. It has been replaced by military specification (MIL).

American wire gauge (AWG) *n*: a standard that assigns a number (a gauge) to electrical wires of various diameters, or sizes. The largest AWG wire is listed as 0000, which is often written as 4/0, and has a diameter of 460 mils, or 0.46 inches. (A mil is 0.001 inches.) A wire often used in small household appliances, such as small table lamps, is 16 gauge, which is 51 mils, or 0.051 inches, in diameter. The smallest is 40-gauge wire, which is 3.1 mils, or 0.0031 inches in diameter. As the AWG number increases, the gauge (thickness) of the wire decreases.

ammeter *n*: an instrument for measuring electric current in amperes.

amp *abbr*: ampere.

ampere (A) *n*: the fundamental unit of electrical current; 1 ampere = 6.28×10^{18} electrons passing through the circuit per second. One ampere delivers 1 coulomb in 1 second.

ampere-hour *n*: a unit of electricity equal to the amount produced in 1 hour by a flow of 1 ampere. See *ampere*.

ampere-hour rating *n*: a measure of a battery's ability to produce current in terms of ampere-hours. For example, if a battery has an ampere-hour rating of 30, the battery could produce 1 ampere of current for a 30-hour period.

amplifier *n*: an electronic device for increasing the magnitude of a quantity such as an electrical measurement signal. Amplifiers may be used to increase a transmitted and received measurement signal.

amplitude *n*: a measure of the extent of a vibratory movement, or an oscillation. It expresses the maximum deviation that a periodically varying quantity takes from its normal value of zero.

amplitude modulation (AM) *n*: the varying (modulating) of a wave in which the height (amplitude) of the wave is the characteristic varied in accordance with the intelligence to be transmitted. Compare *frequency modulation.*

analog *adj*: 1. of or pertaining to an instrument or equipment that measures a continuous variable that is proportional to another variable over a given range. For example, temperature can be represented or measured as voltage, its analog. 2. pertaining to devices, data, circuits, or systems that operate with variables that are represented by continuously measured voltages or other quantities. Compare *digital.*

AND gate *n*: a circuit that has two or more input-signal ports that deliver an output only if and when every input port is simultaneously energized.

angle of declination *n*: the difference, in angular degrees, between magnetic north and true north.

annunciator *n*: a signaling device that operates electromechanically and serves to indicate visually or visually and audibly whether current is flowing, has flowed, or has changed direction of flow in one or more circuits.

anode *n*: in a semiconductor diode, the terminal toward which forward current flows from an external circuit. See *diode.*

apparent power *n*: the vector sum of the power in watts plus the reactive power in volt-ampere reactive (VAR) in a circuit. See *reactive power.*

aquadag *n*: the graphite coating on the inside of certain cathode-ray tubes. It collects secondary electrons emitted by the face of the tube.

arc fault circuit interrupter (AFCI) *n*: a special type of circuit breaker that has a coil and special electronic circuitry that can sense the small fault currents created by arcing between components and that trips the circuit to prevent fires caused by arcing.

arithmetic converter *n*: in a computer, a circuit or component that changes (converts) electronic signals into units that additional circuitry can add, subtract, multiply, divide, or compare.

armature *n*: a part made of coils of wire placed around a metal core, in which electric current is induced in a generator, or in which input current interacts with a magnetic field to produce torque in a motor.

arsenic (As) *n*: a chemical element that occurs as a brittle, steel-gray hexagonal mineral and that is added as an impurity to semiconductors to give them a negative charge.

artificial magnet *n*: a bar or other shape of soft iron or hard steel alloy that is magnetized by placing the bar or other shape in a magnetic field. Magnetite (lodestone) is a natural magnet; other magnets are thus artificial.

As *sym*: arsenic.

ASCII *abbr*: American standard code for information interchange.

aspect ratio *n*: the ratio of a screen's (or frame's) width to the screen's height. For television sets in the U.S. and Great Britain, for example, the TV screen's aspect ratio is 3 to 4, which means that if the screen's height is 3 inches, its width is 4 inches; or, if its height is 12 inches, its width is 16 inches.

atom *n*: the smallest quantity of an element capable of either entering into a chemical combination or existing alone.

atomic particle *n*: one of the small bits of matter that makes up an atom, such as an electron, proton, or neutron.

attenuating probe *n*: in electronics, a small device (a probe) that is put into contact with or inserted into a circuit to make measurements of the circuit and that reduces (attenuates) the strength of the value (such as voltage) being measured in the circuit. The amount of attenuation is usually expressed as a ratio—for example, a 10-to-1 probe reduces the value's strength by 10.

audio transformer *n*: an iron-core transformer used to bring together (to electronically couple) circuits that carry audio frequencies.

augend *n*: a quantity to which another quantity is added.

average current rating *n*: in a solid-state diode, the amount of current that the diode is designed to handle over a range of operating current in a circuit, from a minimum to a maximum amount.

AWG *abbr*: American wire gauge.

AWS *abbr*: American War Standard.

axial lead (pronounced "leed") *n*: in an electronic component such as a composition resistor, a single wire (a lead) that exits each end of the component from the component's center and to which the electrical connections are made to the circuit in which the component is installed. Compare *radial lead.*

B

back electromotive force (emf) *n*: 1. voltage induced in the primary winding of a transformer that opposes the incoming line voltage. 2. voltage induced in the armature of a motor by the magnetic field in which it is turning. Also called counter emf and countervoltage. See *electromotive force.*

back emf *abbr*: back electromotive force.

bar magnet *n*: a piece of hard steel in the shape of a rectangular-shaped rod (a bar) that has been strongly magnetized and holds its magnetism. It is therefore a type of permanent magnet. See *permanent magnet.*

base *n*: 1. a substance capable of reacting with an acid to form a salt. A typical base is sodium hydroxide (caustic), with the chemical formula NaOH. For example, sodium hydroxide combines with hydrochloric acid to form sodium chloride (a salt) and water. This reaction is written chemically as $NaOH + HCl \rightarrow NaCl + H_2O$. 2. in a junction transistor, the region that lies between the transistor's emitter and collector and into which minority carriers are injected. The base is the transistor's control terminal and receives current from an external source into its region. See *collector; emitter.*

battery *n*: an electricity storage device consisting of two or more galvanic cells. See *galvanic cell.*

battery charger *n*: an electrical device (a rectifier) that converts alternating current into direct current and is used to reestablish the battery's ability to produce current by reversing the chemical reaction that occurs in a secondary (wet cell) battery when it discharges. See *rectifier.*

BCD *abbr*: binary coded decimal.

bias *n*: direct-current voltage applied to a transistor control electrode to establish a desired operating point.

binary coded decimal (BCD) code *n*: a system of number representation in which a binary number represents each digit of a decimal number.

binary number *n*: a method of representing numbers using only the digits 0 and 1 in which successive digits are interpreted as coefficients of successive powers to the base 2. For example, the decimal number 3 is 10 in binary numbers and the decimal number 4 is 11 in binary numbers.

binary point *n*: the character, or the location of an implied symbol, that separates the integral part of the numerical expression from its fractional part in binary notation.

bipolar *adj*: of a material capable of assuming a positive or negative electric charge, such as a bipolar transistor.

bistable *adj*: capable of assuming either of two stable states; (pronounced bi-stable).

bistable mutivibrator *n*: see *flip-flop circuit.*

blanking pulse *n*: in an oscilloscope, a positive or negative square-wave variation in electrical power (a pulse) that switches off a part of the oscilloscope's screen for a predetermined length of time.

blocking diode *n*: a diode that restricts (blocks) the flow of current in an undesirable direction in a circuit. See *diode.*

Boolean logic *n*: a system of symbolic logic based on algebraic symbols representing such logical operations as AND, OR, and NOR.

borrowing *n*: in the mathematical operation of subtraction, the act of taking the number 1 from the tens, hundreds, or other column to the left of the ones column, and placing it to the left of the number in the ones column, to raise the number in the ones column to the next higher order digit. This action reduces the number that was borrowed from by 1. Borrowing then continues, if necessary, in each subsequent column to the left until the solution is achieved. For example, to subtract 679 from 831, 1 is borrowed from 3 in the tens column of 831, which leaves 2. Nine is then subtracted from 11 to leave 2. Next, 7 is subtracted from 12, because 1 was borrowed from the hundreds column to make 12. This action leaves 5. Finally, 6 is subtracted from 7 in the hundreds column, which was reduced by 1 when it was borrowed from earlier. This action leaves 1. So, the answer is 152.

braided wire *n*: see *stranded wire.*

breakaway torque *n*: see *starting torque.*

breakdown torque *n*: see *pullout torque.*

British thermal unit (Btu) *n*: a measure of heat energy equivalent to the amount of heat needed to raise 1 pound of water 1°F at or near its point of maximum density (39.1°F). Equivalent to 0.252 kilogram calories or 1,055 joules.

brush *n*: a carbon block used to make an electrical connection between the rotor of a generator or a motor and a circuit.

Btu *abbr*: British thermal unit.

buffer *n*: 1. in solid-state electronics, an isolating circuit in a computer that prevents the action of a driven circuit from affecting the corresponding driving circuit. 2. an electric circuit or component that prevents an undesirable electrical interaction between two circuits or components.

bus *n*: an assembly of electrical conductors for collecting current from several sources and distributing it to feeder lines so that it will be available where needed. Also called bus bar.

bus bar *n*: see *bus.*

C

C *sym*: 1. carbon 2. coulomb.

cable *n*: in electronics, a wire made of a conducting material such as copper or aluminum. Usually, cable is braided from several single strands of wire into a single conductor that is easier to bend than a solid wire of the same gauge.

can-type capacitor *n*: a capacitor whose components are housed inside a small, can-shaped cylinder. See *capacitor*.

capacitive coupling *n*: in electronics, the transfer of energy from one circuit to another through a capacitor.

capacitive reactance *n*: the impedance of current flow in an alternating current circuit (the reactance) caused by a capacitor in the circuit.

capacitor *n*: an electrical device that, when wired in the line of an electrical circuit, stores a charge of electricity and returns the charge to the line when certain electrical conditions occur. Also called a condenser.

carbon (C) *n*: a naturally abundant nonmetallic element that occurs in many inorganic and in all organic compounds, exists freely as graphite and diamond and as a constituent of coal, limestone, and petroleum, and is capable of chemical self-bonding to form an enormous number of chemically, biologically, and commercially important molecules.

carry *v*: in the mathematical operation of addition, to perform the necessary actions to obtain a correct answer when the sum of the digits in a given position equals or exceeds the base of the number system. For example, when adding 352 and 49 in the decimal system, 9 is added to 2 in the ones column, which equals 11. In this case, 1 is written below the ones column and the other 1 is added (carried) to the 5 in the tens column of 352, making the 5 a 6. Then, 6 is added to 4 and a 0 placed below the tens column. The 1 is carried to the 3 in the hundreds column, making it 4. The answer is 401.

cathode *n*: the terminal of a semiconductor diode that is negative with respect to the other terminal when the diode is biased in the forward direction. See *diode*.

cathode ray *n*: a stream of electrons, such as those emitted by a heated filament in a tube, or those emitted by the cathode of a gas-discharge tube when the cathode is bombarded by positive ions.

cathode-ray tube (CRT) *n*: an electron tube in which a beam of electrons can be focused to a small area and varied in position and intensity on a surface.

CCA *abbr*: cold cranking ampere.

cell *n*: a device that delivers an electric current as the result of a chemical reaction; a receptacle, such as a jar, tube, or box, that contains two dissimilar metals immersed in or surrounded by a liquid or solid chemical (an electrolyte) that reacts with both metals.

channel *n*: the main current path between the source and drain electrodes in a field-effect transistor.

chassis *n*: the framework to which the components of a radio, television, or other electronic equipment are attached.

chemical reaction *n*: a change in which a substance or substances is changed into one or more new substances.

chip *n*: 1. the shaped and processed semiconductor die that is mounted on a substrate to form a transistor, diode, or other semiconductor device. 2. an integrated microcircuit performing a significant number of functions and constituting a subsystem.

choke *n*: in electronics, an inductance used in a circuit to present high impedance to frequencies above a specified frequency range without appreciably limiting the flow of direct current.

circuit *n*: a complete electrical path from one terminal of a source of electricity to the other, usually connected to a load. When the circuit is closed, electric current flows through it; when it is opened, the current flow stops.

circuit diagram *n*: see *schematic*.

circular mil (cmil) *n*: the cross-sectional area of a wire or cable expressed in thousandths of an inch (a milli-inch).

cladding *n*: in electronics, braided or solid aluminum or other material that surrounds insulated wires or cables. Cladding protects the wire or cable from being damaged by blows or forces from external sources, such as dropped objects, foot traffic, and the like.

clamp-on meter *n*: an AC voltmeter-ammeter that has a magnetic core in the form of hinged jaws that can be clamped around a current-carrying wire to measure AC voltage or amperage. Also called a snap-on meter.

clip *v*: in electronics, to prevent the amplitude of a waveform from exceeding a specified level while preserving the shape of the waveform at amplitudes less than the specified level.

clipper *n*: an electronic circuit used to prevent the amplitude of a waveform from exceeding a specified level while preserving the shape of the waveform at amplitudes less than the specified level. Also called a limiter.

clipping *n*: in electronics, the act of preventing the amplitude of a waveform from exceeding a specified level while preserving the shape of the waveform. Also called limiting.

clock *n*: see *timing circuit.*

cmil *abbr*: circular mil.

Co *sym*: cobalt.

coax *abbr*: shortened form of coaxial cable.

coaxial cable (coax) *n*: electrical conducting wire (cable) that has a center wire covered by polyethylene insulation. A second conductor of braided copper (or other conductor such as aluminum) fits over the polyethylene insulation. Finally, a vinyl cover protects the braided conductor. The braided conductor protects, or shields, the center conductor from extraneous and undesirable electrical interference that may affect the signal being carried by the center conductor.

cobalt (Co) *n*: a metallic element used chiefly in alloys.

coefficient *n*: a number that serves as a measure of some property or characteristic and that is commonly used as a factor in computations.

coil *n*: see *solenoid.*

cold cranking ampere (CCA) *n*: a measure of a battery's ability to provide current to start (crank) an engine when the engine is started cold—that is, when the engine has not been running for several hours, such as overnight.

collector *n*: in a junction transistor, a semiconductive region through which the primary flow of charge carriers leaves the base of the transistor. The collector is the transistor's output terminal and is connected to the positive polarity of voltage from the supply through a resistor. See *base* (def. 2), *emitter.*

color code *n*: in electronics, the indication of resistor and capacitor values by the application of 12 different colors and no color. Each color, or lack thereof, is assigned a numerical value and also indicates a multiplier as well as the power rating. By correctly reading the color sequence, the resistance or capacity, as well as the power rating, of a resistor or a capacitor can be determined.

commutator *n*: a series of bars connected to the armature coils of an electric motor or generator. As the commutator rotates in contact with fixed brushes, the direction of flow of current to or from the armature is in one direction only.

composition *n*: a special material made of conductive and nonconductive substances, which, when blended and formed into a resistor, give the resistor a specific resistance.

composition resistor *n*: a resistor made of conductive and nonconductive materials, which are combined and formed into a relatively small cylinder. Axial or radial leads in the resistor provide a way to connect the resistor to a circuit.

compound *n*: a substance formed by the chemical union of two or more elements in definite proportions; the smallest particle of a chemical compound is a molecule.

condenser *n*: see *capacitor.*

conductivity *n*: the ability to transmit or convey (as heat or electricity); the opposite of resistivity.

conductor *n*: a material with a high concentration of free electrons. It readily carries (conducts) an electric current.

conduit *n*: solid or flexible metal or other tubing through which insulated electric wires are run.

conservation of energy *n*: see *law of conservation of energy.*

Constantan *n*: an alloy of 55 percent copper and 45 percent nickel; often used in making precision wire-wound resistors.

contactor *n*: a switch used to open or close an electric circuit.

converter *n*: a device for changing alternating current to direct current or direct current to alternating current; usually, if the device changes AC to DC, it is called a rectifier.

copper *n*: a ductile, malleable, reddish-brown metallic element that is an excellent conductor of heat and electricity and is widely used for electrical wiring, water piping, and corrosion-resistant parts, either pure or in alloys such as brass and bronze. Copper's chemical symbol is Cu.

core *n*: in electricity, the interior and central part of a coil; it may be air or a solid material such as iron. See *solenoid.*

core loss *n*: in a transformer, the reduction (loss) in electrical energy that occurs in the core of the transformer's windings. See *core.*

corona effect *n*: a phenomenon that occurs when high voltage is transmitted through uninsulated conductors. At very high voltages, the voltage ionizes the surrounding atmosphere and the atmosphere also begins to conduct electricity. See *ionization.*

cos *abbr*: cosine.

cosec *abbr*: cosecant.

cosecant (cosec) *n*: in a right triangle, the ratio of the hypotenuse to the side opposite a given angle.

cosine (cos) *n*: in a right triangle, the ratio of the length of the side adjacent to an acute angle to the length of the hypotenuse.

cotan *abbr*: cotangent.

cotangent *n*: in a right triangle, the ratio of the side adjacent to a given angle to the side opposite.

coul *abbr*: coulomb.

coulomb (C, coul) *n*: the amount of electric charge that crosses a surface in 1 second when a steady current of 1 ampere is flowing across the surface.

counter electromotive force *n*: see *back electromotive force*.

counter emf *n*: see *back electromotive force*.

countervoltage *n*: see *back emf*.

coupling *n*: in electronics, the relation between two circuits that permits energy transfer from one to the other, usually through a wire, resistor, transformer, capacitor, or other device.

covalent bond *n*: see *electron-pair bond*.

cross-multiply *v*: in mathematics, to multiply the numerator of each of two fractions by the denominator of the other to eliminate the fractions from the equation. For example, in the two fractions $\frac{2}{3} \times \frac{1}{2}$, the 2 in the numerator of the first fraction and the 2 in the denominator of the second fraction are cross-multiplied to cancel each other.

CRT *abbr*: cathode-ray tube.

crystal *n*: a substance formed by the solidification of a chemical element, a compound, or a mixture that has a regularly repeating internal arrangement of its atoms. For example, quartz and silicon are crystals.

current (I) *n*: the flow of electric charge or the rate of such flow, measured in amperes.

current gain *n*: that part of the current flowing into the emitter of a transistor that flows through the base region and out the collector.

current surge rating *n*: a measure of a solid-state diode's ability to operate properly when the circuit in which the diode is installed experiences momentary and intermittent surges in current flowing in the circuit.

cycle *n*: in an alternating current generator (an alternator), one positive and one negative alternation constitutes 1 cycle. As a generator's coil rotates through a magnetic flux to induce voltage in the coil, the magnetic flux induces positive voltage as it rotates from 0 to 180 degrees. It also induces negative voltage as the coil rotates from 180 to 360 degrees. Each positive and negative voltage is one alternation and the two constitute a cycle. See *alternation*.

cycle counter *n*: see *frequency counter*.

cycles per second *n*: see *hertz (Hz)*.

D

d'Arsonval movement *n*: an electronic assembly used in a meter that consists of a magnet, a moving coil, a jewel pivot bearing, and a pointer. The pointer sweeps across a dial that can be marked with numerical values expressing current, voltage, resistance, or whatever is appropriate to the application of the meter. It is named after one of its inventors, the French physiologist Jacque-Arsène d'Arsonval.

DC *abbr*: direct current.

DC inductor *n*: see *choke*.

decade *n*: a group or set of ten, as a period of ten years or a set of ten resistor values.

decimal number *n*: a base 10 numbering system that includes 10 numbers ranging from 0 to 9; numbers in a system wherein place values are read in powers of 10.

decimal point *n*: a dot written either on or slightly above the line that is used as the point at which place values change from positive to negative powers of 10 in the decimal numbering system.

declination *n*: 1. the angle, variable with geographic position, between the direction in which a magnetic needle points and the true meridian. 2. a turning aside or swerving.

decoder *n*: 1. an electronic device that converts binary numbers to decimal numbers. Compare *encoder*. 2. a demodulator. See *demodulate*.

delta connection *n*: in electronics, a combination of three components connected in series to form a triangle similar to the Greek letter delta. Also known as a mesh connection.

demodulate *v*: in electronics, to recover the information transmitted by a modulated wave. See *modulate*.

demodulator probe *n*: in electronics, a small device (a probe) that is placed on or inserted in a circuit and that detects (demodulates) a modulated carrier wave and allows the demodulated signal to be displayed on an oscilloscope's screen for analysis.

denominator *n*: in a mathematical fraction, the term or number that divides the other term or number (called the numerator), and is written below the line. For example, in the fraction 7⁄1∞; 15 is the denominator.

detector *n*: in an electronic receiver, the stage in the receiver that recovers (demodulates) the modulated signal the receiver originally gets (receives). A modulated signal is a signal that is varied by its amplitude, frequency, or phase. See *modulate*.

detector probe *n*: see *demodulator probe*.

dielectric *n*: a substance that is an insulator, or nonconductor, of electricity.

dielectric constant *n*: the value of dielectricity assigned to a substance. A substance that is a good insulator has a high dielectric constant, whereas a poor insulator has a low one.

dielectric strength *n*: the ability of a dielectric material to withstand the effects of voltage applied to it. It is usually measured in terms of the amount of voltage per mil, or thousandths of an inch, required to break down the material.

differential signal *n*: in a circuit, an indicator (a signal) that is the voltage difference between two junctions (nodes), neither of which is at ground potential.

differentiating circuit *n*: a circuit whose output voltage is proportional to the rate of change of the input voltage.

digit *n*: a character that represents a numeric character. See *number*.

digital *adj*: pertaining to data in the form of digits, especially electronic data stored in the form of a binary code. Compare *analog*.

digital logic *n*: 1. a system that describes a statement or condition in a true-false, yes-no, either-or, or on-off manner. Digital logic is used in computer operating systems and in automatic switching and control systems. 2. *adj*: of or pertaining to the various types of solid-state devices such as gates, flip flops, and other on-off circuits used to perform problem-solving functions in the many types of digital computers.

digital multimeter *n*: in electronics, a multimeter that is constructed such that its measurements are displayed digitally rather than as analog readouts. See *multimeter*.

diode *n*: 1. a solid-state electronic device that restricts current flow chiefly in one direction. 2. a radio tube that contains an anode and a cathode.

diode junction voltage *n*: in a circuit in which a diode is used, the voltage generated at the point where the diode (the junction) is wired into the circuit. Junction voltage is 0.7 volts and when multiplied by the current flowing in the circuit results in wattage, which takes the form of heat.

direct coupling *n*: in electronics, the joining (coupling) of two circuits with a nonfrequency-sensitive device, such as a wire, resistor, or battery, so that both direct and alternating current can flow through the coupling path. For an oscilloscope to measure direct current, its circuits must be direct coupled.

direct current (DC) *n*: in an electric circuit, current flow in one direction only. An example of a simple DC circuit is a battery, a wire, and a light bulb, wherein current flows through the wire from one terminal of the battery, through the light bulb, and through another wire back to the battery's other terminal.

discontinuous DC current pulse *n*: the intermittent and irregular fluctuation (a pulse) of direct current flowing in a circuit. DC current pulses create heat in the circuit and can affect the operation of solid-state diodes whose temperature rating is not sufficient to withstand the heat created by such pulses.

distribution transformer *n*: an electrical device designed to reduce voltage from primary distribution levels, usually 7,200 or 12,400 volts, to utilization voltages of 480, 240, or 120 volts. See *power transformer* (def. 1).

dividend *n*: in mathematical division, the quantity that is divided by another quantity. For example, in $9 \div 3 = 3$, 9 is the dividend.

divisor *n*: in mathematical division, the quantity by which another quantity is divided. For example, in $12 \div 3 = 4$, 4 is the divisor.

doping *n*: the addition of impurities to a semiconductor to achieve a desired characteristic, such as producing n- or p- type material. See *n-type structure, p-type structure*.

double-pole, double-throw (DPDT) switch *n*: a six terminal electrical switch that simultaneously connects one pair of terminals to either of other pairs of terminals.

double-pole, single-throw (DPST) switch *n*: an electrical switch with four terminals that simultaneously opens or closes two separate circuits or both sides of the same circuit.

DPDT *abbr*: double-pole-double throw.

DPST *abbr*: double-pole, single-throw.

drain *n*: the region into which majority carriers flow in a field-effect transistor. It is comparable to the collector in a bipolar transistor. See *field-effect transistor; majority carrier*.

dry cell *n*: a primary cell, such as a flashlight battery, in which the electrolyte is a paste. The term "dry" is misleading, because moisture is necessary for the electrolyte to function.

dual-trace oscilloscope *n*: an oscilloscope that can compare two waveforms on the face of its single cathode-ray tube.

dwell *n*: in conventional automobile ignition systems (those that are not solid-state), the number of degrees through which a distributor cam rotates from the time that the contact points close to the time that they open again. Also called dwell angle.

dwell angle *n*: see *dwell.*

dwell meter *n*: an electronic measuring device that, when properly connected to an automobile's ignition system, measures the number of degrees through which a distributor cam rotates from the time that the contact points close to the time that they open again (the dwell).

E

E abbr: volt.

eddy current *n*: an electric current induced within the body of a conductor when the conductor either moves through a magnetic field or is in a region where a change occurs in the magnetic lines of force.

eddy-current loss *n*: in the core of a transformer, losses in power caused by eddy currents in the core. Eddy current is induced in the core by the same magnetic lines of force (flux) that induce voltage in the windings.

EEPROM *n*: see *electrically erasable programmable read-only memory.*

EIA *abbr*: Electronics Industries Alliance.

electrically erasable programmable read-only memory (EEPROM) *n*: an integrated-circuit memory chip that has an internal switch to permit a user to erase the contents of the chip and write new contents into it by means of electrical signals.

electrically neutral *n*: a condition in an electrical circuit or component in which no electrical charge exists. The circuit or component is uncharged—that is, it has neither a positive nor a negative charge.

electric circuit *n*: see *circuit.*

electric coil *n*: see *inductor.*

electricity *n*: the physical phenomena arising from the behavior of electrons and protons that is caused by the attraction of particles with opposite charges and the repulsion of particles with the same charge.

electric motor *n*: an electric device that converts electric energy to mechanical energy. Compare *generator.*

electric potential *n*: work that must be done against electric forces to bring a unit charge from a reference point to another point. It is also called electromotive force and voltage.

electric relay *n*: see *relay.*

electrochemical cell *n*: see *galvanic cell.*

electrode *n*: a conductor of electric current as it leaves or enters a medium such as an electrolyte, a gas, or a vacuum.

electrolyte *n*: a chemical that, when dissolved in water, dissociates into positive and negative ions, thus increasing its electrical conductivity.

electrolytic capacitor *n*: a capacitor that consists of two electrodes separated by an electrolyte. A dielectric film is formed on the surface of one electrode.

electromagnet *n*: a magnet consisting of a coil of conducting wire wound around a soft iron or steel core; the core is strongly magnetized when current flows through the coil, and is almost completely demagnetized when current flow ceases.

electromagnetic induction *n*: voltage created when a conductor cuts across, or is cut by, a magnetic field.

electromechanical relay *n*: a device used to open or close electrical circuits (a relay) that is made up of mechanical parts that are actuated by electricity.

electromotive force (emf) *n*: 1. the force that drives electrons and thus produces an electric current. 2. the voltage or electric pressure that causes an electric current to flow along a conductor. 3. a difference of potential, or electrical, flow through a circuit against a resistance.

electron *n*: a particle in an atom that has a negative charge. An atom contains the same number of electrons and protons (which have a positive charge). Electrons orbit the nucleus of the atom.

Electronic Industries Alliance (EIA) *n*: an organization that writes, promulgates, and publishes *Recommended Standards* (*RS*s) for electronic and physical devices and their means of interfacing. For example, RS-232 is a standard that defines a computer's serial port, connector pins, and electrical signaling process. EIA was founded in 1924 as the Radio Manufacturers Association. Address: 2500 Wilson Blvd., Arlington, VA 22201; (703) 907-7500; www.eia.org.

electron pair *n*: a pair of valence electrons that form a nonpolar bond between two neighboring atoms. See *valance electron.*

electron-pair bond *n*: the combining of atoms (the bonding) in which each atom of a bond pair contributes one electron to form a pair of electrons. Also called covalent bond. See *electron pair.*

electron tube *n*: an almost obsolete electronic device in which electrons move through a vacuum or gaseous medium within an airtight glass (or other material) container. Also called radio tube; tube; vacuum tube; or, in Great Britain, valve.

element *n*: one of more than a hundred simple substances that consist of atoms of only one kind and that either singly or in combination make up all matter. For example, the simplest element is hydrogen, and one of the most abundant elements is carbon. Some elements, such as radium and uranium, are radioactive.

emf *abbr*: electromotive force.

emitter *n*: in a junction transistor, the region from which charge carriers that are minority carriers in the transistor's base are injected into the base and control the current flowing through the transistor's collector. Base and collector currents are combined in the emitter region and it serves as the outlet for total current in the transistor. See *base* (def. 2), *collector*.

encoder *n*: an electronic device that converts decimal numbers to binary numbers. Compare *decoder*.

engine *n*: a machine for converting the heat content of fuel into rotary motion that can be used to power other machinery. Compare *motor*.

EPROM *abbr*: erasable programmable read-only memory.

equation *n*: in mathematics, a statement that each of two expressions is the same as (is equal to) the other. For example, $a = b$ is an equation as is $2 + 5 = 3 + 4$.

erasable programmable read-only memory (EPROM) *n*: an electronic integrated circuit whose stored data (memory) is placed there by the manufacturer and that cannot be programmed electrically with permanent information provided by the manufacturer. The data can only be accessed (read) by a user; a user cannot change it.

Esaki tunnel diode *n*: see *tunnel diode*.

exciting current *n*: the back electromotive force that flows through the primary winding of a transformer when no loads are connected to the secondary winding. See *back electromotive force*.

exclusive or (XOR) *n*: in computer science, an instruction that performs its operation on a bit-by-bit basis for its two operand words, usually storing the result in one its operand locations. Also called XOR. See *operand*.

exponent *n*: in mathematics, a number or symbol, such as the 3 in x^3, placed to the right of and above another number, symbol, or expression, denoting the power to which that number, symbol, or expression is raised to. Also called power.

F

F *symbol*: farad.

farad (F) *n*: the unit of electrical capacitance. One farad is equal to the capacitance of a capacitor that has a potential difference of 1 volt between its plates when the charge of one of its plates is 1 coulomb, there being an equal and opposite charge on the other plate. See *coulomb*.

Fe *sym*: iron.

feedback *n*: in an electric circuit, the return of a portion of the output current of the circuit to its input.

FET *abbr*: field-effect transistor.

field effect *n*: the local change from the normal value that an electric field produces in a semiconductor's charge-carrier region. The field effect is used in field-effect transistors.

field-effect transistor (FET) *n*: a transistor in which the resistance of the current path from source to drain is varied (modulated) by applying a transverse electric field between grid or gate electrodes. The electric field varies the thickness of the depletion layer between the gates to reduce conductance.

filament *n*: in an incandescent light bulb, a metallic wire or ribbon through which electric current is passed to make it produce light.

film *n*: a thin skin or membrane. In electronics, it is used in the manufacture of one type of resistor.

film resistor *n*: a resistor made from a thin skin or membrane (a film). Carbon film, metal film, or metallic oxide film may be used to make resistors whose resistances are very accurate, usually within 5 percent of the desired resistance.

filter *n*: in electronics, a device in a circuit that senses and allows desired electronic constituents to pass through it to the remainder of the circuit and at the same time senses and removes or weakens (attenuates) electronic constituents not desired in the circuit beyond the point of the filter's placement in the circuit. *v*: to remove or attenuate (weaken) electronic constituents not desired in the circuit beyond the point in the circuit where a filter is placed.

filter choke coil *n*: a winding (coil) with a laminated iron core used in a power supply's filter system to pass direct current and simultaneously impede the passage of alternating current.

flip-flop circuit *n*: a type of electronic circuit in which either of two active devices may remain conducting, with the other nonconducting, until the application of an external pulse.

flux *n*: the lines of force in a magnetic field.

flux density *n*: the number of lines of force in a square inch of the cross-sectional area of the core in a coil, or solenoid.

flux linkage *n*: the product of the number of turns in a coil and the magnetic flux passing through the coil.

FM *abbr*: frequency modulation.

forward bias *n*: where a p-type and n-type semiconductor are joined, the condition that occurs when the positive terminal of a battery is connected to the lead on the p-type semiconductor and the negative terminal of the battery is connected to the lead on the n-type semiconductor. Forward biasing attracts free electrons in the n-type material across the junction toward the positive terminal, and attracts holes in the p-type across the junction toward the negative terminal. A forward-biased p-n junction offers low resistance and current flows freely across it. Compare *reverse bias*. See *n-type structure*, *p-type structure*.

forward voltage drop *n*: the voltage across a semiconducting diode or special transistor that carries current in a forward direction.

free electron *n*: an electron on the outer shell of an atom that moves readily from one atom to another.

free hole *n*: a space (hole) in an atom that is not bound to an impurity in a semiconductor. A free hole is the opposite of a free electron and is a positive charge. See *free electron*.

frequency *n*: the number of cycles completed by a periodic quantity in a unit time—for example, the number of complete alternations (cycles) per second of alternating electric current.

frequency counter *n*: an electronic device used to measure the frequency of a signal by counting the number of cycles in the signal during a predetermined time interval.

frequency modulation (FM) *n*: the varying (modulating) of a wave in which the instantaneous frequency of the modulated wave differs from the carrier frequency by an amount proportional to the instantaneous value of the modulating wave. Compare *amplitude modulation*.

full adder *n*: a logic element in an electrical circuit that operates on two binary digits and a carry digit from a preceding stage to produce as output a sum digit and a new carry digit.

full load *n*: in electronics, the greatest amount of electric power (the greatest load) that a circuit or device is designed to carry under specified conditions.

function *n*: a mathematical rule between two sets that assigns to each member of the first set exactly one member of the second.

G

Ga *sym*: gallium.

gage *n*: variation of gauge.

gallium (Ga) *n*: a silvery-white metallic element often added as an impurity to semiconductors to create a positive charge in the semiconductor.

galvanic *adj*: of, relating to, or producing a direct current of electricity.

galvanic cell *n*: an electrical device that contains two dissimilar metals suspended in or surrounded by an electrolyte, which is capable of producing voltage and current. Two or more galvanic cells constitute a battery. Often, and erroneously, a galvanic cell is called a battery. A battery consists of two or more galvanic cells.

galvanic corrosion *n*: a type of corrosion that occurs when a small electric current flows from one piece of metal equipment to another. It is particularly prevalent when two dissimilar metals are present in an environment in which electricity can flow (as when two dissimilar conductors in a circuit are connected).

galvanometer *n*: an instrument that detects or measures a small electric current by movements of a magnetic needle or of a coil in a magnetic field. It can be adapted with shunts or resistors to measure larger currents or voltage. See *resistor*, *shunt*.

gate *n*: in solid-state electronics, one of the electrodes in a field-effect transistor. See *field-effect transistor*.

gate-controlled turnoff *n*: in a semiconductor, the action of switching the semiconductor from its nonconducting, or off, state to its conducting, or on state, by applying a negative pulse to its gate terminal.

gate-turnoff silicon-controlled rectifier (GTO-SCR) *n*: a silicon-controlled rectifier that can be turned off by applying current to its gate. GTO-SCRs are mainly used for direct-current switching because turnoff can be achieved in a fraction of a microsecond. See *silicon-controlled rectifier*.

gauge *n*: in electrical circuits, a measure of the diameter (thickness) of the wire or other device that conducts electricity through a circuit or from one point to another. In general, gauges for wire conductors range from 24 to 4/0 (0000). Some common sizes are 6, 12, and 16 with 6-gauge wire being larger in diameter than 16-gauge wire. See *American wire gauge*.

Ge *sym*: germanium.

generator *n*: a machine that changes mechanical energy into electrical energy in the form of direct current. Compare *electric motor*.

geographic pole *n*: either of two points that lie at either end of an imaginary line drawn through the rotational axis of the earth; one is the North Pole and the other is the South Pole. Compare *magnetic pole*.

germanium (Ge) *n*: a brittle, silvery-gray metallic element in the carbon family. It is a rare metal used in semiconductors, alloys, and glass.

GFCI *abbr*: ground fault circuit interrupter.

gigohm (Gohm, GΩ) *n*: 1 billion (1,000,000,000) ohms. Sometimes expressed as 1,000 megohms or 10^9 ohms.

GΩ *abbr*: gigohm, or 1 billon (1,000,000,000) ohms.

Gray code *n*: a modified binary code in which sequential numbers are represented by expressions that differ only in one bit. Because the expressions differ by only one bit, errors are minimized.

ground *n*: a conducting path, intentional or accidental, between an electric circuit or equipment and the earth, or a conducting body serving in place of the earth. *v*: to connect electrical equipment to earth or to a conducting body, which serves in place of the earth.

ground fault circuit interrupter (GFCI) *n*: a fast-acting circuit breaker that senses very small ground fault currents and which breaks the circuit when it detects such faults.

ground fault detector *n*: in an electric circuit, a device that senses very small ground fault currents that could flow through the body of a person standing on damp ground while touching a live conductor.

GTO-SCR *abbr*: gate-controlled silicon-controlled rectifier.

H

H *sym*: 1. hydrogen. 2. henry. 3. magnetizing force.

H_2SO_4 *form*: sulfuric acid.

hairspring *n*: a thin spiraled recoil spring that regulates the motion of a pointer in a meter or a clock.

half adder *n*: a logic element in an electronic circuit that operates on two binary digits, but no carry digits, from a preceding stage to produce as output a sum digit and a carry digit.

henry (H or L) *n*: a unit of inductance in which an induced voltage (electromotive force) of 1 volt occurs in a circuit when the current changes at the rate of 1 ampere per second.

hertz (Hz) *n*: a unit in the metric system used to measure frequency in cycles per second.

hexadecimal number *n*: a number based on powers of 16 that uses the decimal digits 0 to 9 and six more digits represented by the letters A, B, C, D, E, and F. A hexadecimal number is usually written with the subscript 16 to indicate that it is a hexadecimal number. For example, $84D1F_{16}$ is a hexadecimal number.

high-frequency *adj*: in electronics, involving the rapid cycling of alternating current, radio waves, or sound waves.

holding current *n*: the minimum current required to maintain a switching device, such as a silicon-controlled rectifier, in a closed, or conducting, state after it is energized or triggered.

hole *n*: in electronics, the space left in atoms as free electrons flow from one atom to another. Like free electrons, holes also flow from atom to atom but in the opposite direction from electrons. In electronics, conventional current flow is shown as hole flow and not electron flow. Consequently, circuit diagrams for electronic devices show current flowing from the positive terminal to the negative terminal.

hole-electron pair *n*: in a p-type semiconductor, that which occurs when an impurity atom with only three valence electrons is placed in a semiconducting material with four valence electrons. The fourth electron is left without an electron to pair with, which results in a hole. The hole conducts current.

hookup wire *n*: wire or cable used to connect components in an electronic device.

hot-carrier diode *n*: see *Schottky diode*.

hot terminal *n*: see *positive terminal*.

hydrogen (H) *n*: a flammable, colorless, odorless, tasteless gas whose chemical symbol is H and that usually occurs as the diatomic molecule H_2. It is lighter than air; indeed, it is the lightest element in the universe.

hypotenuse *n*: in a right triangle, the side opposite the right angle. See *right triangle*.

hysteresis *n*: the lagging of changes in the magnetization of a substance behind changes in the magnetic field as the magnetic field is varied.

hysteresis loss *n*: in a transformer, the conversion of electrical energy to heat in the core of the transformer's winding because of hysteresis (the lagging of changes in the core's magnetic field as the transformer's magnetic field is varied). See *hysteresis*.

Hz *sym*: hertz.

I

I *sym*: current.

if-then operation *n*: a statement that asserts that if one thing is true then another thing is also true. Mathematically, the operation is, "If *p*, then *q*."

IGBT *abbr*: insulated-gate bipolar transistor.

impedance (Z) *n*: in an alternating current (AC) circuit, the circuit's total opposition to current flow. Impedance is similar to resistance in a direct current (DC) circuit. Just as resistance is measured in ohms, so too is impedance. When connecting certain AC components, such as a microphone to an amplifier, it is important for both components to have the same, or very close to the same, impedance; otherwise, the connected components may not operate properly.

implication *n*: see *if-then operation*.

impressed voltage *n*: in an electrical circuit, voltage applied to the circuit or to a device in the circuit.

induced current *n*: the flow of an electric charge, measured in amperes, that is the result of induced voltage. See *induced emf*.

induced emf *n*: electromotive force (voltage, or electricity) that is developed in a conductor by moving the conductor within a magnetic field or moving the magnetic field in relation to the conductor.

induced voltage *n*: see *induced emf*.

inductance *n*: the property of an electric circuit or of two neighboring circuits wherein an electromotive force is generated by induction in one circuit by a change of current in itself or in the other. See *induction*.

inductance coil *n*: see *inductor*.

induction *n*: the production of an electric charge, magnetism, or electromotive force in an electric conductor, a magnetizable body, or an electric circuit by the proximity of a similarly energized body or by the variation of the magnetic flux without contact.

induction motor *n*: an alternating-current motor in which the primary winding on a part (usually the stator) is connected to a source of power, and the secondary winding (usually on the rotor) carries current induced by the magnetic field of the primary winding.

inductive coupling *n*: in electronics, the transfer of energy from one circuit to another through the inductance provided by a transformer.

inductive filter *n*: see *LC filter*.

inductive reactance *n*: the retarding effect on the passage of alternating current through a circuit because of inductance. See *inductance*.

inductor *n*: in an electrical circuit, a coil of wire that introduces inductance or magnetic flux into the circuit. It may also react mechanically to changing magnetic flux. Also called electric coil, inductance coil.

in phase *n*: in alternating current circuits, the phenomenon that occurs when the voltage and the current have the same frequency and strength at the same instant in the cycle.

inrush current *n*: the heavy current that develops when an induction motor is started at full voltage. It usually decreases gradually as the speed increases and drops sharply when the motor reaches full speed.

insulated-gate bipolar transistor (IGBT) *n*: in semiconductors, a transistor that combines low forward voltage drop, gate-controlled turnoff, and high switching speed. It is used in high-powered converters.

insulator *n*: a material with a very low concentration of free electrons that resists the flow of electric current.

integer *n*: any positive or negative number including 0.

integrated circuit *n*: an interconnected arrangement of active and passive electronic elements incorporated (integrated) with a single semiconductor substrate or deposited on the substrate by a continuous series of compatible processes, and capable of performing at least one complete electronic circuit function.

integrating amplifier *n*: an operational amplifier with a feedback, or shunt, capacitor in which the waveform at the output is the integral (usually over time) of the input.

International System of Units (SI) *n*: a system of units of measurement based on the metric system, adopted and described by the Eleventh General Conference on Weights and Measures. It provides an international standard of measurement to be followed when certain customary units, both U.S. and metric, are eventually phased out of international trade operations. The symbol SI (Le Système International d'Unités) designates the system, which involves seven base units: (1) metre for length, (2) kilogram for mass, (3) second for time, (4) Celsius for temperature, (5) ampere for electric current, (6) candela for luminous intensity, and (7) mole for amount of substance. From these units, others are derived without introducing numerical factors.

inverter *n*: an electronic device that converts direct current into alternating current.

inverter circuit *n*: a logic circuit (also called a NOT circuit) with one input and one output that inverts the input signal at the output—that is, the output signal is a logical 1 if the input signal is a logical 0 and vice versa.

inverting input *n*: in an operational amplifier, a type of input wherein if the input signal is changed, the output is also changed. For example, if the inverting input increases in voltage, the output voltage inverts, or decreases. Also called minus input. Compare *noninverting input*.

ion *n*: an atom or a group of atoms charged either positively (a cation) or negatively (an anion) as a result of losing or gaining electrons.

ionization *n*: the process by which a neutral atom becomes positively or negatively charged through the loss or gain of electrons.

ionize *v*: to give a neutral atom either a positive or negative charge; after being ionized, the neutral atom becomes an ion. See *ion*.

iron (Fe) *n*: 1. a heavy, magnetic, malleable, and ductile element occurring in the earth's crust in many types of ores. 2. a loose term used in the oilfield for any tool or device generally made of metal, whether it contains iron or not.

iron-cored coil *n*: in a coil of wire that carries electric current, the placement of a length of soft iron in the central interior part of the coil.

J

JEDEC *abbr*: JEDEC Solid State Technology Association.

JEDEC Solid State Technology Association (JEDEC) *n*: originally known as the Joint Electron Device Engineering Council (JEDEC), this association is the semiconductor engineering standardization body of the Electronic Industries Alliance (EIA), a trade association that represents all areas of the electronics industry. JEDEC was originally created in 1960 as a joint activity between EIA and the National Electrical Manufacturing Association (NEMA), to cover the standardization of discrete semiconductor devices and later expanded in 1970 to include integrated circuits. In spite of the name change, the association is usually abbreviated as JEDEC. Address: 2500 Wilson Blvd., Arlington, VA 22201; (703) 907-7534; www.jedec.org.

JFET *abbr*: junction-type field-effect transistor.

Joint Electron Device Engineering Council (JEDEC) *n*: a now defunct consortium of manufacturers that devised standards mostly for computer memory modules. Replaced in 1998 by the JEDEC Solid State Technology Association.

junction *n*: in solid-state electronics, a region of transition between two semiconducting regions in a semiconducting device—for example, the region where the p and n materials come into contact.

junction transistor *n*: a transistor in which emitter and collector barriers are formed between the semiconductor regions of opposite conductivity (the region where the p and n materials come into contact). Also called a junction-type field-effect transistor (JFET).

junction-type field-effect transistor (JFET) *n*: see *junction transistor*.

K

K *sym*: dielectric constant.

kcmil *abbr*: one thousand circular mils. See *circular mil*.

keeper *n*: a soft iron bar placed between the poles of a permanent magnet when it is not being used. A keeper protects the magnet's poles from being demagnetized if the magnet is dropped or struck with a hard object.

kilohm (kohm, kΩ) *n*: 1,000 ohms.

kΩ *abbr*: kilohm.

kiloline *n*: in magnetic circuits, an expression of the flux density of a coil's core in thousands of lines; usually expressed as kilolines per square inch.

kilovolt ampere (kVA) *n*: a rating applied to transformers in which a transformer's power is expressed as 1,000 volt-amperes (VA).

kilowatt (kW) *n*: a metric unit of power equal to approximately 1.34 horsepower; 1,000 watts. See *watt*.

kinetic energy *n*: energy possessed by a body because of its motion. It is equal to one-half the mass of the body times the square of its speed.

knee *n*: on a graph that plots magnetizing force versus flux density for coil cores, the point on the curve of a particular material where further increases in flux density fails to result in significant gains in magnetizing force.

kohm *abbr*: kilohm.

kVA *abbr*: kilovolt ampere.

kW *abbr*: kilowatt.

L

L *abbr*: 1. litre. 2. henry.

ladder logic *n*: a method used to describe the arrangement of electromechanical relays in a control system. Usually, a schematic drawing of such relays consists of vertical lines called rails and horizontal lines called rungs. Consequently, the schematic resembles the rails and rungs of a ladder.

latch circuit *n*: an electronic circuit that reverses and maintains its state each time power is applied.

lattice *n*: the regular geometrical arrangement of points or objects over an area in space—for example, the arrangement of atoms in a crystal. See *crystal.*

law of conservation of energy *n*: the principle that energy cannot be created or destroyed, although it can be changed from one form to another, such as mechanical energy to electrical energy.

LCD *abbr*: liquid crystal display.

LC filter *n*: a low-pass filter that smooths the DC output voltage of a rectifier. It consists of one or more sections in series, each section consisting of an inductor on one of the pair of conductors in series with a capacitor between the conductors. Also known as inductive filter, pi (π) filter.

lead (Pb) (pronounced "led") *n*: a chemical element, the symbol of which is Pb, which is a heavy, soft, malleable, and ductile bluish white metal. It is often used in batteries, solder, and radioactivity shields.

lead (pronounced "leed") *n*: in electronics, a wire, usually insulated, which, when placed in an electrical circuit, connects two points in the circuit.

lead-acid battery *n*: a storage battery in which the electrodes are grids of lead and lead peroxide that change in composition during charging and discharging, and the electrolyte is dilute sulfuric acid.

lead-acid cell *n*: an electrical device, often part of a battery, that consists of lead and lead peroxide plates immersed in a liquid solution (an electrolyte) that is usually sulfuric acid. Such a cell produces voltage and current.

leakage current *n*: in an electrical component, such as a capacitor, that employs a dielectric (insulator) to separate conductive elements in the component and prevent the flow of electricity between the conductive elements, the relatively small amount of electricity that flows between the conductors because no dielectric is a perfect insulator. Consequently, some current flows, or leaks, from one conductive element to another.

leakage reactance *n*: in a coil, or winding, in which electrical current is flowing, the current that escapes, or leaks, from it. This leakage, in turn, tends to hold back, or impede, the flow of current through the winding. In this case, the impedance to flow is termed reactance.

LED *abbr*: light-emitting diode.

Lenz's law *n*: a rule that states that an induced electromotive force (voltage) generates a current that induces a counter magnetic field that opposes the magnetic field generating the current.

light-emitting diode (LED) *n*: a semiconductor device that has two electrodes or terminals (a diode) and emits light when voltage is applied to it. LEDs are used in electronic displays such as calculators, wrist watches, and so on.

limiter *n*: see *clipper*.

limiting *n*: see *clipper*.

linearity *n*: in electronics, the relationship between two quantities that occurs when a change in one of the quantities produces a directly proportional change in the other quantity.

linear taper *n*: in a potentiometer with multiple windings, the uniform winding of the resistance wire that makes up the potentiometer's control. A linear taper provides the potentiometer with directly proportional control—that is, turning the potentiometer's control directly varies the potentiometer's resistance. For example, turning the control one-fourth turn decreases or increases the resistance by one fourth. Compare *logarithmic taper*.

line loss *n*: the reduction in (or the loss of) electrical energy that occurs when electricity flows though a conductor (a line).

lines of force *n pl*: see *magnetic lines of force*.

liquid crystal display (LCD) *n*: an electronic digital display that consists of two sheets of glass separated by a sealed-in, normally transparent, liquid crystal material. The outer surface of each glass sheet has a transparent conductive coating etched into character-forming segments that have leads going to the edges of the display. Voltage applied between front and back electrode coatings disrupts the orderly arrangement of molecules, which darkens the liquid enough to form visible characters although no light is generated.

load *n*: in electronics, a device that consumes electric power.

load loss *n*: in a transformer, the reduction (loss) of electrical energy that occurs when the transformer is loaded—that is, when electricity is being drawn from it.

load voltage *n*: see *voltage drop*.

lo-cap probe *n*: in electronics, a probe that attenuates (reduces) the strength of a signal being measured by the probe. See *attenuating probe*.

lodestone *n*: a kind of naturally occurring iron oxide that has magnetic properties and therefore attracts iron or steel. Also called magnetite.

logarithm *n*: the exponent that indicates the power to which a number is raised to produce a given number. For example, the logarithm of 100 to the base 10 is 2.

logarithmic *adj*: of or pertaining to a logarithm, as a logarithmic scale. See *logarithm*.

logarithmic scale *n*: a range of values—a scale—on which actual distances from the origin are proportional to the logarithms of the corresponding scale numbers. Unlike an arithmetic scale, where distances from the origin relate directly to the number, values on a logarithmic scale indicate a proportional relation. For example, the Richter scale of earthquake intensity is logarithmic; thus, an earthquake with an intensity of 4.0 on the Richter scale is much more than twice as intense as an earthquake with an intensity of 2.0.

logarithmic taper *n*: in a potentiometer, the winding of the resistance wire in such a manner that turning the potentiometer's control knob varies the resistance nonproportionately. That is, turning the control knob one-quarter turn increases or decreases the resistance by a factor different from one-fourth—for example, two or four times more. Compare *linear taper*.

logic *n*: 1. the study of the principles of reasoning, especially of the structure of propositions as distinguished from their content; logic also includes the study of method and validity in deductive reasoning. 2. the nonarithmetic operations performed by a computer, such as sorting, comparing, and matching that involve yes-no decisions.

logical AND operation *n*: a process in logical reasoning whereby two related statements, or premises, must be true before a correct conclusion can be drawn. For example, if one premise is "that dog is panting," and the other premise is that "the temperature is 98°," then the conclusion, "when the temperature is 98°, that dog pants," may be correctly drawn.

logical NOT operation *n*: a process in logical reasoning whereby a true statement is made false by making a false statement.

logical OR operation *n*: a process in logical reasoning whereby if either or both of two statements, or premises, are true, then a correct conclusion can be drawn. For example, the statement that "if 50 tons of iron or 50 tons of coal overload a truck, then the truck is overloaded," is a correct conclusion. Either 50 tons of iron or 50 tons of coal overload the truck.

logic gate *n*: an electronic circuit that provides the action of comparing, selecting, making references, matching, sorting, and merging where yes-or-no (logic) quantities are involved.

low-capacitance probe *n*: see *lo-cap probe*.

M

magnet *n*: a metal that has the property of attracting ferrous (iron) and certain other metals to it.

magnetic attraction *n*: the principle that the opposite (north and south) poles of a magnet attract each other.

magnetic circuit *n*: the path magnetic lines of force (flux lines) take in an electrical device that employs magnets in its operation. A magnetic circuit is analogous to direct current flowing in a conductor such as wire or cable.

magnetic deviation *n*: see *declination*.

magnetic field *n*: the space that magnetic lines of force produced by a magnet or a current-carrying conductor occupy.

magnetic flux *n*: see *flux*.

magnetic iron ore *n*: see *magnetite*.

magnetic lines *n pl*: see *magnetic lines of force*.

magnetic lines of force *n pl*: the magnetism that emanates from a magnet's north pole and travels in the atmosphere around the magnet to its south pole.

magnetic north *n*: a region located in the north polar region of the earth toward which magnetic compasses point. Its location is different from true, or geographic, north. Compare *true north*.

magnetic pole *n*: 1. either of two regions that are located in the polar areas of the northern and southern hemispheres and toward which a compass needle points from any direction throughout adjacent regions. Compare *geographic pole*. 2. either of the poles of a magnet.

magnetic repulsion *n*: the phenomenon that like poles (north-north or south-south) of a magnet repel, or repulse, each other.

magnetism *n*: physical phenomena that include a magnet's or lodestone's attraction for iron and the forces associated with moving electricity; magnets and electric current exhibit these phenomena, which are characterized by fields of force.

magnetite *n*: the mineral form of black iron oxide, Fe_3O_4, that often occurs with magnesium, zinc, and manganese and is an important ore of iron. Also called lodestone, magnetic iron ore.

magnetizing current *n*: the part of exciting current that is 90 degrees out of phase with the impressed voltage on the circuit of a transformer and that produces magnetic flux in the transformer's winding. See *exciting current*.

magnetizing force (H) *n*: in an electric coil, or solenoid, the ampere-turns per inch of the length of the coil's core to be magnetized. Just as a voltage drop, or loss of voltage, occurs along the length of a conductor that is carrying current, a magnetomotive drop occurs along the length of the magnetic circuit. The longer the magnetic circuit, the greater is the magnetomotive force required to magnetize the core to the desired amount.

magneto *n*: a device (an alternator) that produces alternating current for distribution to the ignition system of some internal-combustion engines. See *alternator*.

magnetomotive force (mmf) *n*: in a magnetic circuit, the force that creates magnetic lines of force (magnetic flux); it is analogous to electromotive force, or voltage, in an electric circuit.

magnet wire *n*: wire used as windings in electromagnets, transformers, motors, and generators. Its insulation must be thin so that it can be wound several times around a core; thick insulation would not allow it to be wound an adequate number of times. So, instead of a plastic coating, varying grades of enamel or varnish are baked onto magnet wires to provide good insulating properties.

majority carrier *n*: the type of carrier, either electron or hole, that makes up more than half the carriers in a semiconductor.

mass *n*: the quantity of matter a substance contains, independent of such external conditions as the buoyancy of the atmosphere or the acceleration caused by gravity.

master oscillator *n*: an oscillator that establishes the carrier frequency of the output of an amplifier or transmitter. See *oscillator*.

master switch *n*: an electrical component, usually wired in parallel with other switches, that controls the flow of electricity in the circuit. When closed, the switch allows electricity to flow through the circuit. When open, no electricity can flow. When the master switch is open, other switches in the circuit do not operate.

matrix *n*: in mathematics, a rectangular array of numbers from vector space.

matter *n*: anything that occupies space and can be perceived by one or more senses; a physical body, a physical substance, or the universe as a whole; something that has mass and exists as a solid, liquid, gas, or plasma.

maximum working voltage rating *n*: in a capacitor, the largest amount of voltage that can be applied to the capacitor under normal operating conditions without damaging or destroying it.

MC *abbr*: metal clad.

megohms (Mohms, MΩ) *n*: 1,000,000 ohms. Sometimes expressed as 10^6 ohms.

metal-clad (MC) cable *n*: in electricity, a bundle of several wire conductors, each of which is individually insulated, that is surrounded by a strong metal covering.

metallic oxide semiconductor field-effect transistor (MOSFET) *n*: a field-effect transistor whose gate is insulated from the semiconductor's substrate by a thin layer of silicon dioxide.

metallic oxide varistor (MOV) *n*: a semiconductor with two electrodes whose resistance depends on the voltage flowing through it. Its resistance decreases as applied voltage increases. Also called a voltage-dependent resistor.

μF *abbr*: microfarad.

mF *abbr*: microfarad.

MFD *abbr*: manufacturer's abbreviation for mircofarad; often stamped on the body of the capacitor.

mH *abbr*: millihenry.

mica *n*: a silicate mineral characterized by sheet cleavage—i.e., it separates into thin sheets. Biotite is ferromagnesian black mica, and muscovite is potassic white mica. Sometimes mica is used as a lost circulation material in drilling and as a dielectric (insulator) in electrical components such as capacitors.

mica capacitor *n*: a capacitor whose dielectric (insulating material) is made from mica. See *capacitor*; *mica*.

microfarad (mF, μF) *n*: one-millionth (10^{-6}) of a farad. See *farad*.

microprocessor *n*: a single silicon chip on which the arithmetic and logic functions of a computer are placed.

mil *n*: a distance equal to 0.001 ($1/1{,}000$, or 10^{-3}) inches.

MIL *abbr*: military standards.

mil-foot *n*: a measurement of the length of a wire or conductor based on its cross-sectional diameter; a wire that is 1 foot long and has a diameter of 0.001 inches is 1 mil-foot.

military standard (MIL) *n*: specifications for equipment and procedures purchased by U.S. military forces. One source of MILs is Global Engineering Documents, 15 Inverness Way East, Englewood, CO 80112: 303-397-7956 or 800-854-7179; www.global.his.com.

millihenry (mH) *n*: one-thousandth of a henry, or 10^{-3} henrys.

milli-inch *n*: a measure of the diameter of an electrical wire or cable expressed in thousandths of an inch. For example, if a wire's diameter is 0.1 inches, it is also 0.1 × 1,000 or 100 milli-inches.

mil size *n*: see *circular mil.*

minuend *n*: in mathematics, the quantity from which another quantity is to be subtracted. For example, in the expression 38 – 23 = 15, 23 is the minuend.

minus input *n*: see *inverting input.*

mixture *n*: a physical combination of two or more elements or components that maintain their chemical identity; no chemical reaction is involved.

mmf *abbr*: magnetomotive force.

modulate *v*: in electronics, to vary the amplitude, frequency, or phase of a signal's wave. Modulating a wave with electronic information enables the wave to carry the information over a distance. A receiver, when tuned to the correct frequency, can demodulate the wave to make the information available for interpretation and understanding.

Mohm *abbr*: megohm, or 1,000,000 ohms.

MΩ *abbr*: megohm, or 1,000,000 ohms.

molecule *n*: the smallest particle of a substance that retains the properties of the substance. It is composed of one or more atoms.

MOSFET *abbr*: metal oxide semiconductor field-effect transistor.

motor *n*: any of various power units, such as hydraulic, air, or electric devices, which develop energy or impart motion. Compare *engine*. See *electric motor*.

motor slip *n*: see *slip*.

MOV *abbr*: metallic oxide varistor.

multimeter *n*: an electronic testing instrument that features several ranges for measuring voltage, current, and resistance. Also called a volt-ohm-milliammeter.

multiplicand *n*: in mathematics, the quantity that is multiplied by another quantity. For example, if 6 is multiplied by 2, 6 is the multiplicand.

multiplier *n*: 1. in mathematics, the quantity that multiplies another quantity. For example, if 6 is multiplied by 2, 2 is the multiplier. 2. in electronics, an indication that one quantity should be multiplied by another quantity, usually in multiples of ten. For example, if an electronic component is stamped 2k (or, sometimes, 2K), the k (or K) stands for kilo and indicates that 2 should be multiplied by 1,000 to arrive at 2,000. (The lower case k is the preferred abbreviation for kilo, but manufacturers sometimes use K.) An M (short for mega) after a number indicates that the preceding number should be multiplied by 1 million; a G (short for giga) after the number indicates the preceding number should be multiplied by 1 billion. (The capital letters M and G are used to abbreviate mega and giga.)

multitrace recorder *n*: in electronics, an instrument that senses (traces) and records two or more signals generated by a component to which the recorder is attached.

multivibrator oscillator *n*: a special type of oscillator that uses two tubes, transistors, or other electronic devices in which the output of each tube, transistor, or device is coupled to the input of the other with resistors and capacitors (or other devices) to obtain in-phase feedback voltage. Multivibrator oscillators are used in oscilloscopes to produce a large range of sweep frequencies.

N

NAND circuit *n*: a logic circuit whose output signal is a logical 1 if all of its inputs is a logical 0, and whose output signal is a logical 0 if all of its inputs are a logical 1.

nanofarad (nF) *n*: one-billionth (10^{-9}) of a farad. See *farad*.

National Electrical Manufacturers Association (NEMA) *n*: an organization created in 1926 by the merger of the Electric Power Club and the Associated Manufacturers of Electrical Supplies. NEMA provides a forum for the standardization of electrical equipment to ensure the manufacture of safe, effective, and compatible electrical products. NEMA also makes contributions to the electrical industry by shaping public policy development and operating as a central confidential agency for gathering, compiling, and analyzing market statistics and economics data. Address: 1300 North 17th Street, Suite 1847, Rosslyn, VA 22209; (703) 841-3200; www.nema.org.

National Electric Code (NEC) *n*: a document sponsored by the National Fire Protection Agency (NFPA), the National Electric Code (NEC) establishes standards for safe electrical installations. Electricians, electrical contractors, engineers, and inspectors use the NEC as a guide when installing electrical wiring. Most cities and other authorities in the U.S. require that houses, buildings, factories, etc., be wired to meet NEC standards.

National Fire Protection Agency (NFPA) *n*: founded in 1896, an organization that provides fire, electrical, and life safety to the public. The mission of the international nonprofit NFPA is to reduce the worldwide burden of fire and other hazards on the quality of life by providing and advocating scientifically-based consensus codes and standards, research, training, and education. Address: 1 Batterymarch Park, Quincy, MA 02169; www.nfpa.org.

NEC *abbr*: National Electric Code.

needle valve *n*: a device that controls the rate of flow in a line (a valve), which contains a sharp-pointed, needlelike plug that is driven into and out of a cone-shaped seat to control accurately a relatively small rate of fluid flow.

negative charge *n*: a phenomenon that occurs at the atomic level of elements whereby the element has a surplus of electrons—that is, it has a lower electric potential. Compare *positive charge*.

negative feedback *n*: the return of a portion of the output current of a circuit to its input (feedback) in which the portion returned is 180 degrees out of phase with the input signal. This out-of-phase feedback decreases the amount of amplification in the circuit and stabilizes it with respect to time or frequency. Negative feedback reduces distortion and noise in the amplifier.

negative terminal *n*: the terminal of a battery or other voltage source away from which electrons flow and toward which holes flow. See *hole, terminal.*

NEMA *abbr*: National Electrical Manufacturers Association.

neutral *adj*: of or relating to a particle, an object, or a system that has neither a positive nor a negative electric charge; of or relating to a particle, object, or system that has a net electric charge of zero.

neutron *n*: an electrically neutral subatomic particle in the nucleus of an atom, whose mass is about 1,845 times that of the electron. A neutron is stable when bound in an atomic nucleus. It and the proton form nearly the entire mass of atomic nuclei.

nF *abbr*: nanofarad.

NFPA *abbr*: National Fire Protection Agency.

Ni *sym*: nickel.

Nichrome *n*: the trademarked name of a nickel-chromium alloy that has high electrical resistance and the ability to withstand high temperatures; it is often used as the elements in electric heaters.

nickel (Ni) *n*: a silver-gray, ductile, malleable, and tough metal used in alloys, plating, coins, ceramics, and electronic circuits.

no-load loss *n*: in a transformer, losses in electrical energy that occur in the primary, secondary, and core of the transformer; no-load losses occur whether the transformed is loaded or is not loaded.

nonconductor *n*: a material through which little or no free electrons can flow.

noninverting input *n*: in an operational amplifier, a type of input wherein if the input signal is changed, the output is also changed. For example, if the noninverting input increases in voltage, the output voltage also increases. Also called plus input. Compare *inverting input*.

NOR circuit *n*: an electrical circuit in which output voltage appears only when a signal is absent from all its input terminals.

north magnetic pole *n*: a point on the earth's surface near the north polar region to which a magnetic compass points regardless of the compass's location on earth. The north magnetic pole is located several miles from the North Pole (the north geographic pole); consequently, navigators must correct for this difference when plotting courses or precisely locating a particular area.

NOT circuit *n*: see *inverter circuit*.

npn transistor *n*: a transistor in which n-type material is placed on each side of p-type material. One side of the n-type material is an emitter and the other side is a collector. The p-type material sandwiched in between the two n-type materials is the base. The emitter is negative with respect to the base, while the collector is positive with respect to the base.

n-type structure *n*: in a semiconductor material such as silicon, the arrangement of atoms in the material in such a manner that electron flow occurs between the atoms and thus creates a negative charge. Compare *p-type structure*.

nucleus *n*: the positively charged central region of an atom, composed of protons and neutrons and containing almost all of the mass of the atom.

number *n*: 1. in mathematics, a member of the set of positive integers; one of a series of symbols of unique meaning in a fixed order that can be derived by counting. A member of any of the further sets of mathematical objects, such as negative integers and real numbers. 2. a symbol or a word used to represent a number.

numerator *n*: in a mathematical fraction, the term or number that is divided by the other term or number (called the denominator), and is written above the line. For example, in the fraction $8/9$, 8 is the numerator.

O

O *sym*: oxygen.

octal number *n*: a number in a base 8 numbering system that is represented by the digits 0 through 7. Usually, an octal number is written with the subscript 8 to indicate that it is an octal number. For example, 6270_8 indicates an octal number.

ohm (Ω) *n*: a unit for measuring electrical resistance. One ohm is equal to the resistance through which a current of 1 ampere will flow when a potential difference of 1 volt exists across a circuit.

ohmic *adj*: of or relating to the resistance of an object or a material in terms of ohms. See *ohm*.

ohmmeter *n*: an electrical instrument that measures the resistance of a conductor in ohms. It may be calibrated to read in ohms or megohms.

ohms-adjust potentiometer *n*: an adjustable resistor (a potentiometer) on an ohmmeter that, when turned, varies the resistance in the meter's circuit. Varying the resistance adjusts the meter's pointer to indicate the appropriate resistance of the circuit or component being tested.

Ohm's law *n*: a rule that explains the behavior of electrical flow through a conductor. It is stated as $R = E/I$, where R = resistance in ohms, E = voltage in volts, and I = current in amperes. The law states that resistance, voltage, and current are interrelated and that modifying or changing one value in an electrical circuit modifies or changes the others. It is therefore used to calculate resistance, voltage, and current in electrical circuits.

ohms/mil-ft *abbr*: ohms per mil-foot.

ohms per mil-foot (ohms/mil-ft) *n*: in electronics, a measure of the resistance of a wire based on the wire's gauge and length; in general, the heavier (thicker) and longer the wire is, the more resistance it presents. See *mil-foot*.

oil-filled capacitor *n*: a capacitor whose dielectric (insulating material) is paper but which is enclosed by an oil-filled metal container. See *capacitor*.

ones column *n*: in mathematics, the column of numbers written in the far right-hand column in a group of numbers having more than one digit. For example, in the number 3,583, three is in the ones column. It is a way of denoting that the number 3,583 contains 3 ones. In addition, 3,583 contains 8 tens, 5 hundreds, and 3 thousands. Numbers with only one digit have only a ones column.

op-amp *abbr*: operational amplifier.

open circuit *n*: in an electrical circuit, a phenomenon that occurs when a component, such as a wire, breaks or otherwise fails so that the component can no longer carry current. For example, a broken wire in a circuit causes an open circuit.

open-loop amplifier *n*: an electronic device that increases, or amplifies, a signal without feedback from any part of the circuit. See *feedback*.

operand *n*: in computer science, any one of the quantities entering into or arising from an operation. See *operation*.

operation *n*: in computer science, a process or procedure that obtains a unique result from any permissible combination of operands. See *operand*.

operational amplifier (op-amp) *n*: an electronic device that increases the magnitude or power level of an electric current (an amplifier) and that has high direct-current stability and high immunity to oscillation, which is achieved by a high amount of negative feedback in the circuit in which the amplifier is installed. See *negative feedback*.

OR function *n*: an instruction that performs the logical operation "or" on a bit-by-bit basis for its two or more quantities entering into or arising from an operation (its operand), usually storing the result in one of the operand locations. See *operand*.

orifice valve *n*: a device (a valve) that controls the flow of fluid in a line, which contains an opening (an orifice) the size of which is varied by a mechanically, hydraulically, pneumatically, or electrically operated piece that contacts the orifice.

oscillation *n*: any effect that varies periodically between two values.

oscillator *n*: an electronic circuit that produces high frequencies without rotational force being required. Oscillator voltages produce output current that periodically reverses, or oscillates.

oscilloscope *n*: a test instrument that uses a cathode-ray tube to make visible on a fluorescent screen the instantaneous values and waveforms of electrical quantities that rapidly vary as a function of time or other quantity.

oxidation *n*: a chemical process in which oxygen combines with a compound and causes the compound to lose electrons and gain a more positive charge. For example, when exposed to air (which contains about 21 percent oxygen), iron rusts, which means that part of the iron chemically combines with the oxygen in the air and oxidizes, or becomes rusty. The rust is typically a red iron oxide, which gives iron rust its characteristic color.

oxide film *n*: a thin layer of material (a film) that is coated with metallic oxides and is often used as a dielectric in capacitors. See *capacitor*, *dielectric*.

oxygen (O) *n*: a nonmetallic element constituting 21 percent of the atmosphere by volume that occurs as a diatomic gas, O_2, and in many compounds such as water and iron ore. It combines with most elements, is essential for plant and animal respiration, and is required for nearly all combustion.

P

paper capacitor *n*: a capacitor whose dielectric (insulating material) is made from paper. See *capacitor*.

parallel *adj*: of an electrical circuit in which the components are assembled such that current flows through two or more paths. Compare *series*.

parallel circuit *n*: a circuit with two or more paths through which the current flows. The total current equals the sum of the currents through the various paths; the voltage is the same throughout the circuit. Compare *series circuit*.

Pb *sym*: lead.

peak inverse voltage *n*: see *peak reverse voltage*.

peak reverse voltage *n*: the maximum instantaneous anode-to-cathode voltage in the reverse direction that is applied to a diode in an operating circuit.

peak voltage *n*: the instantaneous increase (peak) in voltage that occurs when voltage flows in an alternating current (AC) circuit. In an AC circuit, the voltage not only reverses direction, but also increases for a moment (peaks) at the point when it reverses direction.

permanent magnet *n*: a magnet usually made of hardened steel that is strongly magnetized and retains its magnetism after the removal of the magnetizing force.

permeability *n*: the property of a magnetizable substance that determines the degree to which it modifies the magnetic lines of force (the flux) in the region occupied by the substance in a magnetic field; specifically, it is the ratio of the induction to the magnetizing force in the substance.

permissible operating temperature *n*: in a solid-state diode, a measure of the diode's ability to handle heat that is generated by voltage and current that occurs where the diode is connected, or joined, to a circuit.

pf *abbr*: power factor.

pF *abbr*: picofarad.

PFD *abbr*: manufacturer's abbreviation for picofarad; sometimes imprinted on the body of a capacitor.

phase *n*: in electronics, the part of a period through which the time variable of a periodic quantity, such as alternating current, has moved. Phase is measured at any point in time from an arbitrary time origin. In the waveform of alternating current, where the waveform is sinusoidal in shape and where positive polarity is above a zero line and negative polarity is below the zero line, phase is measured at the last point at which the quantity passed through a zero position from a negative to a positive direction.

phase shift *n*: a change in the phase of a periodic quantity, such as the waveform generated by alternating current. See *phase*.

photovoltaic cell *n*: a device that detects or measures electromagnetic radiation by generating voltage (potential) at the junction between two types of material when radiant energy (as from the sun or other light source) strikes it.

photovoltaic diode (PVD) *n*: a special solid-state diode to which has been added a special impurity that causes the diode to produce voltage when exposed to bright light, such as the sun.

pi (π) filter *n*: see *LC filter*.

picofarad (pF) *n*: 1 trillionth (10^{-12}) of a farad. See *farad*.

pigtail lead *n*: on an electronic component, such as a composition resistor, a single wire (a lead) that exits the component from either end and to which the electrical connection to a circuit are made.

pixel *n*: the basic and smallest part of the composition of an image on a television screen, computer monitor, or similar cathode-ray tube.

PLC *abbr*: programmable logic controller.

plus input *n*: see *noninverting input*.

pnp transistor *n*: a transistor in which p-type material is placed on each side of n-type material. One side of the p-type material is an emitter and the other side is a collector. The n-type material is sandwiched in between the two p-type materials in the base. The emitter is positive with respect to the base, while the collector is negative with respect to the base.

polarity *n*: the quality of being either negative or positive; in magnetism, the north or south pole of a magnet.

pole *n*: 1. either end of an axis through a sphere. 2. either of the regions bordering the extremities of the earth's rotational axis, which are the North Pole and the South Pole. 3. a magnetic pole. 4. in electricity, either of two oppositely charged terminals, as in an electric cell or battery. 5. a long, slender, and usually cylindrical object, such as fishing pole.

polyphase *n*: in electronics, a phenomenon in which a circuit has several alternating electromotive forces (emfs) of the same frequency and sine wave form. A polyphase generator has two or more circuits in the field windings. *adj*: having or utilizing two or more phases of an alternating-current power line. See *phase*.

positive charge *n*: a phenomenon that occurs at the atomic level of elements whereby the element has a deficiency of electrons—that is, it has a higher electric potential. Compare *negative charge*.

positive terminal *n*: the terminal of a battery or other voltage source toward which electrons flow and from which holes flow. See *hole*, *terminal*.

potential *n*: see *electric potential*.

potential difference *n*: see *electromotive force*.

potential hill *n*: in solid-state semiconductors, an energy barrier created by the semiconductor's space-charge region, which must be overcome by the application of external voltage to the semiconductor to make current flow across the junction where an n-type and p-type semiconductor are joined. See *n-type structure*, *p-type structure*, *space-charge region*.

potentiometer *n*: 1. a resistor having a continuously adjustable sliding contact that is usually mounted on a rotating shaft. By rotating the shaft, the voltage flowing through the potentiometer is varied as desired. A potentiometer is mainly a voltage divider. 2. a device that measures an electromotive force by comparing it with a known potential difference. See *resistor*.

power *n*: 1. the source or the means of providing energy. 2. the time rate at which work is done. 3. an exponent, as the 3 in x^3.

power distribution system *n*: an arrangement of many pieces of equipment, such as transformers, conductors, switches, heat sinks, fans, poles, and towers, which serve to deliver (distribute) electrical power from its source to its point of use. Power sources include hydroelectric generation by dams, natural gas (or other fuel) fired steam turbines to drive electrical generators, wind generators, and solar generators.

power factor (pf) *n*: the ratio of true power to apparent power in an alternating current circuit. True power is always less than apparent power. See *apparent power*.

power law *n*: the rule that allows a person to determine the power a circuit is capable of producing. The law consists of three variables: E or V, measured in volts; current, I, measured in amperes; and power, P, measured in watts. In equation form, the power law is $P = EI$ or $P = VI$.

power rating *n*: 1. in electricity, the power available at the output terminals of a component or a piece of equipment that is operated according to the manufacturer's specifications. 2. in electronics, the amount of power a component, such as a resistor, is capable of handling without malfunctioning or failing. Ratings are usually given in watts so that a component with a power rating of 2 watts means that it is capable of handling 2 watts of power.

power supply *n*: a source of electrical energy, such as a battery, an electrical line, or a set of components with a transformer in an electronic device, that furnishes electrical power to the semiconductor devices of an electronic circuit with the proper voltages and currents for their operation.

power transformer *n*: 1. a transformer that takes power from a generating station and steps up the voltage for transmission, or that takes power from the transmission line and steps down the voltage to the primary distribution level of 7,200 or 12,400 volts. 2. a relatively small transformer used in an electronic circuit to obtain the proper voltage required by the circuit to operate. See *distribution transformer*.

power triangle *n*: a right triangle derived from a vector diagram of inductive and capacitive VAR (VAR_H and VAR_C) and watts. A power triangle allows relationships between reactive power and real power to be easily seen. See *reactive power*, *real power*, *VAR*.

preferred resistance values *n*: in electronics, a set of standard values for resistors that manufacturers use as a guide for making resistors of various tolerances. Preferred values range from a low of 1 percent for E96 resistors, to a high of 20 percent for E6 resistors, with values of 5 and 10 percent in between the extremes.

primary *n*: see *primary winding*.

primary cell *n*: a unit that produces electricity from the chemical reaction of conductors of dissimilar metals through an electrolyte. In a primary cell, the chemical energy has its origin within the cell, and the chemical reaction is not reversible. Also, the electrolyte is not liquid; rather, it is a moist solid, often called "dry." See *dry cell*, *wet cell*.

primary loss *n*: in a transformer, the reduction (loss) in electrical energy that occurs in the primary winding of the transformer.

primary maximum capacity *n*: in a transformer, the product of the maximum volts and amperes of the primary.

primary winding *n*: in a transformer, the first of two windings (coils) to which electrical energy flows to induce electrical energy in the second coil, which is called the secondary winding. Compare *secondary winding*.

product *n*: in mathematics, the quantity that results when two or more quantities are multiplied together. For example, in the equation $2 \times 8 = 16$, 16 is the product.

programmable logic controller (PLC) *n*: a device used to manage, or control, another device or devices that govern the operation of a system or process. An operator, using an attached computer, can program the controller to maintain a given set of desirable circumstances and to respond to changes or upsets in the system or process using ladder logic, which is a logic system that operates much like the rungs on a ladder—that is, before the next rung on the ladder can be scaled, the controller must determine that certain conditions are met on the current rung.

proton *n*: the positively charged elementary particle that occurs with the neutron in an atomic nucleus.

p-type structure *n*: in a semiconductor material such as silicon, the arrangement of atoms in the material in such a manner that few electrons are able to flow between the atoms; a positive charge is thus created. Compare *n-type structure.*

pullout torque *n*: the greatest turning force (torque) under which an electric motor can operate without sharply losing speed.

pulse generator *n*: 1. in electronics, a generator that produces repetitive pulses or signal-initated pulses. 2. an electrical device that produces very short surges of high-voltage or high-current power by discharging capacitors in parallel or in series.

PVD *abbr*: photovoltaic diode.

Pythagorean theorem *n*: in mathematics, the principle stating that in a right triangle, the square of the length of the hypotenuse is equal to the sum of the squares of the lengths of the other two sides. See *hypotenuse, right triangle.*

Q

quotient *n*: in mathematical division, the quantity that results when one quantity is divided by another. For example, where $6 \div 2 = 3$, 3 is the quotient.

R

R *abbr*: resistance, rel.

raceway *n*: in electricity, a tube or channel that holds, guides, and protects electric wires.

radial lead (pronounced "leed") *n*: in an electronic component, a wire (a lead) that exits each end of the component and is arranged so that the leads are attached to the component in a radial manner to the end of the component. Compare *axial lead.*

Radio Manufacturers Association (RMA) *n*: a now defunct association that published manufacturing standards for electronic components. RMA evolved into the Electronic Industries Alliance (EIA). See *Electronic Industries Alliance.*

radio tube *n*: see *electron tube.*

random-access memory (RAM) *n*: electronic data stored in such a manner as to allow a user to access and program it, and then read from it, erase it, and reprogram it.

RC circuit *n*: see *resistance-capacitance circuit.*

reactance *n*: in an alternating current circuit, the impedance of current flow caused by capacitive or inductive components.

reactive power *n*: the value of the power in an electric circuit obtained by multiplying the effective value of the current in amperes, the effective value of the voltage in volts, and the sine of the angular phase difference between current and voltage.

read-only memory (ROM) *n*: a device for storing electronic data in permanent, or nonerasable, form—that is, the data cannot be accessed, rewritten, or altered by a user. It is usually a static electronic or magnetic device that allows rapid access to the data.

real power *n*: the component of apparent power that represents true work. Real power is expressed in watts and equals volt-amperes multiplied by the power factor.

receiver *n*: electronic equipment used for receiving modulated radio waves and converting them into the original intelligence, such as into sounds or pictures, or converting to desired useful information as in a radar receiver.

reciprocal equation *n*: in parallel electronic circuits, the equation

$$R_T = \frac{1}{\frac{1}{R_1} + \frac{1}{R_2} + \frac{1}{R_3} + \frac{1}{R_4} + etc.}$$

which states that total resistance (R_T) in the circuit is equal to 1 divided by the sum of 1 divided by the resistance of each resistor (R) in the circuit, which are indicated as R_1, R_2, R_3, R_4, and so on.

recommended standard (RS)-232 *n*: an Electronic Industries Alliance (EIA) standard that describes the requirements for the interface between data processing and data communications equipment. It is widely used to connect computers to peripheral devices.

rectifier *n*: an electrical device that converts alternating current into direct current.

rectify *v*: to change an alternating current to direct current.

rel (R) *n*: a proposed unit to express reluctance in a magnetic circuit. The rel has not, however, been universally accepted.

relay *n*: a device in an electric circuit that senses electrical variations in the circuit and then makes or breaks one or more connections in the same or another electric circuit.

reluctance *n*: in electromagnetics, a measure of the opposition presented to magnetic lines of force (flux) in a magnetic circuit; it is analogous to resistance in an electric circuit.

remainder *n*: in mathematical division, a quantity that results when one quantity does not divide evenly into another quantity. For example, $7 \div 2 = 3$ with a remainder of 1. That is, 2 goes into 7 three times with 1 left over.

remote terminal unit (RTU) *n*: that part of an automated lease that relays signals from the central computer to the end devices and relays status and alarm conditions from the end devices to the computer.

residual magnetism *n*: the relatively small amount of magnetism that virtually all metallic objects that contain iron retain as a result of the magnetic influences. It affects magnetic compasses by attracting the compass needle away from actual magnetic north to give a false reading.

resistance (R) *n*: opposition to the flow of direct current caused by a particular material or device. Resistance is equal to the voltage drop across the circuit divided by the current through the circuit.

resistance-capacitance (RC) circuit *n*: an electrical circuit that has a resistance and a capacitance in series, and in which inductance is negligible.

resistance tolerance *n*: a measure of a resistor's accuracy, usually stated in percent. For example, if a 2,400-ohm resistor has a tolerance of 5 percent, its resistance should vary no more than 5 percent of 2,400 ohms, either up or down; or its resistance should be no less than 2,280 or more than 2,520 ohms.

resistivity *n*: the electrical resistance offered to the passage of current; the opposite of conductivity.

resistor *n*: a component that tends to impede the flow of electric current, usually without any inductive or capacitive effects.

retentiveness *n*: see *retentivity*.

retentivity *n*: in magnetism, the tendency of an unmagnetized iron or steel bar to retain, or hold, its magnetism after it is magnetized. Iron or steel with low retentivity loses its magnetism relatively quickly, while an iron or steel with high retentivity keeps its magnetism relatively longer.

retrace *n*: in an oscilloscope, the brief instant at which the voltage driving the oscilloscope's beam suddenly decreases and drives the beam to the left of the screen.

reverse bias *n*: where a p-type and an n-type semiconductor are joined, a condition in which no current flows in the semiconductors because the negative terminal of a battery is connected to the lead of the p-type material and the positive terminal of the battery is connected to the lead of the n-type material. Compare *forward bias*. See *n-type structure*, *p-type structure*.

reverse breakdown *n*: see *Zener breakdown*.

rheostat *n*: a resistor that is used to vary the electrical current flow in a system.

ribbon line *n*: a kind of conductor in which two wires are separated a short distance from each other by an insulating material. The conductor thus has the appearance of a flat ribbon. Ribbon line is often used to attach a television antenna to the television set and has an impedance of 300 ohms. See *impedance*.

right angle *n*: an angle of 90 degrees.

right-hand rule for coils *n*: the statement of the fact that if the coil is grasped in the right hand with the fingers wrapped around it in the direction of the current flow, the thumb points toward the north pole.

right-hand rule for conductors *n*: the statement of the fact that if the fingers of the right hand are placed around a wire so that the thumb points in the direction of the current flow, the fingers point in the direction of the magnetic field produced by the wire.

right triangle *n*: in mathematics, a triangle in which one of the angles is a right angle. See *right angle*.

ring *n*: see *shell*.

ripple *n*: the alternating current component in the output of a direct current power supply, which arises in the power supply because of incomplete filtering or from a direct current generator.

RMA *abbr*: Radio Manufacturers Association.

RMS value *abbr*: root-mean-square value.

ROM *abbr*: read-only memory.

root *n*: in mathematics, a number which, when raised to some exponent (or power), equals that number. For example, the root of 4 is 2 because $2^2 = 4$.

root-mean-square (RMS) value *n*: the square root of the time average of the square of a quantity. For a periodic quantity, the average is taken over one complete cycle.

rotor *n*: the rotating part of an electric machine or device, such as the rotating armature of a motor or generator, or the rotating plates of a variable capacitor. A rotor is part of an induction-type alternating current electric motor. Compare *stator*.

Rowland's laws *n pl*: several rules originated by Henry A. Rowland, an American physicist working in the mid-to-late 1800s, about the behavior of magnetic circuits. Perhaps his best-known law is the one for magnetic circuits that is analogous to Ohm's law for current flow through wire circuits. Rowland's law for magnetic circuits is $\varnothing = F/R$, where $\varnothing$ = magnetic flux, F = magnetomotive force, and R = reluctance.

RTU *abbr*: remote terminal unit.

S

S&W *abbr*: sediment and water. API Committee on Petroleum Measurement prefers this abbreviation to the older "BS&W."

saturated core *n*: in magnetic circuits, the instant in the core of a coil whereby the flux density reaches a level beyond which little or no additional density can be achieved by further increases in magnetomotive force.

sawtooth oscillator *n*: an electronic circuit that converts energy from a direct-current source to a periodically varying electric output; its waveform has the shape of a sawtooth.

sawtooth sweep generator *n*: an electronic device that creates, or generates, an output voltage that has a waveform in the shape of a sawtooth. Such a generator produces sweep voltage for cathode-ray tubes. See *sweep voltage*.

schematic *n*: a drawing of an electric circuit that uses several standard symbols to represent electrical components and that shows the arrangement and connections of the conductors and the components in the circuit.

Schmitt circuit *n*: see *Schmitt trigger*.

Schmitt trigger *n*: pulse generator capable of assuming either of two stable states (a bistable pulse generator) in which an output pulse of constant amplitude exists only as long as the input voltage exceeds a certain value. Also called a Schmitt circuit. See *bistable*.

Schottky barrier diode *n*: see *Schottky diode*.

Schottky diode *n*: a diode formed by contact between a semiconductor layer and a metal coating. It is used as a special rectifier. Electrons or holes are emitted from the area where the semiconductor and metal coating contact each other (the p-n junction) and move to the metal coating. It provides very fast switching speeds. Also called a hot-carrier diode, Schottky barrier diode.

SCR *abbr*: silicon-controlled rectifier.

sec *abbr*: 1. second. 2. secant.

secant (sec) *n*: in a right triangle, the trigonometric function that is the ratio of the hypotenuse to the side adjacent to a given angle.

secondary *n*: short for secondary winding. See *secondary winding*.

secondary cell *n*: a device that produces electricity from the chemical reaction of conductors of dissimilar metals through an electrolyte. In a secondary cell, the electrolyte is a liquid solution and the chemical reaction is reversible—that is, the secondary cell can be recharged. It is often called a wet cell. See *dry cell, wet cell*.

secondary loss *n*: in a transformer, the reduction (loss) of electrical energy that occurs in the secondary winding of the transformer.

secondary winding *n*: in a transformer, the second of two windings (coils) in which electrical energy is induced by the magnetic energy in the primary winding. Compare *primary winding*.

sediment and water (S&W) *n*: a material coexisting with, yet foreign to, petroleum liquid and requiring a separate measurement for reasons that include sales accounting. This foreign material includes free water and sediment (dynamic measurement) and/or emulsified or suspended water and sediment (static measurement). The quantity of suspended material present is determined by a centrifuge or laboratory testing of a sample of petroleum liquid.

semiconductor *n*: a solid (such as germanium or silicon) whose electrical conductivity lies between that of a conductor and that of an insulator. Semiconductor materials are used in the manufacture of diode rectifiers, transistors, and integrated circuits.

sequential logic device *n*: a component or element that has at least one input channel, at least one output channel, and at least one internal state variable, so designed and constructed that the output signals depend on the past and present states of the inputs.

series *adj*: of or pertaining to an end-to-end arrangement of electrical components in a circuit that forms a single path for current to travel.

series circuit *n*: a circuit that has two or more electrical devices connected to form a single path for the electric current. The current is the same throughout the circuit; the total voltage is the sum of the voltages across each element of the circuit.

shell *n*: a hypothetical spherical surface centered on the nucleus of an atom that contains orbiting electrons.

short *n*: see *short circuit*.

short circuit *n*: a phenomenon that occurs in electrical circuits wherein two components carrying current inadvertently come into contact and interrupt the current path. For example, if two bare wires in a circuit accidentally touch, current flows directly back to the source of electricity without traveling its normal path. Often, because so little resistance exists, current flow is high enough to melt wires and other components.

shunt *n*: a conductor joining two points in an electrical circuit to form a parallel or alternate path through which a portion of the current may pass.

SI *abbr*: International System of Units.

Si *sym*: silicon.

signal generator *n*: an electronic test instrument that delivers a sinusoidal output at an accurately calibrated frequency ranging anywhere from audible to microwave frequencies. The frequency and amplitude of the device's output are adjustable over a wide range. The output may be frequency- or amplitude modulated.

signal integrator *n*: an operational amplifier that senses electronic signals in a circuit and electronically arranges the signals so that they function together in an efficient and logical way. See *operational amplifier*.

silicon (Si) *n*: a nonmetallic element occurring extensively in the earth's crust in the form of silicon dioxide, or sand. Silicon usually occurs as dark-brown crystals. It is used in glass, steel, semiconducting devices, concrete, brick, and other materials.

silicon-controlled rectifier (SCR) *n*: a semiconductor device (a rectifier) that changes alternating current to direct current by means of a silicon control gate. Normally, an SCR acts as an open circuit, but rapidly switches to a conducting state when an appropriate gate signal is applied to the gate terminal. An SCR conducts current in only one direction. Commonly called an SCR or, erroneously, a thyristor. While one type of thyristor is an SCR, another type exists, which is called a triac. See *triac*.

silicon steel *n*: an iron alloy that contains carbon (a steel) and 0.5 to 4.5 percent silicon. Silicon steel is often used in electric transformer coils. The silicon allows the steel to be drawn into thin, flat plates that are ideal for transformer construction. See *silicon*.

simple circuit *n*: an electrical circuit that consists of a single path from one terminal of the source of electromotive force, through one electrical device, and back to the terminal of the source of electromotive force.

sin *abbr*: sine.

sine (sin) *n*: a trigonometric function equal to the ratio of the side opposite a given acute angle to the hypotenuse.

sine wave *n*: a undulating form (a wave) whose amplitude varies as the sine of a linear function of time. See *sine*.

single-phase motor *n*: a motor energized by a single source of alternating voltage.

single-phase voltage *n*: voltage produced by an alternating current circuit that has only two points of entry, or one which, having more than two points of entry, is intended to be so energized that the potential differences between all pairs of points of entry are either in phase or differ by 180 degrees.

single-pole, double-throw (SPDT) switch *n*: an electrical switch with three terminals that connects one terminal to one of the other two terminals.

single-pole, single-throw (SPST) switch *n*: an electrical switch with two terminals that is arranged so that the switch opens or closes one circuit.

sinusoidal *adj*: having a magnitude (size) that varies as the sine of an independent current. See *sine*.

slip *n*: in an electric motor, the difference between the optimal and the actual output of the motor.

slip ring *n*: a conducting ring that gives current to or receives current from the brushes in a generator or a motor. See *brush*.

snap-on meter *n*: see *clamp-on meter*.

solenoid *n*: a cylindrical coil of wire that resembles a bar magnet when it carries a current so that it draws a movable core into the coil when the current flows.

solid-state *adj*: relating to the properties, structure, or reactivity of solid material, especially relating to the arrangement and behavior of ions, molecules, electrons, and holes in the crystals of a substance, such as a semiconductor. The term distinguishes semiconducting electronic components that are composed of and rely on solid materials for operation instead of materials such as evacuated glass tubes.

solid wire *n*: in electronics, a conductor of electricity that is drawn out of a conducting material (such as copper) to form a single cylindrical length of wire. Compare *stranded wire*.

source *n*: the terminal in a field-effect transistor from which majority carriers flow into the conducting channel in the semiconductor's material. See *field-effect transistor*.

south magnetic pole *n*: a point on the earth's surface near the south polar region with which the south pole of a magnetic compass aligns regardless of the compass's location on earth. The south magnetic pole is located several miles from the South Pole (the south geographic pole).

space-charge region *n*: when p-type and n-type semiconductors are joined, the area (the region) on the semiconductors in which free electrons flow from the n-type semiconductor to the holes of the p-type semiconductor. In this region, the n-type material becomes slightly positive, and the p-type material becomes slightly negative. When this space charge has been established, the junction is stabilized. No more electrons can flow across the junction because they are now repelled by the slight negative charge in the p-type material. See *hole, n-type structure, p-type structure.*

SPDT *abbr*: single-pole, double-throw.

speedup capacitor *n*: in a flip-flop circuit, a capacitor that speeds up the switching action in the circuit. Usually, two speedup capacitors are used.

spider *n*: in an electric motor, a structure consisting of a hub, spokes, and a rim that supports the core of the motor's rotor, or armature.

SPST *abbr*: single-pole, single throw.

squirrel-cage induction motor *n*: an electric motor that uses a rotating device (a rotor) containing a secondary winding that turns inside a stationary unit (a stator) that contains the primary winding. Electric current created by induction makes the motor work. See *induction*, *induction motor*, and *squirrel-cage rotor.*

squirrel-cage rotor *n*: in an induction motor, the rotating device (rotor) that consists of several copper bars running longitudinally along the outside of the rotor's iron core. Because the copper bars surround the core longitudinally, the rotor resembles a cage used to exercise squirrels in captivity.

starting torque *n*: in an electric induction motor, the turning force (torque) that is affected by the resistance of the rotor's winding. The greater the rotor winding's resistance, the greater is the starting torque up to the point at which pullout torque occurs. See *pullout torque*.

stator *n*: the stationary (unmoving) part of an induction-type alternating-current electric motor. The stator is the stationary part of a rotating motor, and it contains the stationary parts of the magnetic circuit and associated windings. Compare *rotor*.

steel *n*: a hard, strong, durable, and malleable alloy of iron and carbon, which usually contains between 0.2 and 1.5 percent carbon. Steel often has other constituents such as manganese, chromium, nickel, molybdenum, copper, tungsten, cobalt, or silicon, depending on the desired alloy properties, and widely used as a structural material.

steering diode *n*: a diode used in logic circuits to guide (steer) voltages in the circuits from one point to another. See *diode*.

step-down transformer *n*: a transformer in which applied voltage of one value is decreased to a lower value by means of primary and secondary windings, which are coils of conductive wire wound around the transformer's laminated steel core. In a step-down transformer, more windings are made in the primary than in the secondary, which decreases the voltage in the secondary. Compare *step-up transformer*. See *transformer*.

step-up transformer *n*: a transformer in which applied voltage of one value is increased to a higher value by means of primary and secondary windings, which are coils of conductive wire wound around the transformer's laminated steel core. In a step-up transformer, fewer windings are made in the primary than in the secondary, which increases the voltage in the secondary. Compare *step-down transformer*. See *transformer*.

stranded wire *n*: a conductor of electricity that is made by interweaving several wires of relatively small gauge to obtain the effect of a single wire of large gauge. In large gauge sizes, stranded wire is easier to bend than solid wire, so stranded wire is often used where large gauge wire is required and where it must be bent or curved rather sharply to fit into a particular installation.

subscript *n*: in mathematics, a number or symbol written below and usually to the right of another number or symbol for any of a number of purposes, such as to identify a particular element or elements of a set, to denote a constant value or a variable, or, in a chemical formula, to indicate the number of atoms of a particular kind of molecule.

subtrahend *n*: in mathematics, the quantity that is subtracted from another quantity. For example, in the equation $18 - 8 = 10$, 8 is the subtrahend.

sulfuric acid (H_2SO_4) *n*: a colorless, oily liquid compound of hydrogen, sulfur, and oxygen (H_2SO_4), strongly poisonous and corrosive. It is formed when hydrogen sulfide (H_2S) or sulfur dioxide (SO_2) is mixed with water (H_2O). Also called vitriolic acid.

sum *n*: in mathematics, the quantity that results when two or more quantities are added together. For example, the sum of $2 + 2$ is 4.

supply voltage *n*: in an electric circuit, the source of the electromotive force (voltage) in the circuit. For example, a battery provides supply voltage in some circuits.

sweep voltage *n*: periodically varying voltage applied to the deflection plates of a cathode-ray tube to give a beam displacement that is a function of time, frequency, or other data base.

switch *n*: a manually, mechanically, or electronically operated device for making, breaking, or changing the connections in an electric circuit.

switchgear *n*: the aggregate of electronic devices in a circuit that bring other devices into the circuit into an operating or nonoperating state.

sync amplifier *abbr*: synchronization amplifier.

sync control *abbr*: synchronization control.

synchronization amplifier *n*: in an oscilloscope, a device that receives a small portion of the vertical input signal and feeds it to the horizontal-sweep oscillator as a pulse that triggers the oscillator in step with the vertical input signal.

synchronization control *n*: in an oscilloscope, a device that varies the amount of pulse strength that reaches the oscillator and that provides enough pulse to stabilize the pattern displayed on the oscilloscope's screen.

synchronous speed *n*: in a two-pole induction motor, the maximum speed at which such a motor is capable of operating. In a 60-Hz system, the synchronous speed is 3,600 revolutions per minute (rpm). Synchronous speed can be stated as the equation—

$$\text{synchronous speed (rpm)} = \frac{120 \times \text{frequency (hertz)}}{\text{number of poles}}.$$

T

tan *abbr*: tangent.

tangent *n*: in a right triangle, the ratio of the side opposite a given angle to the side adjacent to the angle.

telemetry *n*: the process of gathering data by electronic or other kinds of sensing devices and transmitting those data to remote points.

temperature coefficient *n*: an expression of the amount resistance increases in a material as the temperature increases. It is stated as degrees per Celsius per ohm.

temporary magnet *n*: a magnet usually made of soft iron that loses its magnetism shortly after the magnetizing force is removed from it.

tens column *n*: in a column of numbers having more than one digit, the column immediately to the left of the ones column. For example, in the number 3,284, 8 is in the tens column. It denotes that the number 3,284 contains 8 tens. Besides 8 tens, the number also contains 3 thousands, 2 hundreds, and 4 ones.

terminal *n*: 1. a screw, soldering lug, or other point to which electric connections can be made. 2. one of the electrical input or output points of a circuit or component.

test voltage *n*: in capacitors, voltage applied to a capacitor that is higher than the capacitor's maximum working voltage rating. Test voltage is applied to ensure that a capacity can perform at its working voltage without failing. See *maximum working voltage rating*.

thermoelectric device *n*: an electrical component that directly converts heat into electrical energy or directly converts electrical energy into heat.

thousand circular mil (kcmil) *n*: a measure of wire diameter stated in thousands of circular mils. See *circular mil*.

three-mode controller *n*: an instrument that continuously measures the value of signals generated by three variable quantities or conditions and acts to control any deviation from a desired preset value.

three-phase voltage *n*: voltage generated in a circuit that is energized by alternating-current voltages that differ in phase by one-third of a cycle, or 120 degrees.

thyratron *n*: a virtually obsolete electrical device, similar to a vacuum tube, that is a hot-cathode gas tube in which one or more control electrodes initiate but do not limit anode current except under certain operating conditions. Solid-state devices that perform the same function have replaced thyratrons.

thyrector *n*: a solid-state silicon diode that behaves as an insulator up to its rated voltage, and as a conductor above its rated voltage. It protects components in AC circuits from voltage surges.

thyristor *n*: a transistor that initiates but does not limit anode current except under certain operating conditions. As collector current is increased to a critical value, a thyristor gives a high-speed triggering action.

timing circuit *n*: a source of accurately timed pulses, used in synchronizing a digital computer or as a time base in a transmission system. Also called a clock.

torque *n*: the turning force that is applied to a shaft or other rotary mechanism to cause it to rotate or tend to do so. Torque is measured in foot-pounds, joules, newton-metres, and so forth.

total power *n*: see *apparent power*.

tractive force *n*: in an electromagnet, the force of attraction that exists between an iron (or other ferrous material) bar placed close to the poles of the electromagnet.

transducer *n*: a device actuated by power from one system and supplying power to another system, usually in a different form. For example, a telephone receiver receives electric power and supplies acoustic power, or a photoelectric cell transforms light energy into electrical energy.

transformer *n*: a device that uses the principle of induction to convert variations in current and voltage in a primary circuit into variations of current and voltage in a secondary circuit. It typically consists of two coils with different numbers of turns around a single iron core.

transistor *n*: a small electronic device containing a small block of semiconducting material to which at least three electrical contacts are made, usually two rectifying contacts and one ohmic, or nonrectifying contact. A transistor is used in circuits as an amplifier, detector, or switch. Physicists at Bell Laboratories, who invented the device in the late 1940s, coined the term by combining the words transfer and resistor—that is, a transistor transfers a signal across a resistor. See *semiconductor*.

triac *n*: a type of thyristor that switches current to high levels from a low-level control signal in AC and DC circuits. While triacs are similar to silicon-controlled rectifiers, triacs, unlike SCRs, are capable of conducting current in both directions in the circuit; SCRs can conduct current only in one direction.

trickle charge *n*: the continuous flow of direct current electricity into a cell or battery at a low rate to either recharge the cell or battery or to maintain a charge as the cell or battery is used.

trigonometric function *n*: in mathematics, the relationship, or function, such as the sine, cosine, secant, cosecant, tangent, and cotangent of an arc or angle expressed as the ratios of pairs of the sides of a right triangle. See *function*.

triode *n*: a virtually obsolete electrical component, which is a three-electrode vacuum tube that contains an anode, a cathode, and a control electrode. Triode transistors have almost completely replaced triode tubes.

true north *n*: the direction of the north geographic pole—as opposed to the north magnetic pole, which is several hundred miles away, toward Hudson Bay.

truth table *n*: a table listing the results obtained from a logical proposition (a truth value) that result from all the possible combinations of the truth values of its components. See *truth value*.

truth value *n*: the result of a logical proposition, which is either true or false.

tube *n*: see *electron tube*.

tunnel diode *n*: a diode with a special impurity that causes the diode to have negative resistance at low voltages in the forward bias direction and a short circuit in the negative bias direction. Tunnel diodes are used as oscillators and as microwave amplifiers. Also known as an Esaki tunnel diode.

turns ratio *n*: in a transformer, a comparison (the ratio) between the number of turns in the primary winding and the number of turns in the secondary winding. See *transformer*.

U

unipolar transistor *n*: a transistor that utilizes charge carriers of only one polarity, such as a field-effect transistor.

unity *n*: in mathematics, the number 1; a quantity regarded as 1.

V

V *sym*: volt.

VA *abbr*: volt-ampere.

VAC *abbr*: volts alternating current.

vacuum tube *n*: see *electron tube*.

valence electron *n*: a single electron or one of two or more electrons in the outer shell of an atom that is responsible for the chemical properties of the atom. Valence electrons of one atom combine with the valence electrons of other atoms to form many compounds. For example, when the valence electrons of sodium and chlorine combine, common table salt, or sodium chloride (NaCl), is formed.

valve *n*: British term for electron tube. See *electron tube*.

VAR *abbr*: volt-ampere reactive.

VAR_C *abbr*: volt-ampere reactive in circuits with capacitors.

VAR_H *abbr*: volt-ampere reactive in circuits with inductors. Also abbreviated var_H.

VAR_L *abbr*: volt-ampere reactive in circuits with inductors. Also abbreviated var_L.

variable capacitor *n*: a capacitor that is constructed of stationary and movable plates. A shaft attached to the movable plate assembly is turned to rotate the movable plates in and out of the stationary plates. As the area of the movable plates increases or decreases within the stationary plates, the capacity varies. See *capacitor*.

vector *n*: the representation of a quantity that gives the quantity both magnitude and direction and the components of which transform from one coordinate system to another in the same manner as the components of a displacement.

vector addition *v*: the adding together of the components that constitute a vector. See *vector*.

volt (*E* or *V*) *n*: the unit of electric potential, voltage, or electromotive force in the metric system. See *electromotive force*.

voltage *n*: potential difference or electromotive force, measured in volts.

voltage-dependent resistor *n*: see *metallic-oxide varistor*.

voltage drop *n*: the voltage developed across an electrical component or conductor as current flows through a component or conductor.

voltage gain *n*: in solid-state electronics, the increase in voltage produced by a transistor.

voltage regulator *n*: an electronic device, including certain solid-state diodes, that maintains a given voltage across a circuit within required limits despite variations in voltage or load.

volt-ampere (VA) *n*: an expression of apparent, or total, electric power in which power is the product of volts times amperes.

volt-ampere reactive (VAR) *n*: the reactive power in a circuit carrying alternating current when the product of the root-mean-square value of the voltage, expressed in volts, by the root-mean-square value of the current, expressed in amperes, and by the sine of the phase angle between voltage and current, equals 1. See *reactive power*.

voltmeter *n*: an instrument used to measure, in volts, the difference of potential in an electrical circuit.

volt-ohm-milliammeter (VOM) *n*: see *multimeter*.

VOM *abbr*: volt-ohm-milliammeter.

W

W *sym*: watt.

watt (W) *n*: the unit of power in the metric system, equal to 1 joule per second, or 0.746 horsepower.

wattless power *n*: see *reactive power*.

wattmeter *n*: an instrument that measures the magnitude of the power in an electric circuit. It may read watts, kilowatts, or megawatts.

wave *n*: in electronics, a disturbance that begins at one point in a medium and goes to other points in the medium without permanently displacing the medium.

waveform *n*: in electronics, the pictorial representation of the shape (form) of a wave, which is obtained by plotting the displacement of the wave as a function of time, at a fixed point in space. See *wave*.

weight *n*: the force with which a body is attracted to the earth or to another celestial body equal to the product of the object's mass and the acceleration of gravity.

wet cell *n*: a secondary cell in which the electrolyte is a dilute acid.

wire-wound resistor *n*: a resistor made from special wire, usually Nichrome, that is wound around a relatively small cylindrical core or tube. Wire-wound resistors can handle large amounts of power and maintain their ohmic value even under high temperatures.

wye connection *n*: in an electric circuit, a component that has three terminals to which conductors are connected to install the component into the circuit. Also called a Y connection.

XYZ

X_C *sym*: capacitive reactance.

XOR *abbr*: exclusive or.

Y connection *n*: see *wye connection*.

yoke *n*: a stationary piece of ferromagnetic material without windings, which permanently connects two or more magnet cores.

Z *sym*: impedance.

Zener breakdown *n*: a phenomenon that occurs in semiconductor diodes when the electric field across the space-charge region becomes high enough to produce a form of field emission that suddenly increases the number of carriers in this region. This breakdown is typically several hundred to several thousand volts, depending on the semiconductor's construction, and is not destructive. Also called reverse breakdown, Zener effect.

Zener diode *n*: a solid-state diode that uses Zener breakdown to regulate voltage in a circuit in which it is installed. It operates on the principle that the voltage that occurs during breakdown is relatively constant and therefore can be used to regulate voltage. See *Zener breakdown*.

Zener effect *n*: see *Zener breakdown*.

zero potential *n*: the actual potential of the surface of the earth taken as a point of reference. See *electric potential*.